Engineering Applications of Computational Methods

Volume 3

Series Editors

Liang Gao, State Key Laboratory of Digital Manufacturing Equipment and Technology, Huazhong University of Science and Technology, Wuhan, Hubei, China
Akhil Garg, Department of Mechatronics Engineering, Shantou University, Shantou, Guangdong, China

The book series Engineering Applications of Computational Methods addresses the numerous applications of mathematical theory and latest computational or numerical methods in various fields of engineering. It emphasizes the practical application of these methods, with possible aspects in programming. New and developing computational methods using big data, machine learning and AI are discussed in this book series, and could be applied to engineering fields, such as manufacturing, industrial engineering, control engineering, civil engineering, energy engineering and material engineering.

The book series Engineering Applications of Computational Methods aims to introduce important computational methods adopted in different engineering projects to researchers and engineers. The individual book volumes in the series are thematic. The goal of each volume is to give readers a comprehensive overview of how the computational methods in a certain engineering area can be used. As a collection, the series provides valuable resources to a wide audience in academia, the engineering research community, industry and anyone else who are looking to expand their knowledge of computational methods.

More information about this series at http://www.springer.com/series/16380

Min Wu · Weihua Cao · Xin Chen · Jinhua She

Intelligent Optimization and Control of Complex Metallurgical Processes

Science Press
Beijing

Springer

Min Wu
China University of Geosciences
Wuhan, China

Hubei Key Laboratory of Advanced
Control and Intelligent Automation
for Complex Systems
Wuhan, China

Weihua Cao
China University of Geosciences
Wuhan, China

Hubei Key Laboratory of Advanced
Control and Intelligent Automation
for Complex Systems
Wuhan, China

Xin Chen
China University of Geosciences
Wuhan, China

Hubei Key Laboratory of Advanced
Control and Intelligent Automation
for Complex Systems
Wuhan, China

Jinhua She
School of Engineering
Tokyo University of Technology
Tokyo, Japan

China University of Geosciences
Wuhan, China

ISSN 2662-3366 ISSN 2662-3374 (electronic)
Engineering Applications of Computational Methods
ISBN 978-981-15-1144-8 ISBN 978-981-15-1145-5 (eBook)
https://doi.org/10.1007/978-981-15-1145-5

Jointly published with Science Press
The print edition is not for sale in China Mainland. Customers from China Mainland please order the print book from: Science Press.

This Springer imprint is published by the registered company Springer Nature Singapore Pte Ltd.
The registered company address is: 152 Beach Road, #21-01/04 Gateway East, Singapore 189721, Singapore

Preface

Iron and steel industry is the basis for the development of a country. Its level of development determines the economic level and comprehensive national power of a country. However, accelerated energy consumption and significant CO_2 emissions have limited the development of the iron and steel industry. Metallurgical automation technologies have been developed to yield the green manufacturing of the iron and steel industry.

The iron and steel industry involves complex processes, such as raw material proportioning, sintering, blast furnace ironmaking, steelmaking, continuous casting, and rolling. They have the common characteristics of strong nonlinearity and a large number of parameters. In order to achieve the green manufacturing of the iron and steel industry, we have to consider the problems of energy saving, emission reduction, environmental protection, etc. These problems involve the modeling, optimization, and control of complex processes. However, conventional methods of modeling, optimization, and control are not suitable for actual applications for complex processes with uncertainties. Advanced intelligent optimization and control technologies have been applied to the iron and steel industry to improve production efficiency and reduce CO_2 emissions.

We summarized our work and experience in the control of complex metallurgical processes over the last two decades in this monograph. We hope that it would be a useful reference for engineers in the field of metallurgical automation and complex process control, and for graduate students interested in computational intelligence and its applications of industrial process control.

The monograph consists of six chapters. Chapter 1 describes the current status of metallurgy processes and explains conventional intelligent algorithms in the aspects of modeling, optimization, and control. Chapter 2 describes the methods of how to optimize the proportioning of a coal blending process in an iron ore sintering process. An expert control strategy is designed based on a combination of back-propagation networks, mathematical models, and rule models to accurately determine and track the target percentages for a coal blending process. An intelligent integrated optimization system with a hierarchical configuration is devised to solve the problem of optimizing the proportioning for an iron ore sintering process.

Chapter 3 focuses on the problems of how to measure and control the coke-oven temperature. An integrated soft-sensing model of the coke-oven temperature and a hybrid hierarchical control system of a coke-oven combustion process are developed to solve these problems. Chapter 4 describes control methods of thermal state parameters in sintering process. An intelligent control of a sintering ignition process is designed to stabilize the ignition temperature. An intelligent control system for the burn-through point is designed to stabilize the point in an actual sintering process. Chapter 5 describes intelligent decoupling control methods for both the gas collection process of multiple asymmetric coke-ovens and a gas mixing-and-pressurization process in an iron and steel company. Chapter 6 focuses on the problem of how to control the rolling temperature and reduce the energy consumption of the process of a reheating furnace. An optimization and control system for the heating furnace is devised to solve this problem.

We are grateful for the support of Natural Science Foundation of China under Grant 61210011, the 111 project under Grant B17040, the Hubei Provincial Natural Science Foundation of China under Grant 2015CFA010, and the National High Technology Research and Development Program of China (863 Program) under Grant 2012AA040307.

We are also grateful for the support of scholars both in domestic and abroad. We would like to thank Emeritus Professor Michio Nakano of Tokyo Institute of Technology; Professors Deyao Shen, Zixing Cai, and Qi Lei of Central South University; Professor Hongye Su of Zhejiang University; Professor Feng Qian of East China University of Science and Technology; Dr. Yanguang Sun of Automation Research and Design Institute of Metallurgical Industry; and Professor Yonghua Xiong, Associate Professor Jianqi An, Drs. Zhentao Liu, and Luefeng Chen of China University of Geosciences. We would also like to express our appreciation for the great efforts of Drs. Xiaoxia Chen, Fang Gao, and Pan Yu of Central South University, Drs. Jie Hu, Sheng Du and Yali Jin, Ms. Ting Wan and Mr. Kexin Zhang of China University of Geosciences.

Wuhan, China — Min Wu
Wuhan, China — Weihua Cao
Wuhan, China — Xin Chen
Tokyo, Japan — Jinhua She

Contents

About the Authors

Min Wu received his B.S. and M.S. degrees in engineering from Central South University, Changsha, China, in 1983 and 1986, respectively, and his Ph.D. degree in engineering from the Tokyo Institute of Technology, Tokyo, Japan, in 1999.

He was a Faculty Member of the School of Information Science and Engineering at Central South University from 1986 to 2014, and was promoted to Professor in 1994. In 2014, he moved to China University of Geosciences, Wuhan, China, where he is a Professor in the School of Automation. He was a visiting scholar at the Department of Electrical Engineering, Tohoku University, Sendai, Japan, from 1989 to 1990, and a visiting research scholar at the Department of Control and Systems Engineering, Tokyo Institute of Technology, from 1996 to 1999. He was a visiting professor at the School of Mechanical, Materials, Manufacturing Engineering and Management, University of Nottingham, Nottingham, UK, from 2001 to 2002. His current research interests include process control, robust control, and intelligent systems.

He has published 6 monographs and more than 260 SCI-indexed papers. He has been included in the list of Thomson Reuters Highly Cited Researchers from 2014 to 2016. Due to his contributions on control theory and its engineering applications, he received numerous honors including the IFAC Control Engineering Practice Prize Paper Award in 1999, the second class prize of China's National Natural Science Award in 2013, the second class prize of China's National Science and Technology Progress Award in 2004, and the Academic Contribution Award of Chinese Process Control in 2009.

Professor Wu is a Fellow of IEEE, Vice-President of the Technical Committee on Control Theory, China Association of Automation, and an active member of the Technical Committee on Process Control, China Association of Automation. Moreover, he served as an Associate Editor of the editorial board of Control Engineering Practice, Information Sciences, International Journal of Automation and Computing, Control Theory and Applications, and Information and Control.

Weihua Cao received his B.S., M.S., and Ph.D. degrees in engineering from Central South University, Changsha, China, in 1994, 1997, and 2007, respectively.

He was a Faculty Member of the School of Information Science and Engineering at Central South University from 1997 to 2014, and was promoted to Professor in 2009. In 2014, he moved to China University of Geosciences, Wuhan, China, where he is a Professor in the School of Automation. He was a Visiting Student in the Department of Engineering, Kanazawa University, Japan, from 1996 to 1997, and was a Visiting Scholar in the Department of Electrical and Computer Engineering, University of Alberta, Canada, during the 2007–2008 academic year. His research interests include process control, robots, and intelligent systems. He is a member of the Chinese Association of Automation and a member of the Chinese Association for Artificial Intelligence.

Xin Chen received his B.S. and M.S. degrees in engineering from Central South University, Changsha, China, in 1999 and 2002, respectively, and the Ph.D. degree in engineering from University of Macau, China, in 2007.

He was a Faculty Member of the School of Information Science and Engineering at Central South University from 1999 to 2014. In 2011, he finished his postdoctoral research on control science and engineering at Central South University. In 2014, he moved to China University of Geosciences, Wuhan, China, where he is currently a Professor with the School of Automation. His research interests include intelligent control, process control, multi-agent systems, and robotics.

Jinhua She received his B.S. degree in engineering from Central South University, Changsha, China, in 1983, and his M.S. and Ph.D. degrees in engineering from Tokyo Institute of Technology, Tokyo, Japan, in 1990 and 1993, respectively.

In 1993, he joined the School of Engineering, Tokyo University of Technology, Tokyo, where he is currently a Professor. His research interests include control theory and applications, repetitive control, process control, Internet-based engineering education, and rehabilitation support devices.

Dr. She is a member of the Society of Instrument and Control Engineers (SICE), the Institute of Electrical Engineers of Japan (IEEJ), the Japan Society of Mechanical Engineers (JSME), the Architectural Institute of Japan (AIJ), the Asian Control Association (ACA), and the Institute of Electrical and Electronics Engineers (IEEE). He was the recipient of the International Federation of Automatic Control (IFAC) Control Engineering Practice Prize Paper Award in 1999 (jointly with M. Wu and M. Nakano).

Abbreviations

ACO	Ant colony algorithm
ADO	ActiveX data object
AI	Artificial intelligence
ANN	Artificial neural network
ATL	ActiveX template library
BIBO	Bounded-input, bounded-output
BP	Backpropagation
BPNN	Backpropagation neural network
BTP	Burn-through point
C/S	Client/server
CIQPM	Cascade integrated quality-prediction model
COT	Coke-oven temperature
COTC	Coke-oven temperature on the coke side
COTM	Coke-oven temperature on the machine side
CSP	Compact strip production
CVPD	Calorific value and pressure decoupling
DCS	Distributed control system
DDE	Dynamic data exchange
DOB	Disturbance observer
ECS	Expert control system
EGT	Exhaust gas temperature
EGTRP	Exhaust gas temperature rise point
EID	Equivalent input disturbance
ES	Expert system
FCM	Fuzzy C mean
FL	Fuzzy logic
FNNDC	Fuzzy-neural-network decoupling controller
FNNIM	Fuzzy-neural-network inference machine
GA	Genetic algorithm
GAP-EKF	Growing and pruning with an extended Kalman filter

HIC	Hierarchical intelligent control
HMI	Human–machine interface
HPSO	Hybrid particle swarm optimization
IIO	Intelligent integrated optimizer
IIOS	Intelligent integrated optimization system
JSP	Job scheduling problem
LP	Linear programming
LR	Linear regression
LTC	Longitudinal temperature on the coke side
LTM	Longitudinal temperature on the machine side
MFC	Microsoft foundation classes
MIMO	Multiple-input, multiple-output
NN	Neural network
OLEDB	Object linking and embedding database
OPC	OLE for process control
PID	Proportional–integral–differential
PLC	Programmable logic controller
PSO	Particle swarm optimization
QAP	Quadratic assignment problem
RBF	Radial basis function
RBF-RNN	Radial-base-function-based recurrent neural network
SA	Simulated annealing
SDF	Satisfactory degree function
SDNN	Supervised distributed neural network
SIP	Sintering ignition process
SSP	Satisfactory solution principle
TDF	Two-degrees-of-freedom
TSP	Traveling salesman problem
TTR	Temperature at the top of a regenerating chamber
TTRC	Temperature at the top of a regenerating chamber on the coke side
TTRM	Temperature at the top of a regenerating chamber on the machine side
VRP	Vehicle routing problem
WinCC	Windows control center

Chapter 1
Introduction

The iron and steel industry is the base for the development of national economy. It directly influences the industries of construction, machinery, shipbuilding, automobile, household electrical appliances, etc. The complexity and large uncertainties of the processes in the iron and steel industry make it difficult to establish accurate mathematical models and to control the processes using conventional control methods. On the other hand, since intelligent control does not require an accurate mathematics model, and its control algorithm has self-learning and adaptive ability, this control method is playing an increasingly important role in the iron and steel industry.

1.1 Complex Metallurgical Processes

The production flow of the iron and steel industry (Fig. 1.1) contains the processes of preparation of furnace charge before ironmaking, blast furnace ironmaking, steelmaking, continuous casting, rolling, etc.

First, iron ore, limestone, dolomite, coking coal, and returned sinter are mixed according to a specified dosing scheme to produce a raw mix. The raw mix is mixed with water to form granules of different sizes. A roller feeder pours the granules onto a moving pallet to form a sintering bed. As the coke in the bed burns, the heat produced induces chemical reactions and physical changes in the bed. This process produces sinter with a certain chemical composition, permeability, and strength. Next, coke and sinter are sent to a blast furnace where the ore material is gradually reduced while falling, melting into molten iron and iron slag. Then, molten iron is sent to converter and refining furnace. Removal of nonmetallic impurities, like sulfur, from molten iron is carried out through degassing, hardening and tempering, and temperature adjusting. After that, molten steel is cast into billet by a continuous casting machine. Finally, the steel rolling process is applied to obtain varieties of steel products. A utility system provides a basis for the production of the iron and steel industry. It contains water, gas, power, and heat supply. For example, coal–gas supply is a typical utility system in a complex metallurgical process.

M. Wu et al., *Intelligent Optimization and Control of Complex Metallurgical Processes*, Engineering Applications of Computational Methods 3,
https://doi.org/10.1007/978-981-15-1145-5_1

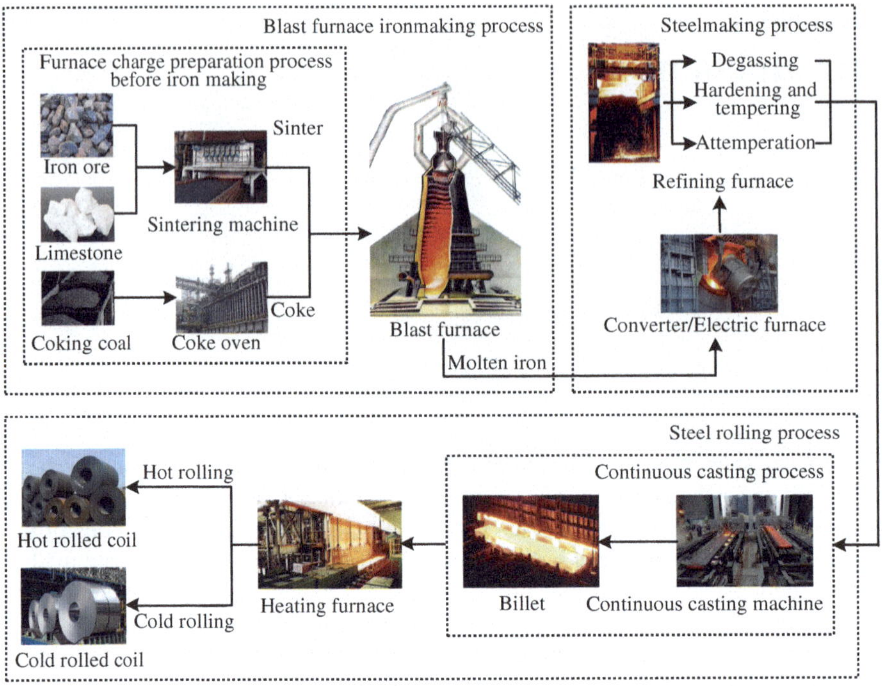

Fig. 1.1 Production flow of iron and steel industry

This book considers the problems of modeling, control, and optimization in the processes of coking, sintering, blast furnace, and heating furnace, and explores modeling and optimal control methods for a complex system.

Along with an increase in the scale of production, the demand on product quality also becomes more and more severe. This imposes strict requirements for process control. The main problem for the control and optimization of the processes is how to determine and track optimal values of process variables in order to achieve high quality, energy saving, and cost reduction.

A complex metallurgical process is a process that involves mass transfer, heat transfer, and complex chemical reactions. Due to an increasing demand on steel and the complexity of the process, it is urgent to improve the production efficiency and product quality, and to reduce product cost. This chapter reviews the methods of modeling, control, and optimization of the complex metallurgical process; explains their limitations; and shows how to use the methods of intelligent optimization and control to solve these problems.

1.2 Modeling, Control, and Optimization of Complex Metallurgical Processes

How to utilize available process information to build a precise process model, to perform online optimization of process variables, and to design a high-efficiency controller are the main problems that the control of the complex metallurgical processes is facing.

1.2.1 Modeling

Building a process model is the basis for advanced control, soft sensing, optimization, and management of the metallurgical process. Modeling of industrial processes has undergone two stages: conventional modeling and intelligent modeling (Table 1.1). When a metallurgical process becomes large, comprehensive, and complex, the difficulties of modeling such a process increase because it involves not only complex mechanisms that contain nonlinearities, uncertainties, time delays, distributed parameters, and time-varying characteristics, but also complicated external environments and mixed human factors.

Conventional modeling methods of metallurgical processes mainly consist of mechanism modeling and system identification. Mechanism models are built based on the balance of mass and energy, and the theory of dynamics and thermodynamics. The accuracy of the models mainly depends on the understanding of the physical changes and chemical reactions of a process. However, since an actual process is complex and has uncertainties, building a precise model is difficult and needs a large amount of time and huge cost.

System identification is a method that measures the behavior of the system and the external influences (inputs to the system) and determines a mathematical relation between them without going into details of what actually happens inside the system. The keys to system identification are the determination of a model type and the selection of the method for parameter estimation. The commonly used methods of parameter estimation are the maximum likelihood method, the least-squares method, the cross-correlation method, the instrumental variable technique, and the stochastic approximation method.

Table 1.1 Modeling stages of complex metallurgical processes

Stage	Theory	Method	Requirement	Form
1	Conventional	Mechanism modeling, system identification	Simple, accurate	Simple
2	Intelligent	Computational intelligence	Intelligent, integrated, efficient	Rich

In accordance with the improvement of computer processing capacity, intelligent modeling methods have gradually been becoming the main modeling methods for a complex metallurgical process. The neural network (NN) and fuzzy modeling methods are the most commonly used intelligent modeling methods for a complex metallurgical process.

An NN has the ability to map any nonlinearities and has a good learning ability. It has been widely used in metallurgical processes. For example, a backpropagation (BP) NN (BPNN) was built to estimate the particle size of sinter, taking the state and operating parameters as inputs [1]. However, an NN is a black model. It is built based only on the training of historical input–output data. Moreover, an NN has the disadvantage of being easily trapped in a local minimum. An improved NN using optimization methods, such as genetic algorithm (GA) and particle swarm optimization (PSO), to optimize the initial settings of the network. This greatly increases the prediction speed and the prediction precision of the model [2].

Fuzzy models use fuzzy logic to describe a process. They make the most of experts' experience using the fuzzy reasoning method to handle uncertain information. A Takagi–Sugeno (T-S) fuzzy model was built to predict the position and the temperature of a burn-through point (BTP) for a lead–zinc sintering process and to solve the problem that the temperature of the BTP cannot be measured directly [3]. It is usually not easy to acquire the experience of a process. An improved modeling method that combines a fuzzy model with an NN enables a model to have both the ability of handling fuzzy information and high prediction speed and precision. This method was used to build a fuzzy NN to estimate the energy consumption of a casting process. This method was also used to solve the problem of being trapped in a local minimum [4].

A complex metallurgy process has the characteristics of multiple parameters, nonlinearities, a large time delay, etc. This makes conventional modeling methods hard to describe the process precisely. Incorporating the knowledge of mechanisms of processes, operating experience, and historical data to build an intelligent model is an effective way to solve the problem and enable us to establish a practically useful model for the process.

1.2.2 Control

The development of the control of a complex metallurgical process is closely related with the development of control theory, computer communication and network, instruments, etc. It has gone through three stages: conventional control, advanced control, and comprehensive control (Table 1.2).

Conventional control theory was used at the first stage (before the 1970s) to design a controller based on the information from conventional instruments to ensure the safety and stability of a process.

The development of computer and the appearance of distributed control computer systems provided the basis for advanced control of the complex metallurgy processed

Table 1.2 Three stages of control of complex metallurgical processes

Stage	Control theory	Control tool	Control requirements	Control level
1	Conventional	Conventional instrument	Safety, smoothness	Low
2	Advanced	Distributed control system	Quality, high yield, low energy consumption	High
3	Comprehensive	Computer network	Intelligence, integration, high efficient	Advanced

at the second stage (from the 1970s to the 1990s). The implementation of model predictive heuristic control, dynamic matrix control, and multivariable predictive control pushed up the level of process control and shortened the distance to the target of high quality and quantity, and low energy consumption. The theory of robust control, nonlinear control, and predictive control has made a big breakthrough at this stage. However, there was still a big gap between the control theory and its applications, and the control theory could not meet actual requirements adequately.

The interdisciplinary integration of different disciplines, such as signal processing, database, and computer network and communication, has provided the basis for the third stage (after the 1990s). Product quality, energy consumption, cost, and emissions are the objects to be optimized. The characteristics of this stage are as follows:

(1) Advanced control theory: Distributed control systems were applied after the 1970s. But due to the limitations of the control theory and technology, simple control strategies, for example, single-loop proportional–integral–differential (PID) control, were mainly used to improve efficiency. But those strategies could not meet the needs of high flexibility and high efficiency. Advanced control theory was then applied to process control. For example, an intelligent decoupling method was devised to eliminate the effect of the coupling of gas pipe pressure of a gas mixing-and-pressurization process [5]. A fuzzy temperature controller was devised to control an ignition process and showed the effectiveness in improving the quality of combustion and utilization rate of gas [6]. Successful applications have attracted considerable attention from the industrial field.

(2) Development of an open system: Distributed control systems produced by different manufacturers were not compatible. After fierce competition, main manufacturers of the distributed control systems have jointly published a standard for real-time distributed control, fieldbus. It is an open, distributed, and intelligent industrial network system. This system has been widely used in complex metallurgical processes such as a blast furnace, a coke-oven, and sintering.

(3) Comprehensive automation system: Keen competition in the global market and the requirements of environmental protection accelerated the reformation of companies to strengthen their competitive power by saving energy, reducing cost, and improving the quantity and quality of products. The trend of current control system is to build a comprehensive automation system that integrates conventional control, advanced control, process optimization, production scheduling, business management, and operating decision.

1.2.3 Optimization

Solving an optimization problem is to find the best or most effective scheme from all alternative schemes. An optimization algorithm provides a theoretical support to solve the problem. The hot topics of the current research on the optimization of complex metallurgical processes are as follows:

(1) Theory and technology of optimizing region for operating state: Conventional optimization theory, for example, linear programming, tries to find a global optimum. However, a metallurgical process is too complex to achieve such a goal. And a local optimum is good enough in many engineering problems. The main research contents of region optimization are the selection of an evaluation function, determination of region for optimization, investigation of optimization algorithms, and convergence analysis.
(2) Random search optimization algorithm [7]: The most commonly used algorithms are the simulated annealing (SA) algorithm, the GA, the evolution computing, and the chemotaxis algorithm. These algorithms have been using to explore optimal solutions to actual optimization problems. For example, an SA algorithm was used to find an optimal dosing scheme for a coking process and to carry out optimal control of a coal blending process.
(3) Intelligent optimization [8–10]: An optimal problem usually has multiple optimization objectives and constraints for a complex metallurgical process. Since ordinary optimization algorithms are easily trapped in combinatorial explosion for this kind of problem, intelligent optimization algorithms, which imitate the behavior of animals and the mechanism of human brains, have been proposed to solve this kind of optimization problem. The NNs, the GA, the PSO, etc., are the commonly used ones.

1.3 Intelligent Control and Optimization Methods

An increase in complexity and scale of an industrial process imposes strict requirements for process control. Strong uncertainties in a process make mathematical model-based conventional control methods no longer applicable. Since knowledge-based intelligent control methods are flexible in dealing with complexity and uncertainties of an industrial process, they have received considerable attention.

1.3.1 Neural Network Modeling

An artificial NN (ANN, simply NN in the rest of this monograph) usually has multiple layers. Signals travel from the first layer to the last layer. The first layer is called the input layer that handles input signals, and the last layer is called the output layer that prepares outputs. All other layers in between these two layers are called hidden layers. NN models attempt to use an organizing principle believed to be used in human brain. An NN is an information processing system that imitates the structure and function of a brain, and human way of thinking. It has very complex dynamic characteristics; parallel processing capability; and the ability to learn, associate, and remember.

NNs are divided into feedforward networks and feedback networks depending on the difference of topological structure. They are also divided into continuous and discrete types of networks depending on the difference of network performance, are divided into supervised and unsupervised networks based on the difference of learning methods, and are divided into first-order linear networks and high-order nonlinear networks based on the property of synaptic junctions.

An NN not only has the similarities with a general nonlinear system, but also has the following characteristics:

(1) The capability of parallel processing: An NN has the high-level parallel structure, which accelerates the speed of information processing. It suits real-time control and dynamic control.
(2) Nonlinear mapping ability: An NN has the ability to map a nonlinearity with a prescribed precision. This makes it possible to solve the problem of effectively modeling a nonlinear system.
(3) Learn through training: An NN is built based on the training of historical data. So, it is applicable to complex processes for which mathematical models are hard to be built.
(4) Self-adaption and integration: The diversity of neurons and plasticity of connecting weights make an NN have strong adaptability. Moreover, the capability of information fusion ensures that an NN can handle a large number of control signals of different categories and can implement the integration and fusion of information.

Commonly used NNs are BPNNs [11, 12], radial basis function (RBF) NNs (RBFNNs) [13, 14], and Hopfield NNs [15, 16].

1. BPNN

A BPNN uses a backpropagation method to train the network. The configuration of a typical three-layer NN is shown in Fig. 1.2.

Suppose that the numbers of input neurons for the input layer, the hidden layer, and the output layer are n, l, and m, respectively. And let $In_j^{(i)}$ and $Out_j^{(i)}$ be the input and the output of the jth neuron in the ith layer, respectively. The relationships between the inputs and the outputs in each layer are

$$Out_j^{(1)} = In_j^{(1)} = x_j, \quad j = 1, 2, \ldots, n, \tag{1.1a}$$

$$In_k^{(2)} = \sum_{k=1}^{l} w_{jk}^{(1)} Out_j^{(1)} + b_k, \quad Out_k^{(2)} = f(In_k^{(2)}), \quad k = 1, 2, \ldots, l, \tag{1.1b}$$

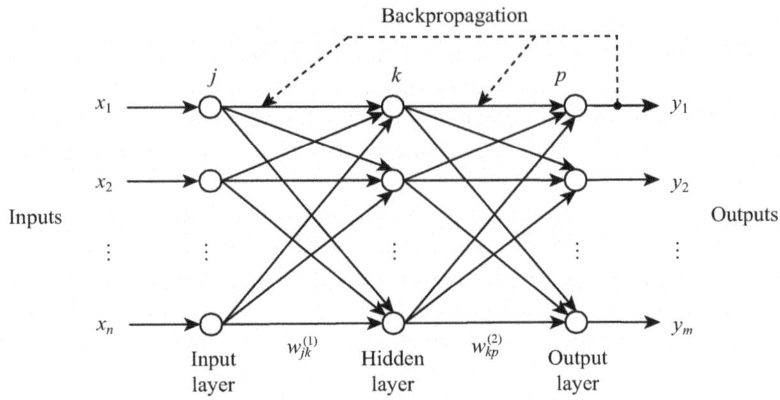

Fig. 1.2 Configuration of BPNN

$$In_p^{(3)} = \sum_{p=1}^{m} w_{kp}^{(2)} Out_k^{(2)} + b_p, \quad Out_p^{(3)} = \varphi(In_p^{(3)}), \quad p = 1, 2, \ldots, m, \tag{1.1c}$$

where $w_{jk}^{(1)}$ is the weight connecting the input neurons to the hidden neurons, $w_{kp}^{(2)}$ is the weight connecting the hidden neurons to the output neurons, $f(\cdot)$ is the activation function of the hidden neurons, $\varphi(\cdot)$ is the activation function of the output layer, b_k is the bias of the kth hidden neuron, and b_p is the bias of the pth output neuron.

The mathematical explanation of the BPNN is

$$y_k = \varphi\left[\sum_{p=1}^{m} w_{kp}^{(2)} f\left(\sum_{k=1}^{l} w_{jk}^{(1)} x_j + b_k\right) + b_p\right], \quad p = 1, 2, \ldots, m. \tag{1.2}$$

The goal of training is to minimize the performance index

$$J = \frac{1}{2}\sum_{q=1}^{h} \left\| \hat{Y}_q - Y_q \right\|^2 = \sum_{q=1}^{h} J_q, \quad J_q = \frac{1}{2}\left\| \hat{Y}_q - Y_q \right\|^2, \tag{1.3}$$

where h is the number of training samples, $\hat{Y}_q$ is the predicted output, and Y_q is the target output.

The weights are updated using the following rule:

$$w(t+1) = w(t) - \eta \frac{\partial J}{\partial w(t)}, \tag{1.4}$$

where η is a learning rate, $w(t)$ is the weight at time t, and $w(t+1)$ is the weight at time $t + 1$. For the three-layer BPNN in Fig. 1.2, Eq. (1.3) becomes

$$w_{jk}^{(1)}(t+1) = w_{jk}^{(1)}(t) - \eta_1 \frac{\partial J}{\partial w_{jk}^{(1)}(t)}, \tag{1.5a}$$

$$w_{kp}^{(2)}(t+1) = w_{kp}^{(2)}(t) - \eta_2 \frac{\partial J}{\partial w_{kp}^{(2)}(t)}, \tag{1.5b}$$

where

$$\frac{\partial J}{\partial w_{jk}^{(1)}} = \sum_{q=1}^{h} \frac{\partial J_q}{\partial w_{jk}^{(1)}} = \sum_{q=1}^{h} \sum_{p} \frac{\partial J_q}{\partial Y_{qp}} \frac{\partial Y_{qp}}{\partial Out_k^{(2)}} \frac{\partial Out_k^{(2)}}{\partial In_k^{(2)}} \frac{\partial In_k^{(2)}}{\partial w_{jk}^{(1)}}, \tag{1.6}$$

$$\frac{\partial J}{\partial w_{kp}^{(2)}} = \sum_{q=1}^{h} \frac{\partial J_q}{\partial w_{kp}^{(2)}} = \sum_{q=1}^{h} \frac{\partial J_q}{\partial Y_{qp}} \frac{\partial Y_{qp}}{\partial In_p^{(3)}} \frac{\partial In_p^{(3)}}{\partial w_{kp}^{(2)}}. \tag{1.7}$$

In Eqs. (1.6) and (1.7), Y_{qp} is the actual output of the pth neuron, i.e., $Out_p^{(3)}$, with respect to the impact of the qth sample.

The BP updating algorithm is summarized as follows:

Step 1: Choose a stop criterion for error, ε. Prepare q $(q = 1, 2, \ldots, h)$ samples $(\hat{X}_q, \hat{Y}_q)$. Choose $\hat{X}_q$ as the input for the network.

Step 2: Calculate the objective function Eq. (1.3). If $J < \varepsilon$, stop; otherwise, go to Step 3.

Step 3: Calculate Eqs. (1.6) and (1.7).

Step 4: Calculate $\partial J/\partial w = \sum_{q=1}^{h} \partial J_q/\partial w$.

Step 5: Adjust weights using Eqs. (1.5a) and (1.5b). Go to Step 1.

2. RBFNN

An RBFNN (Fig. 1.3) is a kind of local approximation of an NN. Since only a few weights for a local input space affect the output, it has a high learning speed.

An RBFNN uses radial basis functions as activation functions. Its output is a linear combination of radial basis functions of the inputs and neuron parameters:

$$y_p = \sum_{p=1}^{m} w_{kp} \mu_k(x), \tag{1.8}$$

where m is the number of neurons on the output layer, w_{kp} is the weight connecting the kth neuron on the hidden layer to the pth neuron on the output layer, and $\mu_k(x)$ is the output of the kth neuron on the hidden layer:

$$\mu_k(x) = \mathrm{e}^{-\frac{(x-c_k)^{\mathrm{T}}(x-c_k)}{2\sigma_k^2}}, \quad k = 1, 2, \ldots, l, \tag{1.9}$$

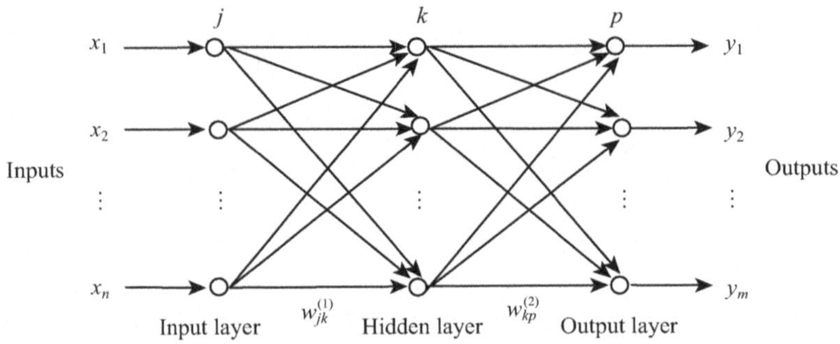

Fig. 1.3 Configuration of RBFNN

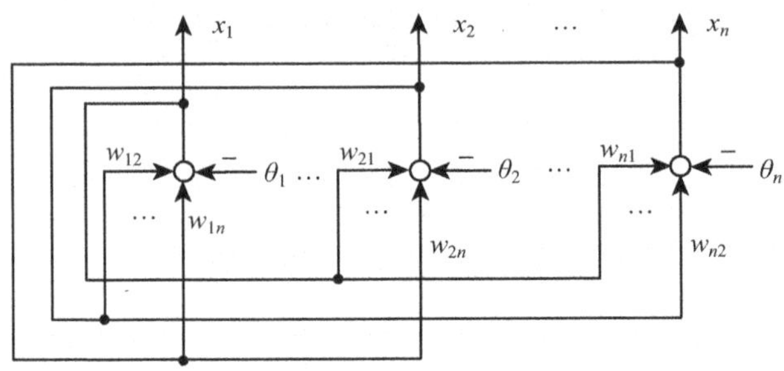

Fig. 1.4 Configuration of Hopfield neural network

where $x = [x_1, x_2, \ldots, x_n]^{\mathrm{T}}$ is the input sample, c_k is the kth center of the Gauss function, σ_k is the kth scale factor of the Gauss function, and l is the number of neurons on the hidden layer.

3. Hopfield NN

A Hopfield NN is a recurrent NN. It has two types: continuous and discrete. A discrete Hopfield NN (Fig. 1.4) is a single-layer network and has n units. The output of each unit is input to all the other units. Each unit does not have self-feedback. For each unit:

$$\begin{cases} s_i = \sum_{j=1}^{n} w_{ij} x_j - \theta_i, \\ x_i = f(s_i). \end{cases} \tag{1.10}$$

The update of a Hopfield network can be performed in asynchronous or synchronous ways. In an asynchronous way, only one unit is updated at a time, the state of other units remains unchanged. It is possible to set the updating order randomly or in a predefined scheme. The process is described by

$$\begin{cases} x_i(k+1) = f\left(\sum_{j=1}^{n} w_{ij} x_j(k) - \theta_i\right), \\ x_j(k+1) = x_j(k), \quad j \neq i. \end{cases} \tag{1.11}$$

In a synchronous way, all units are updated at the same time. This requires a central clock to the system to maintain synchronization. This process is described by

$$x_i(k+1) = f\left(\sum_{j=1}^{n} w_{ij} x_j(k) - \theta_i\right), \quad i = 1, 2, \ldots, n, \tag{1.12}$$

or in the following matrix form:

$$X(k+1) = f(WX(k) - \theta) = F(s), \tag{1.13}$$

where $X = [x_1, x_2, \ldots, x_n]^{\mathrm{T}}$ and $\theta = [\theta_1, \theta_2, \ldots, \theta_n]^{\mathrm{T}}$ are vectors, W is an $n \times n$ matrix that takes w_{ij} as its elements, and $F(s) = [f(s_1), f(s_2), \ldots, f(s_n)]^{\mathrm{T}}$ is a vector function with its elements being

$$f(s) = \begin{cases} 1, & s \geqslant 0, \\ -1, & s < 0. \end{cases}$$

1.3.2 Fuzzy Control

Fuzzy logic was first proposed by Lotfi A. Zadeh of the University of California at Berkeley in 1965 [17]. It has the advantage that a solution to a problem is described in a human way. This makes it possible to apply human experience in the design of a controller.

Fuzzy control is based on fuzzy logic. It uses fuzzy inference to make a control decision. The control system does not need a precise mathematical model and provides a control rule based on fuzzy relationships in a system. For this reason, it has been widely used to solve control problems in engineering practice [18, 19].

The key to a fuzzy control system is a fuzzy controller. There are four main parts in a fuzzy controller: a fuzzifier, an inference engine, a knowledge base, and a defuzzifier (Fig. 1.5).

A fuzzy set is established to quantify an element belonging to a set with a degree of membership. Fuzzifier determines the membership degrees of the inputs based on an antecedent fuzzy set. While an increase in the number of inputs can improve the prediction speed of control, it increases computational expense. In a typical fuzzy control system, the number of membership functions is 2–10, and the number of the elements in the fuzzy domain is 5–30.

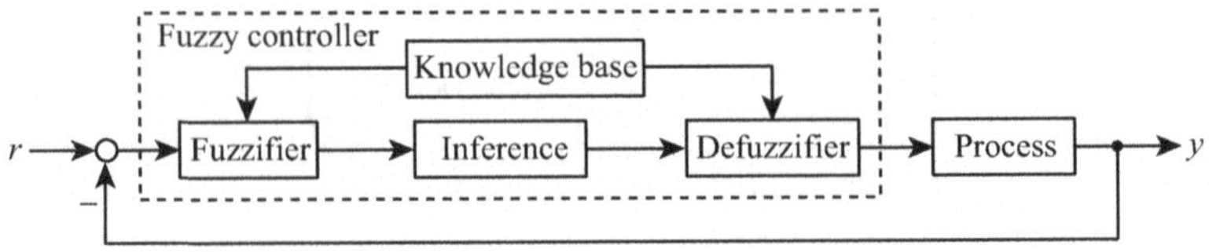

Fig. 1.5 Configuration of fuzzy control system

Quantification connects the mapping of the inputs to the fuzzy domain. Suppose the range for an actual input is $[a, b]$, its fuzzy domain is $\{-n, -n+1, \ldots, -1, 0, 1, \ldots, n-1, n\}$. The quantification factor is given by

$$K_{\mathrm{e}} = \frac{n-(-n)}{b-a}. \tag{1.14}$$

If the input is e_i, the corresponding element in the fuzzy domain is

$$E = \mathrm{INT}\left[K_{\mathrm{e}}\left(e_i - \frac{a+b}{2}\right) \pm 0.5\right]. \tag{1.15}$$

The knowledge base consists of a database and a rulebase. The database stores the vectors of the membership functions of fuzzy subsets for all input and output variables, and the rulebase stores fuzzy control rules for the inference engine.

The inference engine makes a fuzzy decision for each of the input based on the knowledge stored in the rules. The commonly used inference methods are Mamdani method, Zadch method, Baldwin method, Yager method, and Tsukamoto method. The decisions produced by the inference engine are integrated and sent to the defuzzifier.

The defuzzifier calculates the value of the crisp signal from the outputs of the fuzzy controller. The commonly used methods are the maximum-membership-degree method, the centroid method, the maximum-height method, the maximum-average-value method, and the area method.

1.3.3 Expert Control

An expert system (ES) is a knowledge-based system that emulates the decision-making ability of a human expert. The field of ES is one of the most important application areas of the artificial intelligence (AI). It was developed by the AI community in the mid-1960s [20]. Several decades of development of ES pushes up the theory and technology greatly and ensures its wide applications. Expert control is a typical application of the ES in control area. It uses the reasoning engine of an ES to flexibly select control methods, and to yield the combination of parsing rules

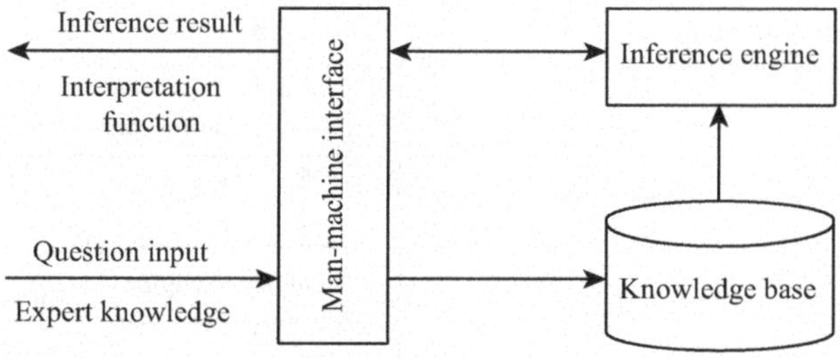

Fig. 1.6 Configuration of ES

and heuristics, and the combination of knowledge models and control models. This makes it possible to achieve satisfactory control performance.

An ES contains a large number of specialized knowledge and experience. It solves problems by imitating a human expert.

An ES (Fig. 1.6) has three parts: a knowledge base, an inference engine, and a man–machine interface. Compared with other systems, the ES has some remarkable features. For example, it has the ability to deal with ill-structured problems; it is mainly based on knowledge processing; and it is of great transparency, flexibility, and expandability.

The knowledge base is the core of an ES. It has two parts: one is data information related to the problems, and the other is general knowledge and experience used in the reasoning processing. The knowledge and experience are represented as

$$\textbf{IF} \;\; \text{condition} \;\; \textbf{THEN} \;\; \text{action}, \tag{1.16}$$

where the condition part is called the antecedent, which represents the state of the system, and the action part is called the consequent, which represents the conclusion. When an ES is constructed based on production rules, it is called a rule-based system.

In order to build a knowledge base, we have to solve the problems of knowledge acquisition and knowledge representation. Knowledge acquisition refers to how to acquire specialized knowledge and experience from experts, and knowledge representation is how to express and store knowledge in a form that computers can be understand.

The inference engine applies logical rules to the knowledge and experience in the knowledge base, and deduces a conclusion.

The man–machine interface is the interface between human and the system. On the one hand, an operator of an ES needs to identify the input problem, data, and other information. And the information has to be transformed into an internal representation of the system. On the other hand, the operator is required to explain the results and the process of reasoning in an easy-to-understand way. The interface ensures the communication between people and the system.

An expert control system (ECS) uses the concepts and techniques of an ES to build a control system by imitating control experts' knowledge and experience. It is

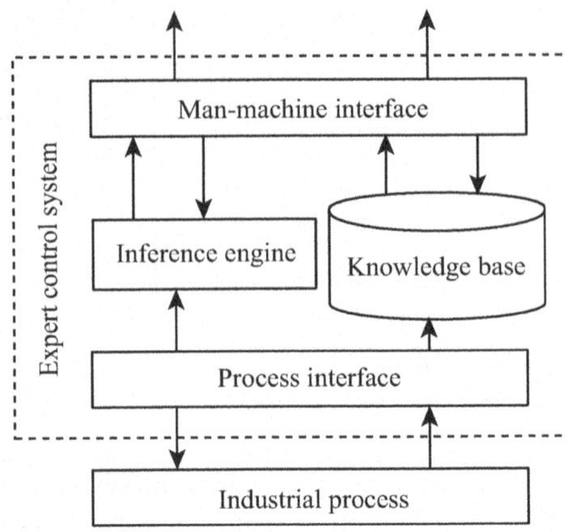

Fig. 1.7 Simplified structure of ECS for industrial process

quite different from a general ES for the following reasons. First, while a general ES usually works in an offline fashion to inquire and/or interpret problems in special fields, an ECS needs to collect online information and to carry out control in a real-time fashion. Second, a general ES performs reasoning based on knowledge and presents a conclusion in a natural language. But an ECS usually performs reasoning based on models and carries out control using an automatically selected control law.

A simplified structure of an ECS for an industrial process is shown in Fig. 1.7. It is the extension of an ES. In addition to a knowledge base, an inference engine, and a man–machine interface, it contains a knowledge base and a process interface. The database stores process data, contains a working memory, and saves the information of process state from the process interface for the "IF" parts in the "IF-THEN" rules. The process interface has two kinds: measurement interface and control interface. The measurement interface provides the ECS the process state and data, and the control interface transforms a reasoning result to a control action.

An ECS can be simplified as an expert controller so as to meet the requirements of control performance, reliability, and real time. It can also be built with a hierarchical structure for a complex process. According to the function of an ECS in a control system, we can divide ECSs into two categories: a direct ECS (Fig. 1.8) and an indirect ECS (Fig. 1.9) as follows:

(1) Direct ECS: An expert controller directly produces an input to control a plant. It carries out inference using the inference engine based on the measured process information and the rules in the knowledge base, and produces a control signal at each sampling time. Clearly, the expert controller is directly contained in a control loop. A control signal is generated by the expert controller at each sampling time so as to guarantee normal running.
(2) Indirect ECS: A distinguishing feature of an indirect ECS is that it separates a basic controller or a control device from an expert controller. The basic controller may use PID control, fuzzy control, pole assignment, and model reference adaptive control. And the expert controller performs self-correction and/or self-

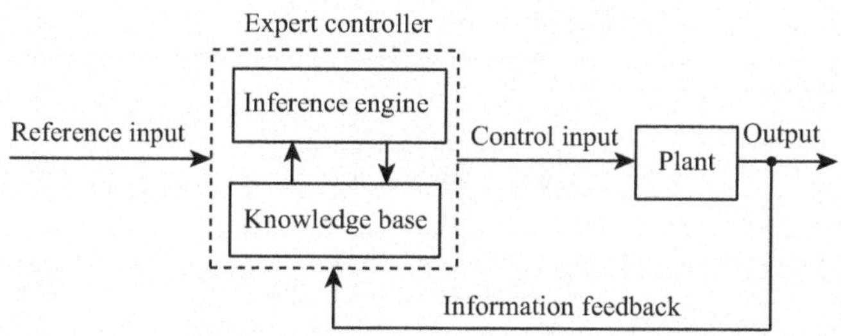

Fig. 1.8 Configuration of direct ECS

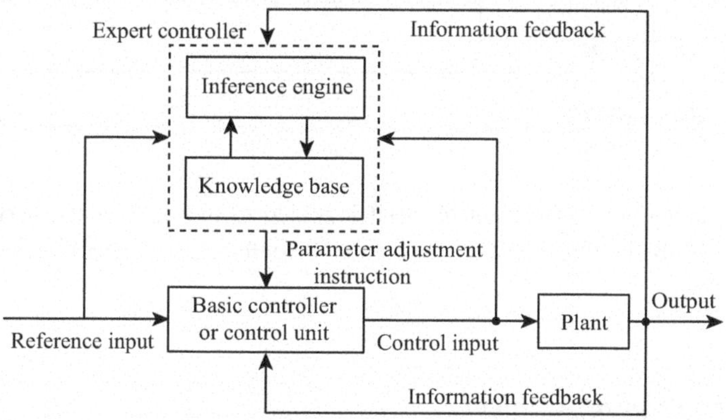

Fig. 1.9 Configuration of indirect ECS

regulating for the adjustment of parameters. The ECS only uses some heuristic knowledge-based rules to ensure the normal conditions of functions and algorithms so as to meet control requirements.

A typical example of an indirect ECS is an expert PID control system. It combines the ECS technology and the conventional PID control. An expert controller is used to adjust a conventional PID controller according to the state of a plant. This ensures that the system has a self-adaptive ability for the change of environment and yields good control performance. Except for a database and an inference engine, the expert controller also contains a mechanism for information acquisition and processing and a mechanism for control. The former guarantees the collection of online data on request, and the latter ensures effective control of the plant and timely adjustment of the parameters of the controller according to reasoning results.

A model-based ECS (Fig. 1.10) makes the best use of conventional control methods by incorporating experts' knowledge and experience.

The inference process of a model-based ECS for producing an appropriate control input is as follows. First, it collects the information of the process state. Then, it produces a control input using reasoning rules and knowledge in the knowledge base. Finally, it predicts the control performance of the plant for the control input

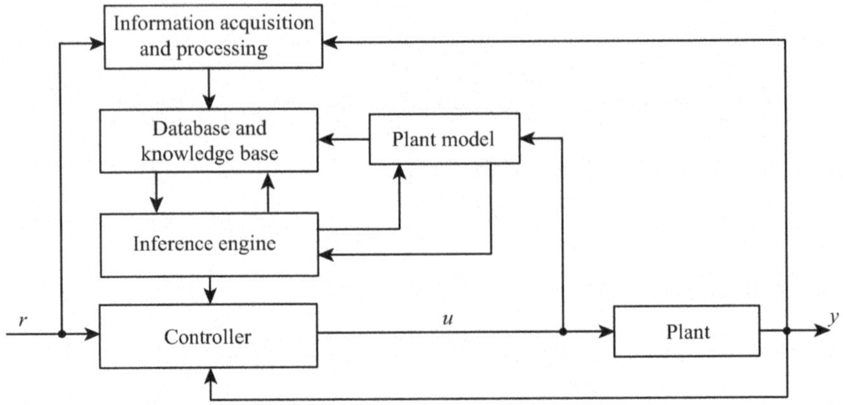

Fig. 1.10 Model-based ECS

using the plant model. The control input is added to the plant if the performance meets requirements. Otherwise, it performs reasoning again to yield a satisfactory control input.

1.3.4 Decoupling Control

One of the main characteristics of a multivariable system is coupling, i.e., a change in one variable affects other variables [21]. This causes coupling between control loops in a multi-input, multi-output (MIMO) control system. It is a widely used strategy to design a decoupling controller to decouple interactions between control loops. This allows us to design a single-input, single-output (SISO) control system for each single loop to achieve good control performance.

Take the control of a two-input, two-output (TITO) control system (Fig. 1.11) as an example. Coupling between the two control loops often results in oscillations for conventional control without decoupling. This hinders us from settling the system quickly.

The transfer function matrix of the TITO plant is

$$G(s) = \begin{bmatrix} G_{11}(s) & G_{12}(s) \\ G_{21}(s) & G_{22}(s) \end{bmatrix}. \tag{1.17}$$

The relationship between the inputs and the outputs of the plant is

$$\begin{bmatrix} Y_1(s) \\ Y_2(s) \end{bmatrix} = G(s) \begin{bmatrix} U_1(s) \\ U_2(s) \end{bmatrix}. \tag{1.18}$$

Note that $X(s)$ is the Laplace transform of $x(t)$, and

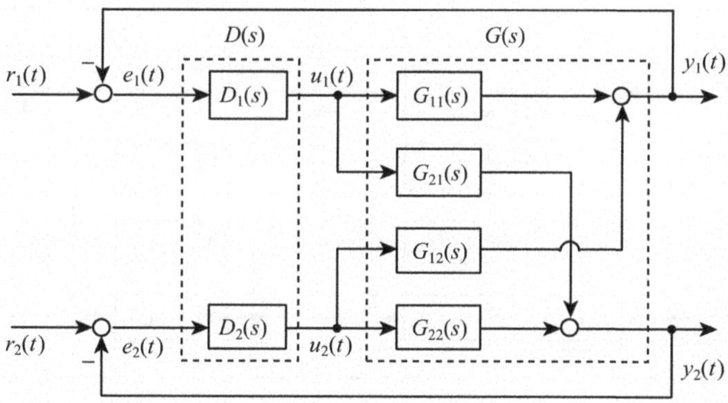

Fig. 1.11 Control system for TITO plant without decoupling

$$\begin{bmatrix} U_1(s) \\ U_2(s) \end{bmatrix} = \begin{bmatrix} D_1(s) & 0 \\ 0 & D_2(s) \end{bmatrix} \begin{bmatrix} E_1(s) \\ E_2(s) \end{bmatrix} = D(s) \begin{bmatrix} E_1(s) \\ E_2(s) \end{bmatrix}, \tag{1.19}$$

where

$$D(s) = \begin{bmatrix} D_1(s) & 0 \\ 0 & D_2(s) \end{bmatrix}.$$

The open-loop transfer function matrix of the system is

$$G_k(s) = G(s)D(s), \tag{1.20}$$

and the closed-loop transfer function matrix of the system is

$$\Phi(s) = [I + G_k(s)]^{-1} G_k(s), \tag{1.21}$$

where I is an identity matrix.

The goal of decoupling control is to make the closed-loop control system matrix becomes a diagonal matrix

$$\Phi(s) = \begin{bmatrix} \Phi_{11}(s) & 0 & \cdots & 0 \\ 0 & \Phi_{22}(s) & \cdots & 0 \\ \vdots & \vdots & & \vdots \\ 0 & 0 & \cdots & \Phi_{nn}(s) \end{bmatrix}, \tag{1.22}$$

so that the control loops are independent with each other. In order to achieve this, we need to insert a decoupling compensator, $F(s)$ (Fig. 1.12), on the inputs of the plant. Equation (1.21) shows that decoupling is achieved when the open-loop transfer function matrix $G_k(s)$ is diagonal.

After inserting a decoupling compensator on the inputs of the plant, the open-loop transfer function matrix becomes

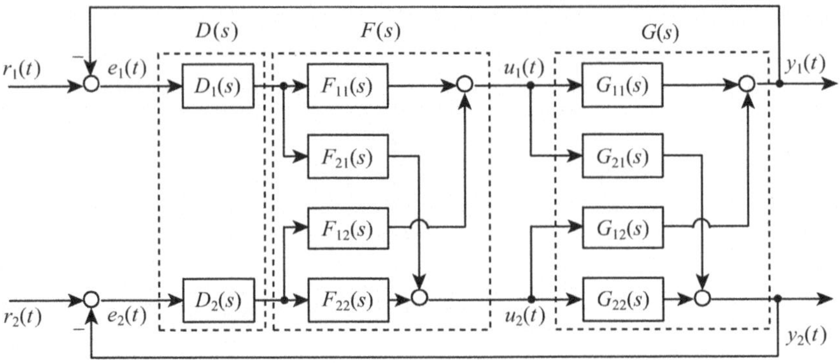

Fig. 1.12 Configuration of decoupling control system

$$G_{kF}(s) = G(s)F(s)D(s), \tag{1.23}$$

where

$$F(s) = \begin{bmatrix} F_{11}(s) & F_{12}(s) \\ F_{21}(s) & F_{22}(s) \end{bmatrix}.$$

Letting

$$\begin{bmatrix} G_{11}(s) & G_{12}(s) \\ G_{21}(s) & G_{22}(s) \end{bmatrix} \begin{bmatrix} F_{11}(s) & F_{12}(s) \\ F_{21}(s) & F_{22}(s) \end{bmatrix} = \begin{bmatrix} \bar{G}_{11}(s) & 0 \\ 0 & \bar{G}_{22}(s) \end{bmatrix} \tag{1.24}$$

gives the decoupling compensator

$$\begin{bmatrix} F_{11}(s) & F_{12}(s) \\ F_{21}(s) & F_{22}(s) \end{bmatrix} = \begin{bmatrix} G_{11}(s) & G_{12}(s) \\ G_{21}(s) & G_{22}(s) \end{bmatrix}^{-1} \begin{bmatrix} \bar{G}_{11}(s) & 0 \\ 0 & \bar{G}_{22}(s) \end{bmatrix}. \tag{1.25}$$

This gives a diagonal matrix synthesis method for decoupling (Fig. 1.12).

Nowadays, there are mainly three types of decoupling control methods:

(1) Conventional decoupling method: It contains frequency-domain and time-domain methods, and has mainly applied to linear MIMO systems. Typical methods are the inverse-matrix method, the relative-gain-array method, the diagonally dominant-matrix method, and the state-feedback method.
(2) Adaptive decoupling method: It combines the adaptive control and decoupling control for an MIMO system to perform adaptive decoupling control. The method tries to make the transfer function of the closed-loop system a diagonal matrix and handles coupling signals as disturbances. The idea of optimal control is used in adaptive decoupling. This is the essential difference between the adaptive decoupling control and conventional decoupling control.
(3) Intelligent decoupling method: It mainly means decoupling control based on an NN or fuzzy control. Since an NN can approximate any functions with arbitrary precision and has the self-learning ability, it is mainly used for time-varying,

nonlinear, and unknown plants. Fuzzy decoupling control is usually used when there is no explicit mapping relationship between the inputs and outputs of a plant [22].

The conventional and adaptive decoupling methods are relatively simple and are easy to implement. Therefore, they are suitable for a linear system with an accurate mathematical model. On the other hand, the NN decoupling method does not need an accurate mathematical model. Some results have been reported for a class of nonlinear systems, but most of them were carried out by trial and error, and only simulation results were presented. The NN decoupling control usually has three forms: ① an NN decoupling compensator is placed between a plant and a controller; ② an NN decoupling compensator is placed before a controller; and ③ an NN decoupling compensator is placed in a feedback loop. An NN decoupling compensator usually is a feedforward NN with three layers and is trained by the BP algorithm. A learning algorithm is used to improve the convergence speed. And a GA is used to avoid local minimum. However, the complexity of implementation and strong direction restricts its applications.

Similar to the NN decoupling, fuzzy decoupling does not need an accurate mathematical model of a plant as well. It only requires preparing a fuzzy decoupling rule table, or in other words, an ES. The table contains rules acquired from experts for decoupling. When decoupling control is carried out, current measured data, facts, and evidence are stored in a database. A rule-application model (inference engine) selects and implements rules, and performs fuzzy reasoning and decoupling. Fuzzy decoupling has been successfully applied in many MIMO systems. The key to success is the simple design of a decoupling controller and the avoidance of difficult theoretical derivation. Since it solves the problem using fuzzy rules, the design of the rules should take into account the coupling among variables carefully.

1.3.5 Hierarchical Intelligent Control

Hierarchical intelligent control (HIC) was developed based on AI, adaptive and self-organizing control, operational research, and others. It is one of the earliest theories in intelligent control. The HIC theory has two kinds. One was proposed by Saridis. It is based on the principle of increasing precision with decreasing intelligence (IPDI) and has three levels. The other was presented by Villa. It is based on knowledge description and mathematical analysis, and has two levels.

An HIC system has many controllers, but each controller at the lowest level controls only one subsystem. A controller receives information from a controller or a decision unit at the upper level to regulate a controller or a subsystem at the lower level.

Conflict between controllers is coordinated by a controller or a coordinator at the upper level. The task of hierarchical coordination control is to coordinate, cooperate, and reinforce with, and to restrict to each other so as to fulfill subtasks and achieve subgoals for subsystems. This ensures the achievement of the overall goal of the whole system.

An example of a three-level HIC is shown in Fig. 1.13. It has an organization level, a coordinate level, and an executive level. C is input commands, U is the classified input signal, f_1 is an offline feedback signal from the coordinate level to the organization level, and f_2 is an online feedback signal from the executive level to the coordinate level.

1. Organization Level

The organization level is the highest level of the HIC system. It is the brain of the system and imitates human behavior and function. The organization level arranges tasks according to incomplete description of a user and the relevant information of the actual process and environment, and offers appropriate control modes to the lower level to achieve the desired control objectives. Its main functions are as follows:

(1) Reasoning: Build a connection between different basic actions with received input commands by inference rules. Produce control goals and relevant activities. And evaluate each action based on the probability theory.
(2) Planning: Sort actions. And use entropy to calculate the uncertainties of activities.
(3) Decision-making: Select a plan with the maximum possibility, or in other words, with the minimum total entropy. The plan is a fully ordered sequence of activities that are tasks or orders assigned to the coordinate level.
(4) Update: Select information feedback from the lower level after each task is executed and evaluated, and the related probability is updated by a kind of learning algorithm. Also, update the contents in a long-term memory.

2. Coordination Level

The coordination level is the second level of the HIC system. The task of this level is to coordinate the control actions of controllers and the execution of subtasks. This level can be further divided into two layers: a control management layer and a control supervision layer. The control management layer decides how to completes tasks given by the organization level based on the information collected from the lower level, and generates control commands to the lower level. The control management layer ensures the normal operations of the controllers in the executive level and locally performs parameter tuning and performance optimization.

The coordinate level usually has a number of coordinate controllers. Each coordinate controller receives commands from organization level and, at the same time, is responsible for the coordination of multiple controllers at the executive level. It is an interface between the organization level and the executive level. While its operation accuracy is relatively low, it has a high decision-making and learning ability.

3. Executive Level

The executive level is the lowest level in the HIC system. It is composed of many hardware controllers. This level only executes specified control tasks. It does not need the ability of decision, reasoning, and learning. While it sends control inputs to

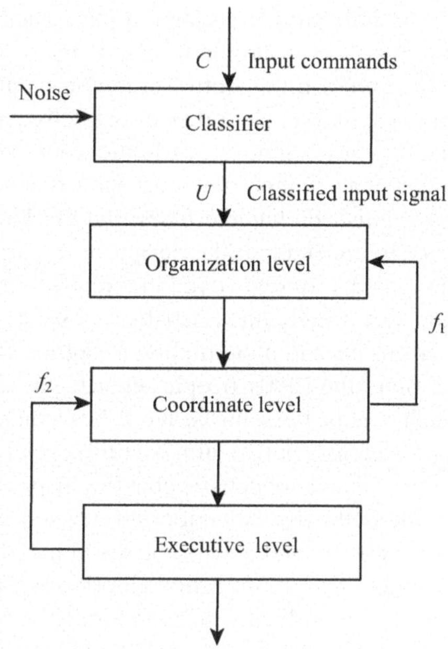

Fig. 1.13 Structure of HIC

a plant, it sends the information detected from sensors back to the higher level in a real-time fashion.

Entropy is used as a measure at all levels in a Saridis' HIC system. It is used at the organization level to measure the requirement of knowledge; at the coordination level to measure the uncertainty of coordination; and at the executive level to calculate the cost of control actions, which is equivalent to the energy consumption of the system. The total entropy is the sum of the three levels. It can be used to represent the total cost of control. The principle of designing and constructing an HIC system is to optimize the total entropy.

In summary, the basic procedure of building an HIC system is as follows. First, build a structure according to the principle of IPDI. Then, design an intelligent controller that has a proper sequence of decision and control, and optimize the total entropy of the system. In this way, the HIC system yields the optimal control of a complex and uncertain system under the guidance of the organization level.

1.3.6 Intelligent Optimization Algorithms

Optimization is an important subject in the fields of scientific research, engineering technology, and economic management. For example, how to select parameters to meet requirements in engineering, as well the cost could be reduced, or how to allocate limited resources for a complex engineering process as good economic benefits is guaranteed.

Optimization is a classic topic. There are two development directions to solve an optimization problem. One is based on functional analysis. It gives strict theoretical proof for an optimization problem, and provides a sophisticated algorithm to find a solution. The other uses intelligent algorithms that are devised based on the intelligent phenomena of biotic population in the natural world. While intelligent algorithms cannot guarantee the existence of an optimal solution, they have some distinguished characteristics: the mechanisms of the algorithms are plain, they are easy to understand; the design of the algorithms is simple, there are no special requirements for objective functions, it is easy to write programs for those algorithms, and they usually find a satisfactory solution within an acceptable time span. For these reasons, research on intelligent optimization is widely discussed lately.

Since the 1990s, people began to pay close attention to the facts that social species, such as ants, bees, birds, and fishes, complete tasks of foraging, cleaning, transportation, defense, build solid, beautiful, rigorous nests, and perform many amazing tasks with efficient collective behavior. Inspired by this, people simulated the mechanisms of the collective behavior and devised intelligent optimization algorithms to solve conventional and new problems in practical applications.

Intelligent optimization algorithms have unique advantage for combinatorial optimization problems that can hardly be handled by conventional optimization techniques. Now, common intelligent optimization algorithms include the GA, the PSO, and the ant colony optimization (ACO). These algorithms have greatly enriched modern optimization technology and have provided a practical solution for the optimization problems that conventional optimization technologies are difficult to deal with.

Since complex metallurgical processes have strong nonlinearity, time-varying parameters, and coupling between operating states, conventional optimization methods are difficult to meet optimization requirements. Intelligent optimization methods have gradually become a powerful tool for the optimization of complex metallurgical processes in recent years. We mainly summarize GAs, ACO, and fuzzy satisficing degree.

1. Genetic Algorithm

A GA is a randomized adaptive search method for global optimization. It mimics some of the biological processes observed in natural evolution [23]. The algorithm was first proposed by Professor John H. Holland of the University of Michigan in a monograph entitled *Adaptation in Natural and Artificial Systems* in 1975 [24]. He and his students conducted a large number of numerical experiments on cellular automata in the 1970s. After that, Dr. David E. Goldberg summarized a series of the research work and presented the basic framework of GA.

The basic procedure of GAs is as follows.

First, we express decision variables in a problem space as a chromosome in a genetic space using a certain coding method and convert the value of the corresponding target function of a decision variable to the value of fitness. Since biological evolution takes groups as the main body, we group individuals in a GA to a generation. Similar to a natural evolutionary process of a biological generation, the operation of a GA is an iterative process. In the process, genetic and evolutionary operation, which is based on the rule of survival of the fittest, generates the next generation of the population that has higher fitness than the present one. Repeating the process finally produces a good individual, which is an optimal or suboptimal solution.

The evolutionary process of organisms is carried out mainly through crossover between and mutation of chromosomes. The search process of an optimal solution using a GA also mimics this process and uses three genetic operators (selection, crossover, mutation) to form a new generation. The selection operator implements the principle of survival of the fittest, that is, an individual in the current generation is selected to be one for later breeding by a probability in proportion to its fitness. This forms a mating pool (an intermediate group between current and next generations). It improves the average fitness of the next generation. Note that this operation does not improve the fitness of the best individual in the group or produce new individuals. The crossover operator produces new individuals. It first selects two parents randomly from the mating pool. Then, it combines the genes of parents in a way to produce a child. The mutation operator alters one or more gene values in a chromosome according to a low mutation probability. New individuals are mainly produced by the crossover operator, which determines the global search ability of a GA. But some new individuals are produced by the mutation operator, which is only an auxiliary method to maintain genetic diversity. This operator is essential because it determines the local search ability of a GA. Crossover and mutation cooperate to complete the global and local search in a search space.

The steps of a GA are as follows:

Step 1: Problem description: Describe an optimization problem using chromosomal representation.

Step 2: Initialization: Set stop criteria of the algorithm. Generate M individuals for the initial generation, $P(0)$, randomly.

Step 3: Evaluation of individuals: Calculate the fitness of each individual in $P(t)$.

Step 4: Judgement: If the current solution satisfies the stop criteria, finish; otherwise, go to Step 5.

Step 5: Selection: Apply the selection operation to the generation $P(t)$.

Step 6: Crossover: Apply the crossover operation to the generation $P(t)$.

Step 7: Mutation: Apply the mutation operation to the generation $P(t)$ and produce the next generation $P(t+1)$.

Step 8: Go to Step 3.

Compared with conventional optimization algorithms, GAs have the following advantages: Start search from not a single point, but a group of points. In the process of searching an optimal solution, only the information of fitness, which is converted from an objective function, is needed; and the other information, such as derivatives, is not needed. The search process is not easy to fall into a local minimum.

Although GAs have above advantages, they are essentially a randomized heuristic search method. So, it also has its own limitations, such as premature convergence, weak local search capability, and the nondirectional problem of genetic operators.

In general, GAs have two characteristics: implicit parallelism and global search ability. As a powerful and widely used method of random search and optimization,

GAs are considered to be one of the most widely affected evolutionary computational methods today. Over the last decades, GAs have produced convincing results in the applications of complex industrial optimization problems. Successful application examples include function optimization, machine learning, combinatorial optimization, training of an NN, an expert system for automatic programming, job scheduling, reliability design, vehicle routing and grouping, and equipment arrangement and allocation.

2. Ant Colony Optimization

Job-shop scheduling, process scheduling, and/or fault diagnosis are often required in the production processes of metallurgical industry. They can be settled by ACO.

ACO is also a randomized heuristic search method of optimization. Inspired by the research on the foraging behavior of real ant colonies in the 1950s, Dr. Dorigo et al. initially proposed this method [25, 26]. It mimics collaboration in an ant colony to lay down pheromone trails for a solution, improves the quality of the solution by leaving and exchanging information on the trails, and finally finds an optimal solution.

The basic principle of ACO is as follows:

Ants wander randomly to search for food. When an ant finds a source of food, it walks back to the colony and lays down a pheromone trail showing the path to food. However, the pheromone on trails evaporates over time. When other ants come across pheromone trails, they follow a path with higher pheromone density at a higher probability. Also, a shorter path is usually followed more frequently, and thus has higher pheromone density. The positive-feedback mechanism eventually leads to that all ants follow a single path, that is, the convergence to an optimal path.

ACO can be used to solve combinatorial optimization problems. There are basically two kinds of combinatorial optimization problems [27]:

(1) Static combinatorial optimization problems: They are the problems in which the cost function does not change during the execution of a solving algorithm. Those problems include the traveling salesman problem (TSP), the quadratic assignment problem, the job-shop scheduling problem, the vehicle routing problem, the sequencing problem, and the graph coloring problem.
(2) Dynamic combinatorial optimization problem: They are the problems in which the cost function changes during the execution of a solving algorithm. Those problems include route planning in transportation networks and routing in ad hoc wireless networks.

A job-shop scheduling problem in metallurgical processes can be converted to a TSP. So, we take this problem as an example to introduce a mathematical model of the basic ACO as follows.

Let $C = \{c_1, c_2, \ldots, c_n\}$ be a set of n cities; n be the size of TSP; and $L = \{l_{ij} | c_i, c_j\}$ be a set of edges l_{ij} $(i, j = 1, 2, \ldots, n, i \neq j)$ between two cities in C, which is the Euclidean distance between Cities i and j

$$l_{ij} = \sqrt{(x_i - x_j)^2 + (y_i - y_j)^2}. \tag{1.26}$$

Let $b_i(t)$ be the number of ants in City i at time t; m be the total number of ants, which is given by $m = \sum_{i=1}^{n} b_i(t)$; and $\Gamma = \{\tau_{ij}(t)|c_i, c_j \subset C\}$ be the residual pheromone on l_{ij} connected Cities i and j in C at time t, which is the same to that on the edge at initial time.

When Ant k $(k = 1, 2, \ldots, m)$ wanders, it selects its direction based on the pheromone density on the edges. We use a tabu list, $tuba_k$ $(k = 1, 2, \ldots, m)$, to record the cities that Ant k has traveled.

Let the heuristic information of edge l_{ij} be

$$\eta_{ij} = \frac{1}{l_{ij}}. \tag{1.27}$$

Then, Ant k moves from City i to City j at time t with the probability

$$p_{ij}^{(k)}(t) = \begin{cases} \dfrac{[\tau_{ij}(t)]^{\alpha}[\eta_{ij}]^{\beta}}{\sum\limits_{j \in allowed_k} [\tau_{ij}(t)]^{\alpha}[\eta_{ij}]^{\beta}}, & j \in allowed_k, \\ 0, & j \notin allowed_k, \end{cases} \tag{1.28}$$

where $\tau_{ij}^{(k)}(t)$ is the pheromone on the edge between (i, j), α is a parameter to control the influence of τ_{ij}, β is a parameter to control the influence of η_{ij}, and $allowed_k$ is the cities allowed to be visited by Ant k in the next step.

The pheromone is updated after an ant has moves one step and also has completed the travel of all the n cities by applying evaporation. For example, the amount of pheromone on the edge (i, j) at time $t + n$ is

$$\tau_{ij}(t + n) = (1 - \rho) \cdot \tau_{ij}(t) + \rho \Delta \tau_{ij}(t), \tag{1.29}$$

$$\Delta \tau_{ij}(t, t + n) = \sum_{k=1}^{m} \Delta \tau_{ij}^{(k)}(t, t + n), \tag{1.30}$$

where ρ is an evaporation rate and satisfies $\rho \subset [0, 1)$, $1 - \rho$ is a residual factor, $\Delta \tau_{ij}^{(k)}(t)$ is the amount of pheromone laid on edge (i, j) by Ant k, and $\Delta \tau_{ij}^{(k)}(t, t + n)$ is the pheromone increment on the edge (i, j) in one complete tour.

Dorigo proposed three different models for ACO: ant-cycle, ant-quantity, and ant-density. The only difference between them is the way of updating $\Delta \tau_{ij}^{(k)}(t)$.

In the ant-cycle model,

$$\Delta \tau_{ij}^{(k)}(t, t + n) = \begin{cases} \dfrac{Q}{L^{(k)}}, & \text{if Ant } k \text{ goes from } i \text{ to } j \text{ in this tour,} \\ 0, & \text{otherwise,} \end{cases} \tag{1.31}$$

where Q is the intensity of pheromone that affects the convergence speed of the algorithm in some degree, and $L^{(k)}$ represents the total length of the path that Ant k traveled in this tour.

In the ant-quantity model,

$$\Delta\tau_{ij}^{(k)}(t) = \begin{cases} \dfrac{Q}{d_{ij}}, & \text{if Ant } k \text{ goes from } i \text{ to } j \text{ between } t \text{ and } t+1, \\ 0, & \text{otherwise.} \end{cases} \quad (1.32)$$

In the ant-density model,

$$\Delta\tau_{ij}^{k}(t) = \begin{cases} Q, & \text{if Ant } k \text{ goes from } i \text{ to } j \text{ between } t \text{ and } t+1, \\ 0, & \text{otherwise.} \end{cases} \quad (1.33)$$

Note that, while the ant-quantity and ant-density models use the local information to update local pheromone trails after one step, the ant-cycle model uses the overall information to update the entire pheromone trails after a complete tour.

The implementation steps of the ACO are as follows:

Step 1: Initialization. Let $t = 0$ and the iteration number $nc = 0$. Select a maximum iteration number $nc_{\max}$. Place $b_i(t)$ ants on Node (Cite) i for $i = 1, 2, \ldots, n$. Assign a constant to each $\tau_{ij}(t)$ for every i and j. And let $\Delta\tau_{ij}(t) = 0$.

Step 2: Update the iteration number: $nc = nc + 1$.

Step 3: Set the index of the tabu list to be $k = 1$.

Step 4: Set the number of ants to be $k = k + 1$.

Step 5: Calculate the probability of an ant that it moves to Node (City) j ($j \in \{C - tuba_k\}$) and move ahead.

Step 6: Update the tabu list by moving the ant to Node j and moving Node j to the tabu list of the ant.

Step 7: If $k < m$, that is, the nodes in C are not completely traveled, then go to Step 5. Otherwise, go to Step 8.

Step 8: Update the amount of pheromone on each edge.

Step 9: If $nc = nc_{\max}$, output the result. Otherwise, clear the tabu list and go to Step 3.

ACO is a distributed positive-feedback parallel algorithm. It has the advantages of strong convergence and robustness, and is easy to be combined with other heuristic algorithms. However, ACO has some disadvantages, mainly including long searching time and stagnation. Convergence performance is also sensitive to the setting of initial parameters. Stagnation is a phenomenon that all solutions become the identical one after searching to a certain extent. This makes it impossible to continue searching for a better solution.

As a relatively perfect and systematic tool, optimization theory has made great contribution to the development of human society. However, the theory also has limitations and is difficult to solve those like high-cost problems, problems without requiring exact solutions, and some real-time problems. Complex metallurgical processes are difficult to establish precise mathematical models. Various disturbances in industrial environment give big influence on production. And along with the increase in the scale of complex industrial processes, it is impossible to control all parameters in the whole production processes. Instead, we try to establish performance indexes according to production requirements, and to optimize these indexes in dynamic environment as much as possible.

3. Fuzzy Satisfying Degree

One of the Nobel Prize winners, H. A. Simon, presented the idea of satisficing in his study of decision-making in 1978. He thought that a satisficing solution should be used instead of a maximum of the conventional meaning in some cases. The proposition of satisficing guides us from pure rational thinking to the state of limited rationality, which provides us a new way to solve decision-making problems. Satisficing is the product of the combination of satisfaction theory and optimization theory. Its core idea is that, in the process of finding an optimal solution in an optimization problem, it does not emphasize to find an optimal solution, but to seek a satisfactory solution for a realistic world. Satisficing ensures a solution for the decision-making under circumstances in which an optimal solution cannot be determined.

After the germination and theoretical research stages, satisficing gradually received attention in several areas, including optimization, control, management, decision-making, resource allocation, and task scheduling.

A satisfying degree and a satisfactory solution are the core of satisficing. They have many definitions, mainly based on the definition of fuzzy mathematics, on the definition of linear satisfaction, and on the fuzzy characteristics of neuron calculation. The related concepts of fuzzy satisfying degree (FSD) are explained as follows.

S is a fuzzy set on $u \in [u_{\min L}, u_{\max L}]$. For an element in $[u_{\min L}, u_{\max L}]$, the value $\mu_s(u) \in [0, 1]$ is called the membership degree of u in S. The membership function (that is, satisficing) is denoted by

$$S[u_{\min L}, u_{\max L}] \to [0, 1], \quad u \to S(u). \tag{1.34}$$

Unsatisfying is denoted by

$$\bar{S}(u) = 1 - S(u). \tag{1.35}$$

Main shapes for satisficing are shown in Fig. 1.14.

If a system has q variables, and the satisficing of Variable i is S_i, then the overall satisficing for the system is

$$S = \sum_{i=1}^{q} \gamma_i S_i, \quad \sum_{i=1}^{q} \gamma_i = 1, \tag{1.36}$$

where γ_i is a weight for S_i that reflects the importance of the variable.

A satisfactory solution defined by P. Pirjanian is as follows [28]. A satisficing set of vector optimization problems is a subset of the feasible solutions in which each

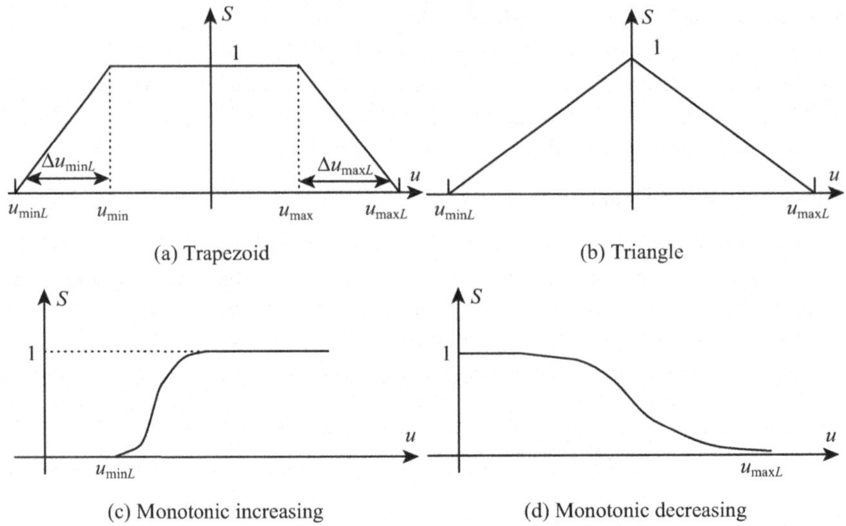

Fig. 1.14 Shapes of satisficing

target is higher than a desired level (or an objective function), where all elements are satisfactory solutions.

The development of satisficing up to present has formed the theory with the concept of satisficing, satisfactory solution, and satisfactory control, and has received large attention in many fields. It has effectively solved optimization problems that could not be solved by conventional optimization theory. However, compared with the theory of optimization and control, it is far from mature and has the following problems:

(1) As a more general theory rather than the conventional optimization theory, the systematization of knowledge of the theory needs to be completed. For example, we need to complement the principle of satisficing and the related concepts, establish an axiomatic system with satisficing theory, and build standards of applying the theory.
(2) The system for solving problems using satisficing is not enough. In particular, how to provide a systematic way for the representation and establishment of satisficing is an urgent task. Satisficing has rich connotations. This gives a variety of design methods of satisficing for different problems. It is important to present a general way to systematically build a reasonable satisficing and to find a satisfactory solution.
(3) Most studies use satisficing on the feasible region of a problem. However, generally speaking, finding feasible solutions itself is a difficult task. And feasible solutions do not fill to the full of the whole solution space for many problems. Sometimes, a feasible region is a non-simply-connected space, or feasible solutions are difficult to describe. Therefore, how to determine a feasible region is the

first problem to be solved. And how to determine a feasible region is particularly important for some specially nonconvex or discontinuous problems.
(4) Satisficing follows humanistic idea to some extent. So, how to reflect human preference in applications is one of the key problems.

Generally speaking, satisficing has the characteristics of universality, fuzziness, intelligence, and relativity. It is suitable to reflect the human nature. For this reason, satisficing not only includes optimal principle, but also is able to be used for an inference engine, knowledge acquisition, artificial intelligence, pattern recognition, management engineering, and reliability engineering.

1.4 Outline of This Book

This book focuses on the intelligent control of complex metallurgical processes. It has six chapters and is organized as follows.

Chapter 1 explains methods of modeling, control, and optimization of complex metallurgical processes, and describes methods of intelligent control and optimization.

Chapter 2 describes a coal blending process in an iron ore sintering process. After the explanation of their characteristics and control requirements, an expert control strategy is designed based on the combination of BPNNs, mathematical models, and rule models to determine and track control objectives accurately for the coal blending process. An intelligent integrated optimization system (IIOS) with a hierarchical configuration is devised that solves the problem of optimizing the proportioning of iron ore in a sintering process.

Chapter 3 first describes a coking process and the configuration of a control system for a combustion process. Next, it presents an integrated method combining linear programming (LP) and NN-based modeling to avoid the drawbacks of mechanism modeling and to create a precise model for the estimation of coke-oven temperature (COT) based on actual data. Then, it explains a hierarchical intelligent integrated control structure for a combustion process based on the determination of an operating state, and also an operation plan and optimal scheduling of a coke-oven. Finally, it presents and comments on the results of actual runs.

Chapter 4 first describes the thermal state parameters of a sintering process. Then, an intelligent control system is designed for the sintering thermal state parameters. Finally, the effectiveness of the intelligent control method is verified in an actual sintering process.

Chapter 5 first analyzes the characteristics of a gas collection process and a gas mixing-and-pressurization process with multiple asymmetric coke-ovens in an iron and steel company. Then, a hierarchical intelligent decoupling control method and an integrated intelligent decoupling control method are presented for the two processes. Finally, the results of actual runs of the processes are presented to demonstrate the validity of the methods.

Chapter 6 analyzes the combustion process of a regenerative reheating furnace and a compact strip production(CSP) soaking furnace. The prediction model of furnace temperature is established based on the analysis of the mechanism. The optimization and control systems of the regenerative heating furnace and the CSP soaking furnace are designed for different control requirements. The effectiveness of the system is tested on an actual production line.

References

1. Li Y, Wu M, Cao WH, Lai XZ, Wang CS (2012) PSO-BP control algorithm of granulation process based on evaluation and optimization of granularity distribution. Acta Autom Sin 38(6):1007–1016 (In Chinese)
2. Hou C, Yu X, Cao Y, Lai C, Cao Y (2018) Prediction of synchronous closing time of permanent magnetic actuator for vacuum circuit breaker based on PSO-BP. IEEE Trans Dielectr Electr Insul 24(6):3321–3326
3. Wu M, Duan P, Cao WH, She JH, Xiang J (2012) An intelligent control system based on prediction of the burn-through point for the sintering process of an iron and steel plant. Expert Syst Appl 39(5):5971–5981
4. Guan L (2006) An intelligent modeling method in Slab's hot rolling process based on the rolling information feedback. Dalian University of Technology, Dalian (In Chinese)
5. Wu M, Cao WH, He CY, She JH (2009) Integrated intelligent control of gas mixing-and-pressurization process. IEEE Trans Control Syst Technol 17(1):68–77
6. Li F, Ma X, Cao WH, Yuan Y (2011) Fuzzy compensation decoupling control of calorific value and pressure in gas mixing process. J Cent South Univ (Science and Technology) 42(1): 94-99 (In Chinese)
7. Deng J, Lai XZ, Wu M, Cao WH (2007) Intelligent optimization method for coal blending based on neural network and simulated annealing algorithm. Metall Ind Autom 3:19–23 (In Chinese)
8. Shang XQ, Lu JG, Sun YX, Liu J, Ying YQ (2010) Data-driven prediction of sintering burn-through point based on novel genetic programming. Int J Iron Steel Res 17(12):1–5
9. Liao YX, She JH, Wu M (2009) Integrated hybrid-PSO and fuzzy-NN decoupling control for temperature of reheating furnace. IEEE Trans Ind Electron 56(7):2704–2714
10. Wang XD, Wang ZF, Liu Y, Du FM, Yao M, Zhang XB (2016) A particle swarm approach for optimization of secondary cooling process in slab continuous casting. Int J Heat Mass Transf 93:250–256
11. McClelland JL, Rumelhart DE (1988) Explorations in parallel distributed processing: a handbook of models, programs, and exercises. MIT Press, Cambridge
12. Simon H (2011) Neural networks and learning machines, 3rd edn. Prentice Hall PTR, Upper Saddle River
13. Fei J, Wang T (2018) Adaptive fuzzy-neural-network based on RBFNN control for active power filter. Int J Mach Learn Cybern. https://doi.org/10.1007/s13042-018-0792-y
14. Hui W, Xie W, Pei J (2015) A pre-radical basis function with deep back propagation neural network research. In: Proceedings of IEEE international conference on signal, pp 1489–1494
15. Yu S, Chen S, Zhang Z, Zhang Y (2018) A novel blind detection algorithm based on double sigmoid hysteretic chaotic hopfield neural network. J Inf Hiding Multimed Signal Process 9(2):452–460
16. Silva HO, Bastos-Filho CJA (2018) Inter-domain routing for communication networks using Hierarchical Hopfield Neural Networks. Eng Appl Artif Intell 70:184–198
17. Zadeh LA (1965) Fuzzy sets, information and control. Inform Control 8(3): 338-353

18. Li H, Wu C, Yin S, Lam HK (2016) Observer-based fuzzy control for nonlinear networked systems under unmeasurable premise variables. IEEE Trans Fuzzy Syst 24(5):1233–1245
19. Oliveira KF, César MB, Gonçalves J (2017) Fuzzy based control of a vehicle suspension system using a MR damper. In: Proceedings of the 12th Portuguese confernece on automatic control, pp 571–581
20. Liao SH (2005) Expert system methodologies and applications–a decade review from 1995 to 2004. Expert Syst Appl 28(1):93–103
21. Chen Q, Liu G, Cai X, Xu G (2018) Decoupling control of five-phase fault-tolerant permanent magnet motor by radial basis function neural network inverse. In: IEEE international magnetics conference, pp 1–2
22. Wu M, Yan J, She JH, Cao WH (2009) Intelligent decoupling control of gas collection process of multiple asymmetric coke ovens. IEEE Trans Ind Electron 56(7):2782–2792
23. Holland JH (1992) Adaptation in natural and artificial systems. MIT Press, Cambridge
24. John HH (1975) Adaptation in natural and artifical systems. MIT Press, Combridge
25. Colorni A, Dorigo M, Maniezzo V (1991) Distributed optimization by ant colonies. In: Proceedings of European conference on artificial life, pp 134–142
26. Dorigo M, Gambardella LM (1997) Ant colony system: a cooperative learning approach to the traveling salesman problem. IEEE Trans Evol Comput 1(1):53–66
27. Chahal P, Singh J (2012) Study of mathematical model and ant colony optimization (ACO). Int J Educ Appl Res 2(1):101–105
28. Pirjanian P (1998) Satisficing action selection. In: SPIE conference on intelligent systems and advanced manufacturing, pp 153–164

Chapter 2
Intelligent Optimization and Control of Raw Material Proportioning Processes

The proportioning of raw materials in a complex metallurgical process contains a coal blending process and an iron ore sintering process.

Different types of coal are blended in the coal blending process in suitable percentages to form a coal blend of required quality. The coal blend is distillated in a coke-oven to make coke for iron smelting in the iron and steel industry. Under certain distillation conditions, the quality of coke is mainly determined by the quality of the coal blend. This means that the quality and percentage of each type of coal to be blended are key factors influencing the quality of coke [1]. To obtain the desired coke, it is important to rigorously control the coal blending process.

Two key problems in the control of this process are to determine the target percentage of each type of coal and to track them. It is especially important to determine the target percentages from the quality requirements of the coke and the quality of each type of coal by predicting the quality of the coal blend and coke. Conventional methods involve constructing mathematical models to predict quality based on measured data for coal blending and distillation, and then determining the target percentages using the models. The models mainly employ linear system identification techniques, such as the least-squares method [2–4]. However, it is difficult to yield accurate percentages by conventional methods because the determination is based solely on mathematical models, which do not describe the exact relationships among the parameters in coal blending and distillation. This requires us to develop a way of constructing highly accurate quality-prediction models and determining the target percentages with a high accuracy.

Recent progresses on ESes and NNs bring us lots of means to effectively control the coal blending process. ESes have been used for process control in the iron and steel industry [5, 6]. Empirical knowledge was used in those systems to solve the control problems [7–11]. NNs are a powerful tool for the modeling and control of complex systems [12, 13]. Especially, a BPNN can approximate large classes of continuous functions and is most widely used in process control applications [14–16]. Since the relationships among the parameters in coal blending and distillation can be expressed through a combination of BPNNs, and mathematical models and

M. Wu et al., *Intelligent Optimization and Control of Complex Metallurgical Processes*, Engineering Applications of Computational Methods 3,
https://doi.org/10.1007/978-981-15-1145-5_2

rule models are based on associated empirical knowledge and statistical data, the ES and NN techniques can be employed to construct highly accurate quality-prediction models and to determine precise target percentages for the coal blending process.

Iron ore sintering is a critical process in iron and steel production. It produces qualified sinter for the blast furnace [17–19]. The process is complex and nonlinear, and many of the process factors are coupled with each other. Proportioning is the first step in the sintering process. The chemical components of the raw materials and their proportions determine the chemical and physical characteristics of the resulting sinter, and thus the quality of the sinter. They are also closely related to energy efficiency, and to the reduction of costs and SO_2 emissions. These things strongly affect the competitiveness of an enterprise.

There are two types of sintering machines: downdraft and updraft. The downdraft Dwight-Lloyd type is now the most common one, and the effective area of a sintering machine has been expanded to more than 450 m^2. A larger effective area means greater efficiency, greater energy savings, and greater environmental benefits. Due to technological constraints, most sintering machines in iron ore sintering plants have an effective area of 360 m^2.

The sintering process consists of proportioning, mixing, feeding, segregation, ignition, breaking, cooling, and screening [20, 21]. Since the chemical components of the raw materials and their proportions determine the chemical and physical characteristics of the sinter, and thus the quality of the sinter, a desired dosing scheme for the proportioning step, is the key to the quality of the sinter.

The proportioning step considered in this study has two phases: the first and second proportionings. In the first phase, an iron ore mix consisting of different kinds of iron ores is prepared. In the second phase, the iron ore mix, coke, limestone, dolomite, and returned sinter are mixed to produce a raw mix for the production of qualified sinter. It is important to design appropriate dosing schemes for these two proportioning phases. Sintering takes about two hours from the loading of the iron ore to the discharge of the sinter. After the materials for the second proportioning phase are mixed according to a prescribed dosing scheme, the components of that phase cannot be changed. For these reasons, two key problems in the proportioning process are the prediction of sinter quality and the design of an adequate dosing scheme.

A large number of studies have attempted to solve the problem of how to predict sinter quality, and many models of the sintering mechanism have been constructed since the 1980s [22]. However, they are difficult to be applied to an actual industrial process for two reasons: ① the difference between an ideal model of the mechanism and the actual process always exists, and ② the measurement technology that the models require does not exist. Time-sequence quality-prediction models have been proposed that are based on an analysis of historical data and that use regression analysis, grey system theory, etc. [23, 24]. These models are applicable to stable, high-quality iron ore sources and a steady sintering process. On the other hand, it is also possible to use the sintering mechanism, a principal component analysis, the calculation of correlation coefficients, grey theory, etc., to build a factor model containing the factors that determine sinter quality. Furthermore, an integrated factor model, which combines a factor model with the methods of computational intelligence [25] (NN, fuzzy system, ES, evolutionary algorithm, etc.) can handle the problem of how to predict sinter quality for variable iron ore sources [26, 27]. As modeling technologies progress, it should become possible to improve the prediction precision and to expand the applicable range by combining these kinds of models and/or by integrating different modeling methods [28–30].

Conventional proportioning optimization is based primarily on practical knowledge and experience of a sintering process. It is widely used to design dosing schemes. It requires a large number of sintering experiments and the accumulation of a large amount of data, and it also strongly depends on the iron ore sources. Moreover, differences in the levels of experience and the psychological states of operators influence the quality of the sinter. Conventional optimization methods, such as linear programming (LP) [31], are often used for proportional optimization; but the diversity of optimization objectives makes it hard to meet the requirements of a sintering process. The global trends in energy, cost, and carbon reduction make it an urgent necessity to devise a new dosing scheme based on the optimization of multiple objectives (cost, emissions, etc.). Intelligent methods of optimal proportioning have been devised that employ the SA, an evolutionary algorithm, the PSO, or the ACO, etc. [33–39]. But they have a common problem in which the convergence of the algorithms is initially very quick but later becomes slow and may even stop unexpectedly. Since combining conventional optimization methods and computational intelligence-based optimization methods can exploit the advantages of both and overcome their weak points, the integration of different optimization methods improves the speed and accuracy of the search for an optimal solution. This is the current trend in proportional optimization for a sintering process [39, 40].

The above studies only considered the proportioning step alone and ignored the relationships between the proportioning step and the other steps in the sintering process. The quality of sinter is closely related to a raw mix produced in the proportioning step and to the process state and operating parameters. So, there is much room for improvement. Most of the optimization objectives focus on cost reduction and do not consider overall performance, that is, cost, emissions, and other indices. So, they cannot yield both optimal economic and environmental performance. Moreover, those optimal proportioning models do not take into account the different characteristics of the first and second proportioning phases. This restricts their area of applications.

This chapter describes an expert control strategy based on a combination of BPNNs, mathematical models, and rule models to determine and track the target percentages accurately for a coal blending process. The architecture of the expert control for the process is first described. Next, highly accurate quality-prediction models that consist of BPNNs and mathematical models are constructed, and rule models are established based on statistical data and empirical knowledge. Then, an expert reasoning method combining the networks and models is devised for determining the target percentages, and a distributed controller is employed to track them. Finally, the results of actual runs are presented. An intelligent integrated optimization system with a hierarchical configuration is presented to optimize the proportioning step by the following procedure. First, the sintering process is analyzed, and the requirements of the proportioning step are specified. Next, an optimization system with two levels (intelligent integrated optimization, basic automation) is built. In the intelligent integrated optimization level, an intelligent integrated optimizer produces an optimal dosing scheme. The intelligent integrated optimizer has three parts: a cascade integrated quality-prediction model, the optimization of the first proportioning, and the optimization of the second proportioning. Computational intelligence methods predict the quality of sinter. Then, the predicted quality indices are fed back to the optimization of the first and second proportionings to find feasible optimal dosing schemes. The system was implemented in an iron and steel plant. Actual runs show that the system reduced production costs by 43.014 CNY/t and SO_2 emissions by 0.001% on average.

2.1 Process Description and System Configuration

This section describes the coal blending process and iron ore sintering process, and the configurations of the intelligent optimization and control system.

2.1.1 Process Description and Characteristic Analysis

A coal blending process is shown in Fig. 2.1. Various kinds of raw coal from different mines are classified according to their properties into seven types. Each type is fed from a hopper through a screw conveyor to a central belt conveyor in the target percentage, where it is blended with the others. The coal blend is pulverized and put in a coke-oven, where destructive distillation produces coke.

An important requirement of the process is to compute the target percentage of each type of coal from the quality requirements of the coke and the quality of coal to be blended. These percentages are tracked by controlling the speeds of the screw conveyors. The coke produced must satisfy given quality requirements. Coal quality is mainly characterized by the caking property index, the volatile matter content, the sulfur content, and the ash content. The quality of coke has to satisfy the general quality requirements that the MICUM strength 40-mm index is less than 76.5, the MICUM strength 10-mm index is larger than 9, the sulfur content is less than 0.7, and the ash content is less than 14.5. Moreover, the lack of some kinds of scarce coal makes the dosing plan of coal blending process unable to be accomplished. So, the proportions of those kinds of scarce coal have to be limited to as low as possible to guarantee continuous production.

In a typical sintering process (Fig. 2.2), different types of iron ore are mixed in the first proportioning phase and sent to the next proportioning phase, where the iron ore mix is combined with coke, limestone, dolomite, and returned sinter. This produces what is called a raw mix. A mixing drum adds water to the raw mix and sends it to a feed bin. A roller feeder discharges the material onto a moving pallet, after which a segregation plate moves the coarse grains to the bottom and the fine grains

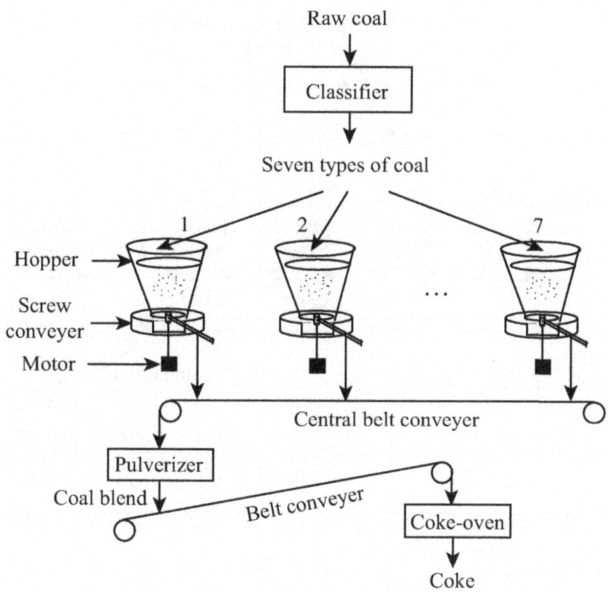

Fig. 2.1 Coal blending process

to the top. This forms a bed on a moving trolley. The surface of the bed is ignited with stationary burners in the ignition hood. The heat from the combustion induces chemical reactions and physical changes in the bed. Air is sucked in through wind boxes situated under the moving trolley, thereby ensuring that the sintering process progresses downward through the bed. This process continues until the sintering front reaches the BTP, which is defined to be the location on the trolley where the oxidation front initially passes all the way through the blended-ball layer. The BTP indicates the completion of sintering. At the end of the sintering machine, the sinter cake is crushed in a sinter breaker, cooled in a sinter cooler, and screened. The qualified sinter is sent to the blast furnace, and the undersized sinter is sent back to the second proportioning phase as returned sinter. Different kinds of iron ore have different grades of iron, S contents, and prices. The iron ore mix, coke, limestone, dolomite, and returned sinter that are combined in the second proportioning phase have different contents of Fe, SiO_2, CaO, MgO, Al_2O_3, S, and P that directly affect the chemical components of the raw mix, as well as the reactions. So, a suitable dosing scheme is a key to producing high-quality sinter with a certain chemical composition, permeability, and strength.

50–70% of the iron ores used in the iron and steel plant considered in this study are imported from Australia, Africa, Brazil, and India. While imported iron ores are of consistent quality and have a low sulfur content, they are expensive. The use of 30–50% domestic iron ores reduces the cost, but these ores suffer from a large variation in quality and a high sulfur in content. So, finding a suitable dosing scheme that

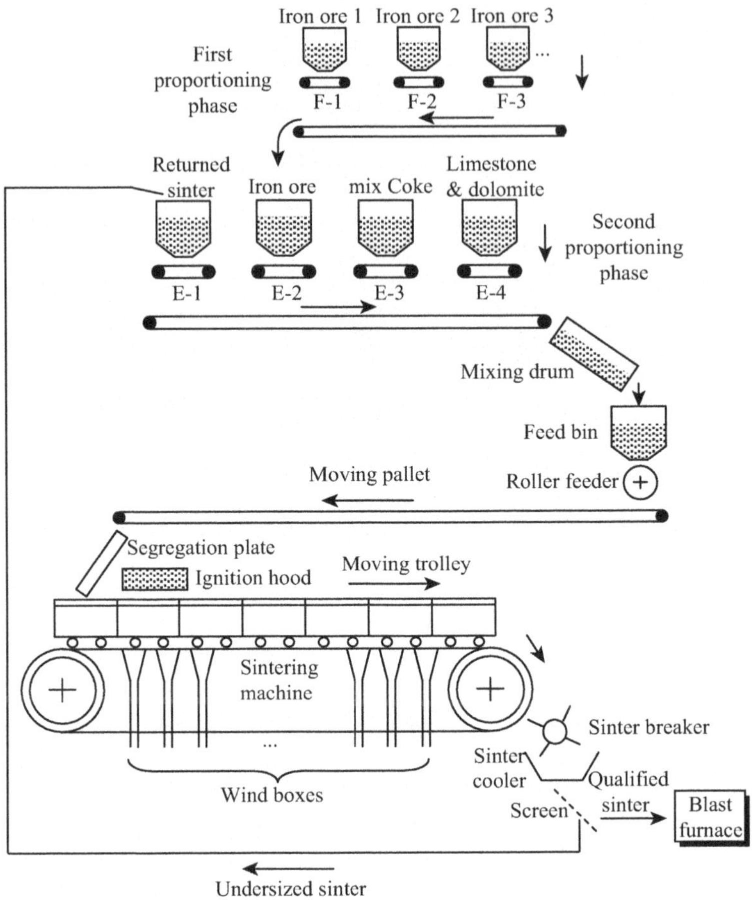

Fig. 2.2 Sintering machine with an area of 360 m^2

blends different kinds of iron ores is important to reduce the cost and SO_2 emissions while assuring quality.

In the rest of this chapter, we used superscripts (1) and (2) to indicate the variables related to the first and the second proportioning phases, respectively.

The first proportioning phase focuses on seven properties of the resulting iron ore mix: the iron grade, and the contents of SiO_2, CaO, MgO, Al_2O_3, S, and P. Fe is the key component of the iron ore mix. S is a harmful source of pollution produced in ferrous metallurgy. The requirements for the iron ore mix in the first proportioning phase are:

(1) The iron grade, $T_{Fe}^{(1)}$, must be greater than 60%.

(2) The SiO_2 content, $y_{SiO_2}^{(1)}$, must be in the range [4, 6%].

(3) The contents of CaO, MgO, and Al_2O_3, $y_{CaO}^{(1)}$, $y_{MgO}^{(1)}$, and $y_{Al_2O_3}^{(1)}$, must be lower than 2%.

(4) The contents of S and P, $y_S^{(1)}$ and $y_P^{(1)}$, must be lower than 0.2%.

In the second proportioning phase, in addition to the seven chemical items examined in the first phase, two other indices are considered: the basicity, R, and the tumbler index, J_t. They are determined from the sinter produced by the process. The requirements for the sinter are as follows:

(1) The iron grade, $T_{Fe}^{(2)}$, must be greater than 55%.

(2) The SiO_2 content, $y_{SiO_2}^{(2)}$, must be in the range [4, 6%].

(3) The CaO content, $y_{CaO}^{(2)}$, must be in the range [8, 11%].

(4) The contents of MgO and Al_2O_3, $y_{MgO}^{(2)}$ and $y_{Al_2O_3}^{(2)}$, must be lower than 2%.

(5) The contents of S and P, $y_S^{(2)}$ and $y_P^{(2)}$, must be lower than 0.1%.

(6) The basicity of the sinter, R, must be in the range [1.75, 2.00].

(7) The tumbler index, J_t, must be greater than 81.5%.

The above specifications are the minimum requirements. Adjustments may be made to suit the actual situation.

The two raw material proportioning processes also have the common characteristics of strong nonlinearities and a large time delay. These things make proportioning a difficult task. The chemical reactions and physical changes involved in the proportioning processes are usually in a dynamic equilibrium state. The components of the raw mix along with the process state and the operating parameters have a strong effect on the quality indices. The process state and the operating parameters vary continuously and generally cannot be determined from online measurements. They are coupled with each other, and the relationships among them are nonlinear. This makes them difficult to estimate. Moreover, the sintering processes, from proportioning to discharge, take a long time. This large time delay makes the data collected online unusable for the real-time control of the proportioning step.

The expert control of the coal blending process is designed based on the characteristics and control requirements of the process. Its hierarchical configuration is shown in Fig. 2.3, which employs an expert controller and a distributed controller.

The expert controller uses an expert reasoning method based on BPNNs, mathematical models, and rule models, and combines forward chaining and model-based reasoning to determine the target percentages from the quality requirements of the coke and the quality of each type of coal so as to obtain the desired coke. Moreover, using the determined target percentages, the total flow rate of the coal blending, and the moisture content of each type of coal, the expert controller calculates the target flow rate of each type of coal and sends it to the distributed controller.

The distributed controller is used for blending different types of coal in accordance with the target percentages. More specifically, the distributed controller generates control actions by using a tracking control algorithm to control the speeds of screw conveyors so as to ensure that the actual flow rates track the target flow rates.

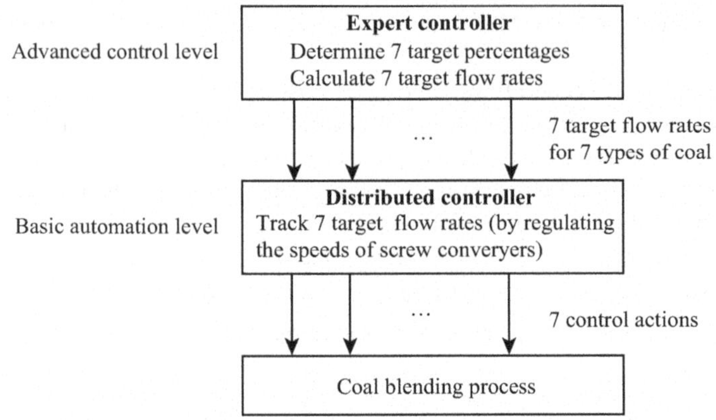

Fig. 2.3 Expert control architecture

2.1.2 Control Architecture

Based on the characteristics and requirements for the proportioning step, we devised an IIOS with a two-level hierarchical configuration: a basic automation level and an intelligent integrated optimization level (Fig. 2.4). The intelligent integrated optimization level contains an intelligent integrated optimizer (IIO) with four parts: optimization of the first proportioning, optimization of the second proportioning, a cascade integrated quality-prediction model (CIQPM), and an LP and GA-PSO algorithm.

Optimization of the first proportioning outputs an optimal dosing scheme for the source iron ores. The inputs are the prices of the iron ores, the constraints imposed by the specifications, and modifications to the constraints produced by the CIQPM. It uses the LP and GA-PSO multi-objective optimization method to produce optimal proportions of iron ores.

The inputs of the second proportioning optimization are the prices of the coke, limestone, and dolomite; constraints imposed by the specifications; modifications to the constraints produced by the CIQPM; and the predicted values for the main chemical components of the iron ore mix based on the optimization of the first proportioning. This step uses the LP and GA-PSO optimization method to produce optimal proportions of raw materials for the second proportioning phase.

The CIQPM predicts the quality of the sinter based on the optimal proportions of raw materials produced by optimization of the second proportioning and historic state parameters. This prediction takes the time delay of the process into account.

The basic automation level contains a distributed control system (DCS) and a communication interface. The DCS carries out process control and collects the operating parameters and the state of the sintering process. The communication interface enables the exchange of information between the IIO and the DCS.

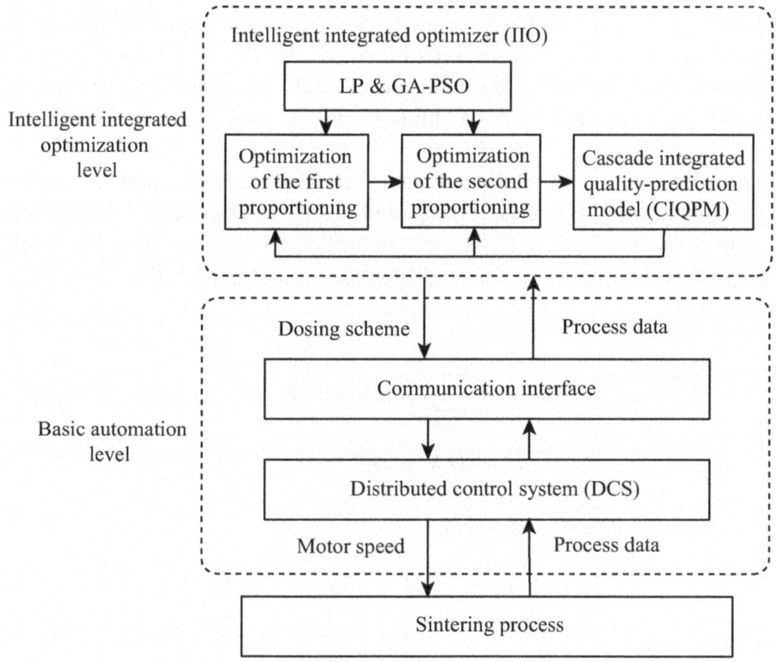

Fig. 2.4 Hierarchical configuration of an IIOS

2.2 Intelligent Optimization and Control of Coal Blending Process

The quality-prediction models for the coal blend and coke consist of BPNNs and mathematical models. Rule models are represented in **IF-THEN** form. These networks and models are based on statistical data and empirical knowledge on coal blending and distillation.

The expert controller is designed to determine the target percentages. It uses an expert reasoning method based on the constructed BPNNs, mathematical models, and rule models, and a combination of forward chaining and model-based reasoning. The reasoning is implemented in two algorithms that determine the allowable values of the quality of the coal blend and the target percentages. The predictions of the coal blend and coke quality are repetitively performed in two algorithms.

2.2.1 Quality-Prediction Models for Coal Blend

In the coal blending process, coal quality is mainly characterized by the caking property index, the volatile matter content, the sulfur content, and the ash content.

Assume that G_i, V_{bi}, S_{bi}, and A_{bi} denote these properties of the ith type of coal; G, V_b, S_b, and A_b are for those of the coal blend; $\hat{G}$, $\hat{V}_b$, $\hat{S}_b$, and $\hat{A}_b$ are the predicted values. x_i is the percentage of the ith type of coal. The quality-prediction models for the coal blend are constructed to obtain $\hat{G}$, $\hat{V}_b$, $\hat{S}_b$, and $\hat{A}_b$ from G_i, V_{bi}, S_{bi}, A_{bi}, and x_i.

Empirical knowledge and statistical data show that G, V_b, S_b, and A_b for the most part depend only on $x_i G_i$, $x_i V_{bi}$, $x_i S_{bi}$, and $x_i A_{bi}$, respectively. In particular, the relationship between G and $x_i G_i$ is more complicated than the other three relationships. To predict the quality of the coal blend accurately, the following expressions are introduced:

$$\hat{G} = \sum_{i=1}^{7} a_i x_i G_i + \Delta G, \tag{2.1}$$

$$\hat{B} = DX + \Delta B, \tag{2.2}$$

where a_i is the correlation coefficient,

$$\hat{B} = \begin{bmatrix} \hat{V}_b \\ \hat{S}_b \\ \hat{A}_b \end{bmatrix}, \quad D = \begin{bmatrix} V_{b1} & V_{b2} & \cdots & V_{b7} \\ S_{b1} & S_{b2} & \cdots & S_{b7} \\ A_{b1} & A_{b2} & \cdots & A_{b7} \end{bmatrix}, \tag{2.3}$$

$$X = \begin{bmatrix} x_1 \\ x_2 \\ \vdots \\ x_7 \end{bmatrix}, \quad \Delta B = \begin{bmatrix} \Delta V_b \\ \Delta S_b \\ \Delta A_b \end{bmatrix}, \tag{2.4}$$

and ΔG, ΔV_b, ΔS_b, and ΔA_b are compensation values that are used to improve the accuracy of the prediction of coal blend quality.

Equation (2.1) describes a BPNN with two layers for predicting G (BP2L-G) that has an input layer with seven neurons and an output layer with one neuron, while a general BPNN usually has an input layer, several hidden layers, and an output layer [14, 16]. In the input layer, both the input and output of the ith neuron are $x_i G_i$, and in the output layer, those of the neuron are $\hat{G}$. a_i is the weight of the signal from the ith neuron of the input layer to the neuron of the output layer and ΔG is the bias of the neuron in the output layer; they are determined by training BP2L-G based on statistical data.

Let B denote the measured value corresponding to $\hat{B}$. $\Delta B(k)$ for the kth blending is given by

$$\Delta B(k) = \sum_{j=1}^{k-1} \left[\hat{B}(j) - B(j) \right] + \Delta B(1), \tag{2.5}$$

where $\Delta B(1)$ is the compensation value for the first blending and is determined by empirical data.

2.2.2 *Quality-Prediction Models for Coke*

The quality of coke is characterized by the MICUM strength 40-mm index, the MICUM strength 10-mm index, the sulfur content, and the ash content. Let M_{40}, M_{10}, S, and A denote these properties; and $\hat{M}_{40}$, $\hat{M}_{10}$, $\hat{S}$, and $\hat{A}$ denote the corresponding predicted values. The quality prediction models for coke are constructed to obtain $\hat{M}_{40}$, $\hat{M}_{10}$, $\hat{S}$, and $\hat{A}$ from M_{40}, M_{10}, S, and A.

M_{40} and M_{10} have been shown to depend mainly on G , V_{b}, and A_{b} under certain distillation conditions. However, these relationships are nonlinear. To predict M_{40} and M_{10} from G, V_{b}, and A_{b} accurately, two BPNNs with three layers, BP3L-M40 and BP3L-M10, are constructed.

BP3L-M40 and BP3L-M10 have the same structure: an input layer with three neurons, a hidden layer with twelve neurons, and an output layer with one neuron. The inputs of the three neurons of the input layer for both BP3L-M40 and BP3L-M10 are G , V_{b}, and A_{b}, and their outputs are the same as the inputs. The input and output of the ith neuron of the hidden layer of BP3L-M40 are defined to be

$$p_i^{\mathrm{H1}} = w_{i,1}^{\mathrm{H1}} G + w_{i,2}^{\mathrm{H1}} V_b + w_{i,3}^{\mathrm{H1}} A_b + b_i^{\mathrm{H1}} \tag{2.6}$$

and

$$y_i^{\mathrm{H1}} = \mathrm{tansig}\left(p_i^{\mathrm{H1}}\right), \tag{2.7}$$

respectively, and those of the neuron of the output layer of BP3L-M40 are defined to be

$$p^{\mathrm{O1}} = \sum_{i=1}^{12} w_i^{\mathrm{O1}} y_i^{\mathrm{H1}} + b^{\mathrm{O1}} \tag{2.8}$$

and

$$\ddot{M}_{40} = p^{\mathrm{O1}}, \tag{2.9}$$

respectively, where tansig$(\cdot)$ denotes the tan-sigmoid transfer function, which has the form

$$\mathrm{tansig}(x) = \frac{2}{1 + \mathrm{e}^{-2x}} - 1 \tag{2.10}$$

and maps the input to the interval $(-1, 1)$ [16]. In Eqs. (2.6) and (2.8), $w_{i,1}^{\mathrm{H1}}$, $w_{i,2}^{\mathrm{H1}}$, and $w_{i,3}^{\mathrm{H1}}$ denote the weight of the signal from the three neuron of the input layer to the ith neuron of the hidden layer, b_i^{H1} denotes the bias of the ith neuron of the hidden layer, w_i^{O1} denotes the weight of the signal from the ith neuron of the hidden layer to the neuron of the output layer, and b^{O1} denotes the bias of the neuron of the output layer.

Equations (2.6)–(2.10) can be written as

$$\hat{M}_{40} = \sum_{i=1}^{12} w_i^{\mathrm{O1}} \mathrm{tansig}\left(w_{i,1}^{\mathrm{H1}} G + w_{i,2}^{\mathrm{H1}} V_{\mathrm{b}} + w_{i,3}^{\mathrm{H1}} A_{\mathrm{b}} + b_i^{\mathrm{H1}}\right) + b^{\mathrm{O1}}. \tag{2.11}$$

Using the same method for M_{10} yields

$$\hat{M}_{10} = \sum_{i=1}^{12} w_i^{\mathrm{O2}} \mathrm{tansig}\left(w_{i,1}^{\mathrm{H2}} G + w_{i,2}^{\mathrm{H2}} V_{\mathrm{b}} + w_{i,3}^{\mathrm{H2}} A_{\mathrm{b}} + b_i^{\mathrm{H2}}\right) + b^{\mathrm{O2}}. \tag{2.12}$$

The weights $w_{i,1}^{\mathrm{H1}}$, $w_{i,2}^{\mathrm{H1}}$, $w_{i,3}^{\mathrm{H1}}$, w_i^{O1}, $w_{i,1}^{\mathrm{H2}}$, $w_{i,2}^{\mathrm{H2}}$, $w_{i,3}^{\mathrm{H2}}$, and w_i^{O2}, and the biases b_i^{H1}, b^{O1}, b_i^{H2}, and b^{O2} are determined by training BP3L-M40 and BP3L-M10 based on statistical data.

From statistical data and empirical knowledge, $\hat{S}$ and $\hat{A}$ are computed from

$$\hat{S} = \frac{\eta_{\mathrm{s}}}{\eta_{\mathrm{c}}} S_{\mathrm{b}} + \Delta S, \tag{2.13}$$

$$\hat{A} = \frac{1}{\eta_{\mathrm{c}}} A_{\mathrm{b}} + \Delta A, \tag{2.14}$$

where η_{s} is the residual coefficient of the sulfur content of the coal blend, and η_{c} is the coking coefficient of the coal blend. Usually, $\eta_{\mathrm{s}} = 0.65$–0.75 and $\eta_{\mathrm{c}} = 0.75$–0.80. The compensation values ΔS and ΔA improve the prediction accuracy, and the values in the kth blending are obtained by a process similar to that for Eq. (2.5):

$$\Delta S(k) = \sum_{j=1}^{k-1} \left[\hat{S}(j) - S(j)\right] + \Delta S(1), \tag{2.15}$$

$$\Delta A(k) = \sum_{j=1}^{k-1} \left[\hat{A}(j) - A(j)\right] + \Delta A(1), \tag{2.16}$$

where $\Delta S(1)$ and $\Delta A(1)$ are the compensation values in the first blending, which are determined from empirical data.

The constructed networks are trained based on statistical data to accurately predict G, M_{40}, and M_{10}. The training requires a set of network inputs and target outputs culled from measured statistical values. More specifically, the network inputs of

BP2L-G are x_1G_1, x_2G_2, ..., x_7G_7, and those of BP3L-M40 and BP3L-M10 are G, V_b, and A_b. The target output of BP2L-G is G, and those of both BP3L-M40 and BP3L-M10 are M_{40} and M_{10}. The performance functions of the networks are the average squared errors between the network outputs and the target outputs. The weights and biases of the networks are iteratively adjusted to minimize the associated performance function during training.

A basic backpropagation training algorithm [14] is used to determine the weights and biases. It employs the gradient of the performance function to adjust the weights and biases and minimize that function during training. The gradient is determined by the backpropagation technique, which involves performing computations backward through the network using the chain rule of calculus. The weights and biases are moved in the direction of the negative gradient. A batch training method is used to implement the gradient descent algorithm. In the training, the weights and biases are updated only after all the training data have been fed to the network. The gradients calculated during each training session are added together to determine the changes in the weights and biases.

The initial weights and biases of BP2L-G, BP3L-M40, and BP3L-M10 are based on statistical data for the last two years. When new statistical data are collected, the weights and biases are updated based on the previous weights and biases, and the new data.

2.2.3 Rule Models

In coal blending and distillation, there may exist several sets of percentages of coal to be blended that satisfy the same quality requirements for coke. It is important to efficiently determine a practical percentage for each type of coal. On the other hand, there are some relationships that cannot be described by BPNNs and mathematical models, but these relationships influence the quality-prediction accuracy and the computational accuracy of the target percentages. In addition, how suitable compensation values are selected in mathematical models is also an important aspect for improving the quality-prediction accuracy. To meet these requirements, we need to construct rule models based on the empirical knowledge of experts and veteran operators.

All rule models use the following production rule form [8]

$$\text{Rule No.}: \textbf{IF}\, condition\, \textbf{THEN}\, action,$$

where Rule No. is the number of the rule model, *condition* is the operating state or a logical combination, and *action* is the conclusion or operation.

Some relationships among the parameters expressing the quality and the percentages can be represented by rule models based on statistical data and empirical knowledge. For example, some typical rule models are listed as follows:

R_1: **IF** G_i increases, **THEN** G increases.
R_2: **IF** V_{bi} increases, **THEN** V_b increases.
R_3: **IF** S_{bi} increases, **THEN** S_b increases.
R_4: **IF** A_{bi} increases, **THEN** A_b increases.
R_5: **IF** (G increases) OR (V_b decreases) OR (A_b decreases), **THEN** (M_{40} increases) AND (M_{10} decreases).
R_6: **IF** S_b increases, **THEN** S increases.
R_7: **IF** A_b increases, **THEN** A increases.

These rule models are also used to examine whether or not the constructed networks and models are correct.

The determination of the target percentages is divided into two steps:

Step 1: Determine the allowable values of the quality of the blend coal from the quality requirements of the coke;

Step 2: Determine the target percentages from both the allowable values and the quality of each type of coal.

The solution of each step is not unique. To obtain a suitable solution in each step quickly, the rule models for determining the target percentages are constructed, which effectively use empirical knowledge and data. About 150 rule models for determining the allowable values of the coal blend and the target percentages were developed.

For instance, if the given quality index of coke is assumed to be M_{40}^{g}, M_{10}^{g}, S^{g}, and A^{g}, and the allowable values of the quality of the coal blend are G^{g}, V_b^{g}, S_b^{g}, and A_b^{g}, then some typical rule models are listed as follows:

R_8: **IF** ($\hat{M}_{40} < M_{40}^{g}$) OR ($\hat{M}_{10} < M_{10}^{g}$), **THEN** (increases G^{g}) AND (decreases V_b^{g}).
R_9: **IF** $\hat{S} > S^{g}$, **THEN** decreases S_b^{g}.
R_{10}: **IF** $\hat{A} > A^{g}$, **THEN** decreases A_b^{g}.
R_{11}: **IF** ($\hat{G} < G^{g}$) AND ($G_i > G^{g}$), **THEN** increases x_i.
R_{12}: **IF** ($\hat{V}_b > V_b^{g}$) AND ($V_{bi} > V_b^{g}$), **THEN** decreases x_i.
R_{13}: **IF** ($\hat{S}_b > S_b^{g}$) AND ($S_{bi} < S_b^{g}$), **THEN** increases x_i.
R_{14}: **IF** ($\hat{A}_b > A_b^{g}$) AND ($A_{bi} < A_b^{g}$), **THEN** decreases x_i.

2.2.4 Determination of Target Percentages Based on Rule Models

The control objective of the coal blending process is to make the quality of the coke satisfy the following quality index requirements:

$$M_{40} \geqslant M_{40}^{g}, \quad M_{10} \leqslant M_{10}^{g}, \quad S \leqslant S^{g}, \quad A \leqslant A^{g}. \tag{2.17}$$

To achieve this, the objective of the first step of the determination is to obtain the allowable values G^{g}, $V_{\mathrm{b}}^{\mathrm{g}}$, $S_{\mathrm{b}}^{\mathrm{g}}$, and $A_{\mathrm{b}}^{\mathrm{g}}$ so that the predicted values $\hat{M}_{40}$, $\hat{M}_{10}$, $\hat{S}$, and $\hat{A}$ for the quality of the coke satisfy

$$M_{40}^{\mathrm{g}} \leqslant \hat{M}_{40}^{\mathrm{g}} \leqslant M_{40}^{\mathrm{g}} + \Delta M_{40}^{\mathrm{e}}, \tag{2.18}$$

$$M_{10}^{\mathrm{g}} \leqslant \hat{M}_{10}^{\mathrm{g}} \leqslant M_{10}^{\mathrm{g}} + \Delta M_{10}^{\mathrm{e}}, \tag{2.19}$$

$$\hat{S} \leqslant S^{\mathrm{g}}, \ \hat{A} \leqslant A^{\mathrm{g}}. \tag{2.20}$$

The objective of the second step of the computation is to obtain the target percentages x_i $(i = 1, 2, \ldots, 7)$ so that the predicted values $\hat{G}$, $\hat{V}_{\mathrm{b}}$, $\hat{S}_{\mathrm{b}}$, and $\hat{A}_{\mathrm{b}}$ for the quality of the coal blend satisfy

$$G^{\mathrm{g}} \leqslant \hat{G} \leqslant G^{\mathrm{g}} + \Delta G^{\mathrm{e}}, \ \hat{V}_{\mathrm{b}} \leqslant V_{\mathrm{b}}^{\mathrm{g}}, \tag{2.21}$$

$$\hat{S}_{\mathrm{b}} \leqslant S_{\mathrm{b}}^{\mathrm{g}}, \ \hat{A}_{\mathrm{b}} \leqslant A_{\mathrm{b}}^{\mathrm{g}}. \tag{2.22}$$

$\Delta M_{40}^{\mathrm{e}}$, $\Delta M_{10}^{\mathrm{e}}$, and ΔG^{e} are empirically determined values that are used to keep the determination from being too conservative. In general, $\Delta M_{40}^{\mathrm{e}} = 2$–5, $\Delta M_{10}^{\mathrm{e}} = 1$–2, and $\Delta G^{\mathrm{e}} = 2$–6.

An expert controller is designed to achieve the objective in each step. It consists of a knowledge base, a database, a working memory, an inference engine, a calculation module, a communication interface, and a man–machine interface to determine the target percentages and the corresponding flow rates.

The knowledge base and database store empirical knowledge and data. When the quality requirements of the coke or the quality of the coal to be blended change, the associated states and data are sent to the working memory. The inference engine receives the empirical knowledge and data from the knowledge base and database, and uses a reasoning strategy combined with forward chaining [8] and model-based reasoning [9] to determine the target percentages. The intermediate results and states are stored in the working memory and employed repetitively. The determined target percentages are stored in the database and sent to the calculation module, which computes the target flow rates from the target percentages, moisture content, and total flow rate of the coal to be blended. The moisture content and total flow rate come from the database. The communication interface is used to send the target flow rates to the distributed controller and to provide process data to the database. The man–machine interface is employed to edit and modify the knowledge base and database, and display data in the database in the form of tables and graphics.

An important feature of the expert controller is that it combines the BPNNs, mathematical models and rule models. The combination ensures accuracy in the determination of the target percentages.

The determination of the target percentages is implemented in two algorithms. One determines the allowable value of the coal blend quality from the quality requirements of the coke, and the other determines the target percentages from the allowable values and the quality of each type of coal. More specifically, the allowable values $S_{\mathrm{b}}^{\mathrm{g}}$ and

A_b^g are obtained from expression [10] as follows:

$$S_b^g = \frac{\eta_c}{\eta_s}(S^g - \Delta S), \tag{2.23}$$

$$A_b^g = \eta_c(A^g - \Delta A). \tag{2.24}$$

The allowable values G_b^g and V_b^g are determined so that $\hat{M}_{40}$ and $\hat{M}_{10}$ obtained from BP3L-M40 and BP3L-M10 satisfy inequalities Eqs. (2.18) and (2.19), respectively. This can be achieved through an iterative algorithm. The following algorithm is developed to determine the allowable value of the coal blend quality.

Algorithm 1 Determination of Allowable Values

Step 1: Compute the allowable values S_b^g and A_b^g from S^g and A^g using Eqs. (2.23) and (2.24), respectively.

Step 2: Select suitable empirical values of G and V_b from the knowledge base as the initial values of G^g and V_b^g, respectively.

Step 3: Compute $\hat{M}_{40}$ and $\hat{M}_{10}$ from BP3L-M40 and BP3L-M10 by replacing G and V_b with G^g and V_b^g, respectively.

Step 4: Check if $\hat{M}_{40}$ and $\hat{M}_{10}$ satisfy Eqs. (2.18) and (2.19), respectively. If not, use the rule models, such as R_8, and adjustment laws, such as

$$G^g = G^g + \alpha_1\left(M_{40}^g - \hat{M}_{40}\right), \tag{2.25}$$

$$V_b^g = V_b^g + \alpha_2\left(\hat{M}_{40} - M_{40}^g\right), \tag{2.26}$$

when $\hat{M}_{40} < M_{40}^g$, to adjust G^g and V_b^g, and return to Step 3. If so, go to Step 5.

Step 5: Check if G^g and V_b^g are in the empirical range. If so, take G^g, V_b^g, S_b^g, and A_b^g obtained in Steps 1–4 to be the allowable value of the coal blend quality, and stop the algorithm. If not, choose other empirical values of G and V_b as the initial values of G^g and V_b^g, and return to Step 3. If suitable G^g and V_b^g cannot be obtained in a given number of iterations, stop the algorithm and report that the allowable values of G and V_b do not exist.

Note that α_1 and α_2 in Eqs. (2.25) and (2.26), respectively, are empirically determined positive values that determine the convergence rate of the iterative algorithm.

The target percentages in Algorithm 1 are determined by the following iterative algorithm.

Algorithm 2 Determination of target percentages

Step 1: Select a set of empirical percentages of coal to be blended from the knowledge base as the initial values of the target percentages x_i $(i = 1, 2, \ldots, 7)$.

Step 2: Predict the coal blend quality, i.e., $\hat{G}$, $\hat{V}_b$, $\hat{S}_b$, and $\hat{A}_b$, from the quality and the selected percentage of each type of coal using BP2L-G and mathematical models Eqs. (2.2) and (2.5).

Step 3: Check if Eq. (2.17) is satisfied. If not, use rule models, such as R_{11}–R_{14}, and adjustment laws, such as

$$x_i = x_i + \beta \left(G^g - \hat{G} \right), \tag{2.27}$$

when $\hat{G} < G^g$ and $G_i > G^g$, to adjust x_i, and return to Step 2. If so, go to Step 4.

Step 4: Check if the percentages obtained in Steps 1–3 are in the empirical range. If so, take the obtained percentages to be the target percentages, and stop the algorithm. If not, choose other empirical percentages from the knowledge base and return to Step 2. If suitable x_i cannot be obtained in a given number of iterations, stop the algorithm and report that useful percentages do not exist.

In Eq. (2.27), β is empirically determined positive value.

The target flow rate of each type of coal is determined from the following expression:

$$Q_i = \frac{x_i}{1 - b_i} Q, \tag{2.28}$$

where Q_i and b_i are the target flow rate and moisture content of the ith type coal, and Q is the total flow rate of the coal blend. The target flow rates are tracked by the distributed controller to obtain the desired coal blend and thus the desired coke.

2.2.5 Determination of Target Percentages Based on Simulated Annealing Algorithm

SA can be used to find target percentages. This is summarized in this subsection.

The objective is to find an optimal dosing scheme that minimizes the production cost and SO_2 emissions. Suppose that there are n kinds of coal, the cost of the jth kind of coal is C_j, and its proportion is x_j. Minimize the production cost can be mathematically expressed as

$$\min f_1(x) = \min \sum_{j=1}^{n} C_j x_j, \tag{2.29}$$

where $x = [x_1, x_2, \ldots, x_n]^T$. Reducing the SO_2 emissions is environmental friendly, and is important to sustainable development of the industry. Suppose there are n kinds of coal, the S content of the jth kind of coal is S_j, its proportion is x_j. Minimize the SO_2 emissions can be mathematically expressed as

$$\min f_2(x) = \min \sum_{j=1}^{n} S_j x_j. \tag{2.30}$$

Let $W = [w_1, w_2, \ldots, w_5]^T$ and $V = [v_1, v_2, \ldots, v_6]^T$. The quality indices of the coke, w_i $(i = 1, 2, \ldots, 5)$ are the moisture content, the ash content, the volatile matter content, the sulfur content, and the adhesive index of the coal blend; and v_i $(i = 1, 2, \ldots, 6)$ are the ash content, the sulfur content, the crushing strength, the scuff resistance, the reactivity index, and the post-reaction strength. They have to be in a certain range:

$$\begin{cases} W_{\min} \leqslant W \leqslant W_{\max}, \\ V_{\min} \leqslant V \leqslant V_{\max}. \end{cases} \tag{2.31}$$

The proportion of a kind of coke has to be limited to a certain range according to the inventory. The proportion of the jth kind of coke, x_j, should not larger than the ratio of the inventory H_j and the planned amount of coal S:

$$x_j \leqslant \frac{H_j}{S}, \quad j = 1, 2, \ldots, n. \tag{2.32}$$

The sum of the proportions of coke has to satisfy

$$\sum_{j=1}^{n} x_j = 100\%. \tag{2.33}$$

So, the determination of the target percentage of coke is equivalent to solve the constraint nonlinear optimization problem:

$$\begin{aligned} &\min \; F(f_1(x_1, x_2, \ldots, x_n), f_2(x_1, x_2, \ldots, x_n)), \\ &\begin{cases} \text{s.t. } L_1(x) = W_{\min} - W \leqslant 0, \\ \quad L_2(x) = W - W_{\max} \leqslant 0, \\ \quad L_3(x) = V_{\min} - V \leqslant 0, \\ \quad L_4(x) = V - V_{\max} \leqslant 0, \\ \quad L_5(x) = x - H/S \leqslant 0, \\ \quad L_6(x_1, x_2, \ldots, x_n) = \sum_{j=1}^{n} x_j - 1 = 0, \\ \qquad j = 1, 2, \ldots, n, \end{cases} \end{aligned} \tag{2.34}$$

where $H = [H_1, H_2, \ldots, H_n]^{\mathrm{T}}$. The objective function of the optimization problem is

$$\begin{aligned} \min \; Z = {} & F(f_1(x_1, x_2, \ldots, x_n), f_2(x_1, x_2, \ldots, x_n)) \\ & + C \Bigg\{ \sum_{j=1}^{5} \max[0, L_{1j}(x)] + \sum_{j=1}^{5} \max[0, L_{2j}(x)] \\ & + \sum_{j=1}^{6} \max[0, L_{3j}(x_j)] + \sum_{j=1}^{6} \max[0, L_{4j}(x_j)] \\ & + \sum_{j=1}^{7} \max[0, L_{5j}(x_j)] + |L_6(x)| \Bigg\}. \end{aligned} \tag{2.35}$$

An SA algorithm is used to solve the optimization problem. The SA is a stochastic search method analogous to a physical annealing process in which an alloy is cooled gradually to ensure that a minimal energy state is achieved. The SA avoids getting stuck in local optima (hill climbing) and finds its global optimal solution.

The SA algorithm based on this idea is given below:

Step 1: Set the starting vector of decision variables x_0 (i.e., the vector of the initial values of decision variables).

Step 2: Generate a new vector x':

$$x' = x + k(0.5 - URN), \tag{2.36}$$

where k is an algorithm variable, and URN is a uniform random number.

Step 3: If the objective function F meets the condition $F(x') < F(x)$, take the new point x' as the best solution and return to Step 2. Otherwise, take the vector x to be the best solution and go to Step 4.

Step 4: Calculate probability p_1 by use of the equation:

$$p_1 = e^{-[F(x')-F(x)]/T}. \tag{2.37}$$

Step 5: Generate a random number, p, within the interval [0, 1] (uniform distribution).

Step 6: If $p_1 > p$, take the new point x' as the best solution and return to Step 2.

Step 7: If $p_1 < p$, reject new point and return to Step 2.

2.2.6 Tracking Control of Target Percentages

The distributed control system is connected to seven control loops so as to ensure that the actual flow rate tracks the target flow rate for each type of coal. The proportional–integral (PI) control algorithm is used for the tracking of target flow rates. Figure 2.5 shows the diagram of the PI-based tracking control of the flow rates. The target flow rates from the expert controller and the measured flow rates from the electronic scales are input into the PI controller. The PI controller outputs the control signals to the inverters, and then to the AC motors and screw conveyors to control the flow rates.

The PI control is a combination of proportional and integral feedback control. The proportional feedback control is

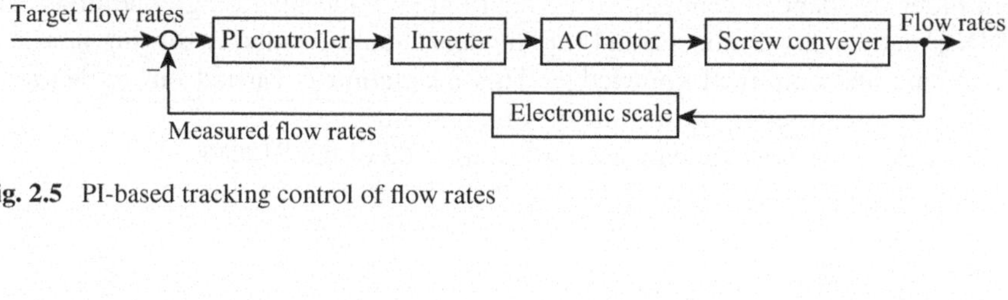

Fig. 2.5 PI-based tracking control of flow rates

$$u(t) = K_{\mathrm{P}} e(t), \tag{2.38}$$

where $e(t)$ is the error between actual and target flow rates, K_{P} is the proportional gain, and $u(t)$ is the control signals. The integral feedback control is

$$u(t) = K_{\mathrm{I}} \int_0^t e(\tau)\mathrm{d}\tau, \tag{2.39}$$

where K_{I} is an integration gain factor. And the PI controller is a given by

$$u(t) = K_{\mathrm{P}} e(t) + K_{\mathrm{I}} \int_0^t e(\tau)\mathrm{d}\tau. \tag{2.40}$$

2.3 System Implementation for Coal Blending Process

The expert control strategy was applied to the real-time control of a coal blending process in an iron and steel plant. The validity of this strategy has been proved by the results of actual runs.

2.3.1 System Configuration and Implementation

The expert control system (Fig. 2.6) consists mainly of an expert control computer system, a distributed control system, and a quality-measurement system. The expert control computer system uses an IPC 810 type computer, and the distributed control system uses an S9000 series controller made by Honeywell Corporation. The expert controller was implemented in the expert control computer system. The S9000 controller is a distributed controller connected to seven control loops so as to ensure that the actual flow rate tracks the target flow rate for each type of coal.

One important issue in the implementation of the expert controller is to determine suitable weights and biases for the BPNNs, and empirical values for the mathematical models. The initial weights and biases were obtained by training the BPNNs based on statistical data collected in 1995 and 1996, and the initial empirical values were culled from the statistical data of those two years. In addition, the rule models, such as R_1–R_7, were also used to check if the initial weights, biases, and empirical values were suitable.

In order to adapt to changes in the environment and operating conditions, the BPNNs, mathematical models, and rule models should be modified using new statistical data and empirical knowledge. This adaptation is carried out by learning functions of the expert control strategy, which mainly:

(1) Renew the statistical data used in the training of the BPNNs.

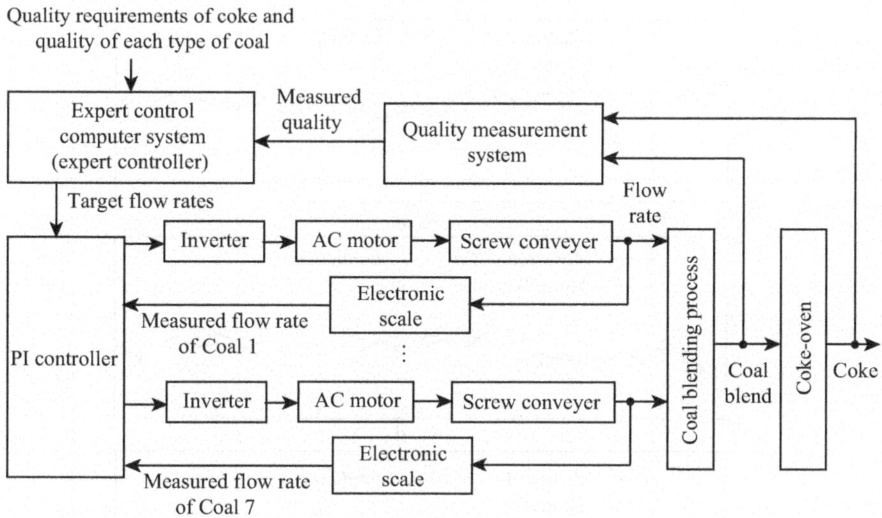

Fig. 2.6 Structure of expert control system for coal blending process

(2) Update the weights and biases of the BPNNs, and the empirical values of the mathematical models and rule models.

(3) Add new rule models and delete unnecessary old rule models.

2.3.2 *Results of Actual Runs of Coal Blending Process*

Figure 2.7 shows the run results during 1 month. The quality index of coke and the empirical values used in determining the target percentages were

$$M_{40}^{g} = 765, \quad M_{10}^{g} = 9.0, \quad S^{g} = 0.7, \quad A^{g} = 14.5, \tag{2.41}$$

$$\Delta M_{40}^{e} = 3.0, \quad \Delta M_{10}^{e} = 2.0, \quad \Delta G^{e} = 5. \tag{2.42}$$

The mean measured values of M_{40}, M_{10}, S, and A are 78.18, 8.27, 0.63, and 13.78, and those mean predicted values are 78.28, 8.24, 0.62, and 13.89, respectively. The results show that the coke quality satisfied the requirements described in expression, and the measured values approach the predicted values very closely.

With the development of modeling technology and the automation of industrial process, intelligent optimization method, such as the SA, have been applied to find the target percentages of the coal blending process. An intelligent optimization system was implemented in an iron and steel plant. The prediction results of crushing strength, scurf resistance, reactivity index, and post-reaction strength are shown in Figs. 2.8, 2.9, 2.10 and 2.11, in which the predicted values fitted the actual values

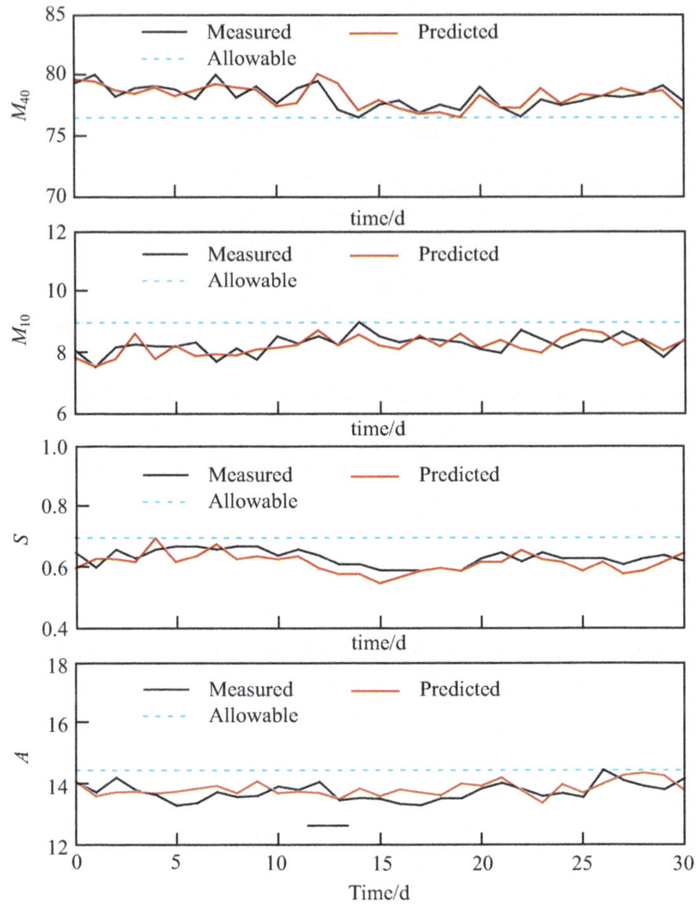

Fig. 2.7 Results of actual runs

with a high precision. It shows the validity of the system. The implementation results have shown that the prediction accuracy of the quality of coke reached 90%, which greatly reduced the production cost of coke.

2.4 Intelligent Integrated Optimization System for Proportioning of Iron Ore in Sintering Process

The intelligent integrated optimization level contains an intelligent integrated optimizer with four parts: optimization of the first proportioning, optimization of the second proportioning, a cascade integrated quality-prediction model, and an LP and GA-PSO algorithm (Fig. 2.4).

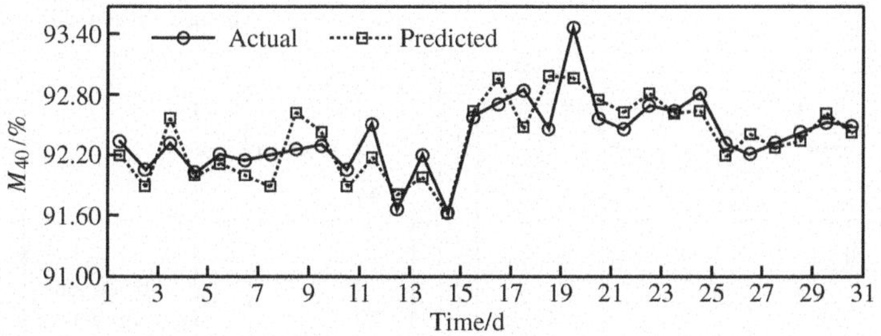

Fig. 2.8 Actual and predicted results of crushing strength

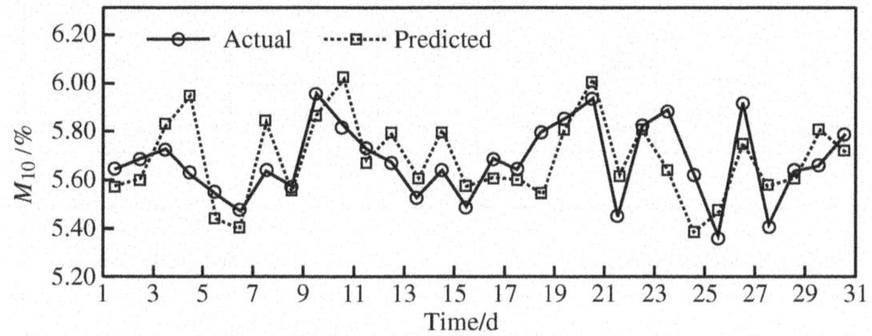

Fig. 2.9 Actual and prediction results of scurf resistance

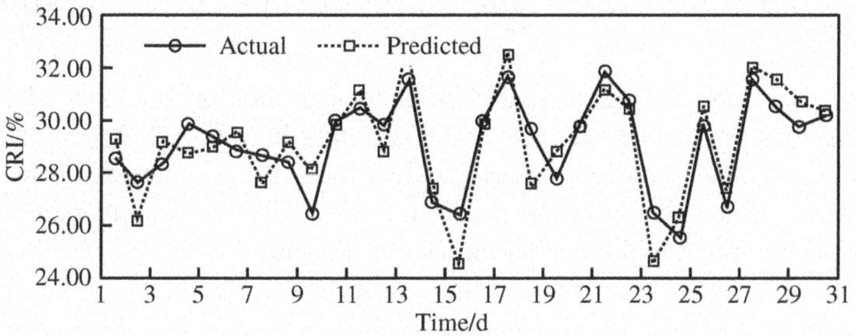

Fig. 2.10 Actual and prediction results of reactivity index

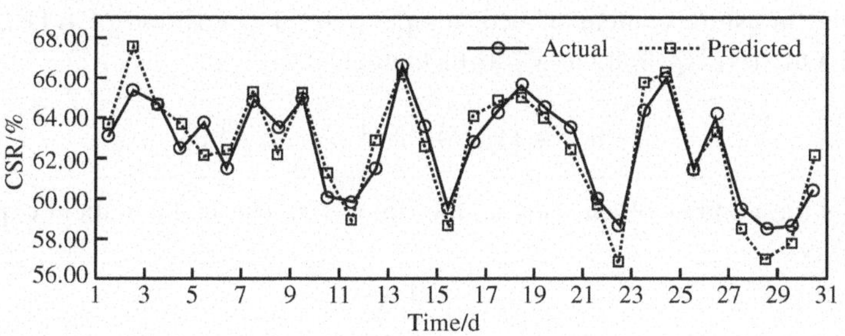

Fig. 2.11 Actual and prediction results of post-reaction strength

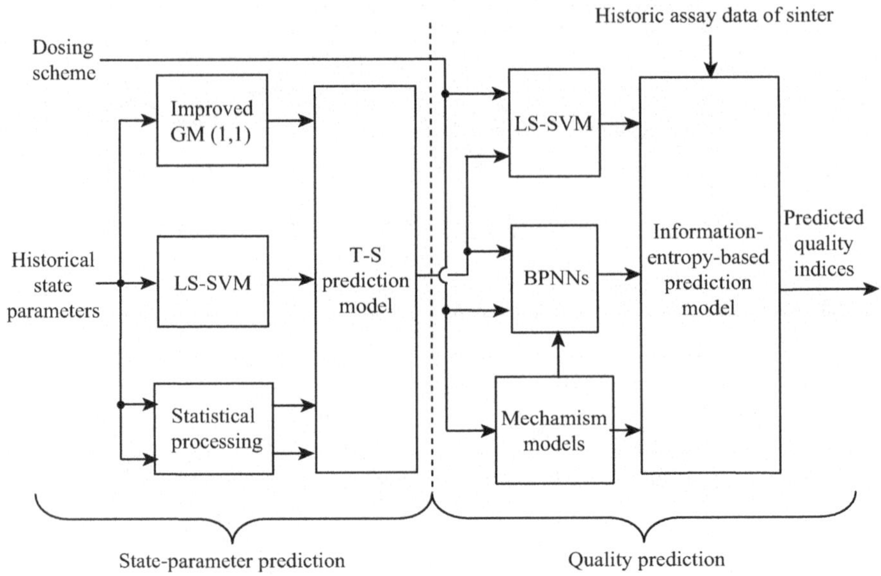

Fig. 2.12 Configuration of cascade integrated quality-prediction model

2.4.1 *Cascade Integrated Quality-Prediction Model for Sinter*

We devised a cascade integrated quality-prediction model for sinter (Fig. 2.12). It has two parts: state-parameter prediction and quality prediction. A T-S prediction model in the state-parameter prediction part predicts the state parameters of the process, and an information entropy-based prediction model in the quality-prediction part estimates the chemical and physical indices of the sinter.

To predict the state parameters, we use an improved GM (1,1) model of the grey system theory and a least-squares support vector machine (LS-SVM) to process time series of the state parameters.

We construct a column vector, y_{sp}, with the entries being the state parameters (height and moisture content of bed, temperature, and position of BTP), and we construct a matrix, Y_{sp}, that has y_{sp} as its elements:

$$Y_{sp}(n) = \left\{ y_{sp}(i) | i = 1, 2, \ldots, n \right\}, \tag{2.43}$$

where n is the number of samples on the time axis. The mathematical expectation for $Y_{sp}(n)$ is

$$\varepsilon(Y_{sp}(n)) = \frac{1}{n} \sum_{i=1}^{n} y_{sp}(i), \tag{2.44}$$

and its standard deviation is

$$\sigma(Y_{\mathrm{sp}}(n)) = \sqrt{\frac{1}{n}\sum_{i=1}^{n}\left[y_{\mathrm{sp}}(i) - \mathcal{E}(Y_{\mathrm{sp}}(n))\right]^2}. \tag{2.45}$$

The relative error of $Y_{\mathrm{sp}}(n)$ is defined to be

$$\epsilon(Y_{\mathrm{sp}}(n)) = \frac{\sigma(Y_{\mathrm{sp}}(n))}{\mathcal{E}(Y_{\mathrm{sp}}(n))} \times 100\%. \tag{2.46}$$

To calculate the change in the current operating conditions, we collect the latest n_{s} elements from $Y_{\mathrm{sp}}(n)$ and construct a historic time series over a short time span:

$$Y_{\mathrm{sp(s)}}(n) = \{y_{\mathrm{sp(s)}}(i) | y_{\mathrm{sp(s)}}(i) = y_{\mathrm{sp}}(n + i - n_{\mathrm{s}}),\ i = 1, 2, \ldots, n_{\mathrm{s}}\}. \tag{2.47}$$

This is input into the improved GM (1,1) and the LS-SVM models.

To estimate the variation in the operating conditions, we also collect the latest n_{l} elements ($n_{\mathrm{l}} \geqslant 3n_{\mathrm{s}}$) of $Y_{\mathrm{sp}}(n)$ and construct a historic time series over a long time span:

$$Y_{\mathrm{sp(l)}}(n) = \left\{y_{\mathrm{sp(l)}}(i) | y_{\mathrm{sp(l)}}(i) = y_{\mathrm{sp}}(n + i - n_{\mathrm{l}}),\ i = 1, 2, \ldots, n_{\mathrm{l}}\right\}. \tag{2.48}$$

To construct the improved GM (1,1) model, we use $Y_{\mathrm{sp(s)}}(n)$ and define

$$\xi_0(i) = \begin{cases} y_{\mathrm{sp(s)}}(i), & i = n - n_{\mathrm{s}} + 1, \\ \beta y_{\mathrm{sp(s)}}(i) + (1 - \beta)\xi_0(i - 1), & i = n - n_{\mathrm{s}} + 2, \ldots, n, \end{cases} \tag{2.49}$$

where β is a smoothing factor in the range [0, 1]. Let

$$\varXi_1 = \left\{ \xi_1(i) \left| \xi_1(i) = \sum_{j=n-n_{\mathrm{s}}+1}^{i} \xi_0(j),\ i = n - n_{\mathrm{s}} + 1, \ldots, n \right. \right\}. \tag{2.50}$$

The GM (1,1) model is

$$\begin{cases} \xi_0(i) + a z_1(i) = b, & i = n - n_{\mathrm{s}} + 2, \ldots, n, \\ z_1(i) = \theta\xi_1(i) + (1 - \theta)\xi_1(i - 1), & i = n - n_{\mathrm{s}} + 2, \ldots, n, \end{cases} \tag{2.51}$$

where a is a development coefficient; b is a grey action quantity; and θ is a coefficient in the range [0, 1], which is often chosen to be 0.5. a and b are obtained by the method of least-squares estimation:

$$\begin{cases} \begin{bmatrix} a \\ b \end{bmatrix} = \left(\Gamma^{\mathrm{T}}\Gamma\right)^{-1}\Gamma^{\mathrm{T}}\Phi, \\ \Gamma = \begin{bmatrix} -z_1(n-n_s+2) & 1 \\ -z_1(n-n_s+3) & 1 \\ \vdots & \vdots \\ -z_1(n) & 1 \end{bmatrix}, \quad \Phi = \begin{bmatrix} \xi_0(n-n_s+2) \\ \xi_0(n-n_s+3) \\ \vdots \\ \xi_0(n) \end{bmatrix}. \end{cases} \tag{2.52}$$

The time response of Eq. (2.51) in discrete form is

$$\tilde{\xi}_1(n+1) = \left[\xi_0(n-n_s+1) - \frac{b}{a}\right]\mathrm{e}^{-an_s} + \frac{b}{a}. \tag{2.53}$$

So, the parameter predicted by the GM (1,1) model is

$$\hat{y}_{\mathrm{spGM}}(n+1) = \tilde{\xi}_0(n+1) = \tilde{\xi}_1(n+1) - \tilde{\xi}_1(n) = \left[\xi_0(n-n_s+1) - \frac{b}{a}\right]\left(1-\mathrm{e}^{a}\right)\mathrm{e}^{-an_s}. \tag{2.54}$$

On the other hand, we also input $Y_{\mathrm{sp(s)}}(n)$ into the LS-SVM model to calculate an estimate of $y_{\mathrm{sp}}(n+1)$. First, we construct l training samples $[(Y_{\mathrm{sp(s)}}(n-1), y_{\mathrm{sp}}(n)), (Y_{\mathrm{sp(s)}}(n-2), y_{\mathrm{sp}}(n-1)), \ldots, (Y_{\mathrm{sp(s)}}(n-l), y_{\mathrm{sp}}(n-l+1)]$ and normalize the training sample set. The LS-SVM problem can be written as

$$\min\ \phi(w,h,e) = \frac{1}{2}w^{\mathrm{T}}w + \frac{1}{2}\gamma\sum_{k=1}^{l}e_k^2, \tag{2.55}$$

$$\text{s.t. } y_{\mathrm{sp}}(n-k+1)\left[w^{\mathrm{T}}\varphi(Y_{\mathrm{sp(s)}}(n-k)) + h\right] = 1 - e_k, \quad k = 1, 2, \ldots, l,$$

where w is a vector of weights, e_k is the fitting error of each training sample ($e = [e_1, e_2, \ldots, e_l]^{\mathrm{T}}$), γ is the penalty factor, h is a constant vector, and $\varphi(x)$ is a nonlinear function. The Lagrange function is defined to be

$$L(w,h,e,\alpha) = \phi(w,h,e) - \sum_{k=1}^{l}\alpha_k\left\{y_{\mathrm{sp}}(n-k+1)\left[w^{\mathrm{T}}\varphi(Y_{\mathrm{sp(s)}}(n-k)) + h\right] - 1 + e_k\right\}, \tag{2.56}$$

where α_k $(k = 1, 2, \ldots, l)$ are Lagrange multipliers ($\alpha = [\alpha_1, \alpha_2, \ldots, \alpha_l]^{\mathrm{T}}$). Letting

$$\frac{\partial L}{\partial w} = 0, \quad \frac{\partial L}{\partial h} = 0, \quad \frac{\partial L}{\partial e} = 0, \quad \frac{\partial L}{\partial \alpha} = 0 \tag{2.57}$$

yields

$$\begin{cases} w = \sum_{k=1}^{l} \alpha_k y_{\rm sp}(n-k+1)\varphi(Y_{\rm sp(s)}(n-k)), \\ \sum_{k=1}^{l} \alpha_k y_{\rm sp}(n-k+1) = 0, \\ \alpha_k = \gamma e_k, \\ y_{\rm sp}(n-k+1)\left[w^{\rm T}\varphi(Y_{\rm sp(s)}(n-k)) + h\right] - 1 + e_k = 0. \end{cases} \tag{2.58}$$

h and α are calculated from the above equations. Thus,

$$f_{\rm LS-SVM}(Y_{\rm sp(s)}(n)) = {\rm sgn}\left[\sum_{k=1}^{l} \alpha_k y_{\rm sp}(n-k+1) K(Y_{\rm sp(s)}(n), Y_{\rm sp(s)}(n-k)) + h\right], \tag{2.59}$$

where $K(Y_{\rm sp(s)}(i), Y_{\rm sp(s)}(j))$ is a kernel function that is chosen to be

$$K(Y_{\rm sp(s)}(i), Y_{\rm sp(s)}(j)) = {\rm e}^{-\frac{\|Y_{\rm sp(s)}(i)-Y_{\rm sp(s)}(j)\|^2}{2\sigma_k^2}} \tag{2.60}$$

with σ_k being a kernel parameter. We obtain $\hat{y}_{\rm spLS\text{-}SVM}(n+1)$ from the reverse normalization $f_{\rm LS\text{-}SVM}(Y_{\rm sp(s)}(n))$.

The improved GM (1,1) is applicable to a time series with a slow rate of change and small variations. However, the prediction accuracy becomes significantly worse when the data of a time series are influenced by a large disturbance or contain many abnormal data. The LS-SVM implements structural risk minimization, and it is good at predicting the state and operating parameters when the data of a time series has large variations.

The T-S prediction model combines the improved GM (1,1) and LS-SVM methods based on the relative error to predict the state parameters. When the variations in the operating conditions are small, that is, when the operating conditions are relatively stable, the integrated model mainly employs the output of the improved GM (1,1) model to produce a precise estimate. Otherwise, when the operating conditions change a great deal, it mainly employs the output of the LS-SVM model to obtain sufficient estimation precision. The membership function of the T-S fuzzy model is based on the relative error of $Y_{\rm sp(s)}(n)$ and $Y_{\rm sp(l)}(n)$, as explained below.

The fuzzy linguistic variables NB, ZO, and PB mean that the variations are negatively big, zero (=very small), and positively big, respectively. Their membership functions are

$$\text{NB}: f_{\rm NB}(x) = \frac{1}{1+{\rm e}^{\frac{x-0.5\mathcal{E}\left(Y_{\rm sp(l)}(n)\right)}{\mathcal{E}\left(Y_{\rm sp(l)}(n)\right)}}}, \tag{2.61}$$

$$\text{ZO}: f_{\rm ZO}(x) = {\rm e}^{-\frac{\left[x-\mathcal{E}(Y_{\rm sp(l)}(n))\right]^2}{2\mathcal{E}^2(Y_{\rm sp(l)}(n))}}, \tag{2.62}$$

$$\text{PB}: f_{\text{PB}}(x) = \frac{1}{1 + e^{-\frac{x - 1.5\mathcal{E}\left(Y_{\text{sp(l)}}(n)\right)}{\mathcal{E}\left(Y_{\text{sp(l)}}(n)\right)}}}. \tag{2.63}$$

On the other hand, we calculate the weighting coefficients using $Y_{\text{sp(s)}}(n)$:

$$w_j = \frac{f_j\left[\mathcal{E}\left(Y_{\text{sp(s)}}(n)\right)\right]}{\sum_j f_j\left[\mathcal{E}\left(Y_{\text{sp(s)}}(n)\right)\right]}, \; j = \text{NB, ZO, PB}. \tag{2.64}$$

Then, we use them to derive estimates of the parameters:

$$\hat{y}_{\text{sp}}(n+1) = \left(w_{\text{NB}} + \frac{w_{\text{ZO}}}{2}\right)\hat{y}_{\text{spGM}}(n+1) + \left(\frac{w_{\text{ZO}}}{2} + w_{\text{PB}}\right)\hat{y}_{\text{spLS-SVM}}(n+1), \tag{2.65}$$

where $\hat{y}_{\text{spGM}}(n+1)$, $\hat{y}_{\text{spLS-SVM}}(n+1)$, and $\hat{y}_{\text{sp}}(n+1)$ are the outputs of the GM (1,1) model, the LS-SVM model, and the T-S prediction model, respectively.

The quality of sinter is determined by $T_{\text{Fe}}^{(2)}$, R, and J_{t}. In this study, we focused on strong generalization and the ability to avoid the local extrema of an LS-SVM, the universal approximation ability of a BPNN, and mechanism models to build an information entropy-based prediction model that integrates them to predict the quality of sinter.

It is possible to build mechanism models for the chemical indices, $T_{\text{Fe}}^{(2)}$ and R, but it is almost impossible to do that for the physical index, J_{t}. For this reason, we need to consider the problem of how to separately predict these two kinds of performance indices, which means that we need different structures to predict them. So, we integrated a mechanism model, a BPNN model, and an LS-SVM model to predict $T_{\text{Fe}}^{(2)}$ and R; and we integrated a BPNN model and an LS-SVM model to predict J_{t}.

Let $d_j^{(2)}$ be the evaporation rate of the jth raw material. The mechanism models used to estimate $T_{\text{FeMech}}^{(2)}$ and R_{Mech} are

$$\hat{T}_{\text{FeMech}}^{(2)} = \frac{\sum_{j=1}^{m} x_j^{(2)} T_{\text{Fe}j}^{(2)}}{\sum_{j=1}^{m}\left(1 - d_j^{(2)}\right)x_j^{(2)}}, \tag{2.66}$$

$$\hat{R}_{\text{Mech}} = \frac{\hat{y}_{\text{CaO}}}{\hat{y}_{\text{SiO}_2}} = \frac{\sum_{j=1}^{m} x_j^{(2)} y_{\text{CaO}j}^{(2)}}{\sum_{j=1}^{m} x_j^{(2)} y_{\text{SiO}_2 j}^{(2)}}. \tag{2.67}$$

Note that $d_j^{(2)}$ $(j = 1, 2, \ldots, m)$ are used in Eq. (2.66). Since we cannot obtain precise values of $d_j^{(2)}$, we use empirical values instead. The mechanism models are sometimes a little conservative.

The BPNN models used to estimate $T_{\rm Fe}^{(2)}$, R, and $J_{\rm t}$ are

$$f(x) = \frac{1}{1 + \mathrm{e}^{-x}}, \tag{2.68}$$

$$\hat{T}_{\rm FeBPNN}^{(2)} = f\left(-\sum_{i=1}^{20} w_{{\rm TO}i} f\left(-\sum_{j=1}^{7} w_{{\rm T}ij} y_j - \beta_{{\rm T}i}\right) - \beta_{\rm TO}\right), \tag{2.69}$$

$$\hat{R}_{\rm BPNN} = f\left(-\sum_{i=1}^{20} w_{{\rm RO}i} f\left(-\sum_{j=1}^{7} w_{{\rm R}ij} y_j - \beta_{{\rm R}i}\right) - \beta_{\rm RO}\right), \tag{2.70}$$

$$\hat{J}_{\rm tBPNN} = f\left(-\sum_{i=1}^{20} w_{{\rm JO}i} f\left(-\sum_{j=1}^{8} w_{{\rm J}ij} y_j - \beta_{{\rm J}i}\right) - \beta_{\rm JO}\right), \tag{2.71}$$

where w_{xij} is the weight of the signal from the jth input neuron to the ith hidden neuron, β_{xi} is the bias of the ith hidden neuron, $w_{x{\rm O}i}$ is the weight of the signal from the ith hidden neuron to the output neuron, and $\beta_{x{\rm O}}$ is the bias of the output neuron, $x = {\rm T, R, J}$. These weights are determined by training the BPNNs.

In order to avoid the problem of overfitting the BPNN model, we first scanned raw data and removed the abnormal ones, and then used a zero-phase-shift filter to smooth them. We chose the number of hidden-layer neurons to be 20 and selected 200 samples for training. The verification results in Sect. 2.4.2 show that overfitting did not occur. These BPNN models all had three layers. The only differences were the input signals and the number of neurons in the input layers. All the models took the proportion of coke, the height and moisture content of the bed, and the temperature and position of BTP as the inputs. $T_{\rm FeMech}^{(2)}$ and $\hat{y}_{\rm SiO_2Mech}^{(2)}$ were additional inputs of the model for $T_{\rm Fe}^{(2)}$; $\hat{y}_{\rm CaOMech}^{(2)}$ and $\hat{y}_{\rm SiO_2Mech}^{(2)}$ were additional inputs of the models for R; and $\hat{y}_{\rm CaOMech}^{(2)}$, $\hat{y}_{\rm SiO_2Mech}^{(2)}$, and $\hat{y}_{\rm MgOMech}^{(2)}$ were additional inputs of the model for $J_{\rm t}$. The models for $T_{\rm Fe}^{(2)}$ and R had seven input neurons, and the model for $J_{\rm t}$ had eight input neurons. As an example, the BPNN model for $T_{\rm Fe}^{(2)}$ is shown in Fig. 2.13.

The concept of the LS-SVM model for the prediction of $T_{\rm Fe}^{(2)}$, R, and $J_{\rm t}$ is the same as that for the prediction of the state parameters and is omitted.

Let $\hat{T}_{{\rm Fe}i}^{(2)}$ and $\hat{R}_i$ be the predicted values of the ith submodel (i = Mech, BPNN, LS-SVM); and let $\hat{J}_{{\rm t}i}$ be the predicted $J_{\rm t}$ of the ith submodel (i = BPNN, LS-SVM). We assume that $x_i(j)$ is the jth predicted x of the ith submodel, $j = 1, 2, \ldots, n$, where n is the number of samples.

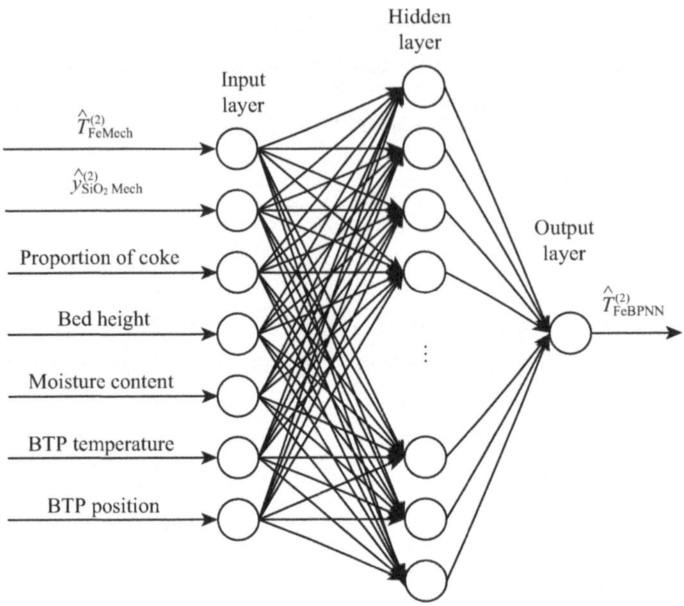

Fig. 2.13 BPNN model for $T^{(2)}_{\mathrm{Fe}}$

The algorithms for the prediction of $T^{(2)}_{\mathrm{Fe}}$, R, and J_{t} are based on information entropy and are almost the same. Here, we use the algorithm for the prediction of R as an example.

Algorithm for the prediction of R:

Step 1: Calculate the relative prediction error

$$\epsilon_{Ri}(j) = \begin{cases} \left|1 - \dfrac{\hat{R}_i(j)}{R(j)}\right|, & 0 \leqslant \left|1 - \dfrac{\hat{R}_i(j)}{R(j)}\right| < 1, \\ 1, & \left|1 - \dfrac{\hat{R}_i(j)}{R(j)}\right| \geqslant 1, \end{cases} \tag{2.72}$$

$$\eta_{Ri}(j) = \frac{\epsilon_{R_i}(j)}{\sum\limits_{i} \epsilon_{R_i}(j)}, \tag{2.73}$$

where $R(j)$ is the value of R measured at the sampling point j.

Step 2: Calculate the entropy of the relative error using the ith submodel

$$E_{Ri} = -\frac{1}{\ln n} \sum_{j=1}^{n} \eta_{R_i}(j) \ln \eta_{R_i}(j). \tag{2.74}$$

Step 3: Determine the weights for the submodels:

$$w_{R\text{Mech}} = \frac{1}{3-1}\left(1 - \frac{1 - E_{R\text{Mech}}}{3 - E_R}\right), \tag{2.75}$$

$$w_{R\text{BPNN}} = \frac{1}{3-1}\left(1 - \frac{1 - E_{R\text{BPNN}}}{3 - E_R}\right), \tag{2.76}$$

$$w_{R\text{LS-SVM}} = \frac{1}{3-1}\left(1 - \frac{1 - E_{R\text{LS-SVM}}}{3 - E_R}\right), \tag{2.77}$$

where

$$E_R = E_{R\text{Mech}} + E_{R\text{BPNN}} + E_{R\text{LS-SVM}}.$$

Step 4: Calculate the predicted value

$$\begin{aligned}\hat{R}(n+1) = {} & w_{R\text{Mech}}\hat{R}_{\text{Mech}}(n+1) + w_{R\text{BPNN}}\hat{R}_{B\text{PNN}}(n+1) \\ & + w_{R\text{LS-SVM}}\hat{R}_{\text{LS-SVM}}(n+1).\end{aligned} \tag{2.78}$$

2.4.2 *Verification of Quality-Prediction Model*

We used actual data collected from a sintering machine with an area of 360 m^2 to verify the effectiveness of the prediction models. A time sequence of measured values is denoted by $y(i)$ $(i = 1, 2, \ldots, n_\text{v})$, and the corresponding predicted sequence is denoted by $\hat{y}(i)$ $(i = 1, 2, \ldots, n_\text{v})$. Five items are employed in the evaluation:

(1) Mean value of residuals:

$$e = \frac{1}{n_\text{v}} \sum_{i=1}^{n_\text{v}} \left|\hat{y}(i) - y(i)\right|. \tag{2.79}$$

(2) Standard deviation:

$$\sigma = \sqrt{\frac{1}{n_\text{v}} \sum_{i=1}^{n_\text{v}} \left[\hat{y}(i) - y(i)\right]^2}. \tag{2.80}$$

(3) Maximum error:

$$e_{\max} = \max_i \left\{\left|\hat{y}(i) - y(i)\right|\right\}. \tag{2.81}$$

(4) Accuracy

$$\rho = P\left(\hat{y}(i) \in [y(i) - \zeta, y(i) + \zeta]\right). \tag{2.82}$$

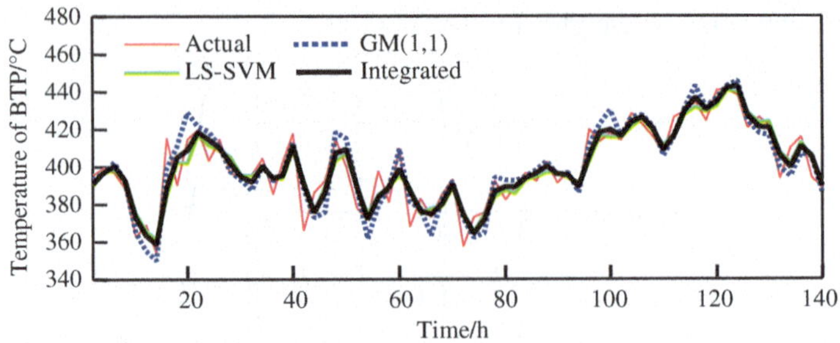

Fig. 2.14 Comparison of prediction results for temperature of BTP

Table 2.1 Prediction results for temperature of BTP for different models

Model	e	σ	$e_{\max}$	$\rho\ (\zeta = 0.5)$	P_r
Improved GM (1,1)	7.4063	9.2238	22.4130	0.9683	0.9769
LS-SVM	6.8540	8.6761	26.00	0.9683	0.9783
Integrated model	6.2336	8.1204	25.1061	0.9841	0.9797

where P is the probability function and ζ is the threshold.

(5) Precision:

$$P_r = \left(1 - \frac{\sigma}{y_{avg}}\right) \times 100\%,\ y_{avg} = \frac{1}{n_v} \sum_{i=1}^{n_v} y(i). \tag{2.83}$$

A larger e means larger variations, and the same is true for σ and $e_{\max}$. The larger ρ and P_r are, the more accurate and more precise the prediction is. The larger σ and P_r are, the better the prediction model is.

First, we used the temperature of the BTP as an example to illustrate the verification of a state-parameter prediction. The verification results are shown in Fig. 2.14.

It is clear from Fig. 2.14 and Table 2.1 that the integration of the improved GM (1,1) and LS-SVM models produced a smaller prediction error than either individual model. So, the integrated model has a higher accuracy and smaller variations than the conventional models do.

We used the prediction of $T_{Fe}^{(2)}$ as an example to illustrate the verification of the CIQPM. Figure 2.15 shows the timewise change in the actual value of $T_{Fe}^{(2)}$ along with the values predicted by the mechanism model, the LS-SVM model, the BPNN model, the one-stage integrated model, and the CIQPM. Note that the one-stage integrated model used only the quality-prediction part; instead of using the state-parameter prediction part, it used $y(j-1)$ to approximate $y(j)$ so as to predict the quality.

The statistical data in Table 2.2 show that the prediction error is the largest for the one-stage integrated model and the smallest for the CIQPM. The LS-SVM, BPNN, and mechanism models combined with state-parameter prediction produced smaller

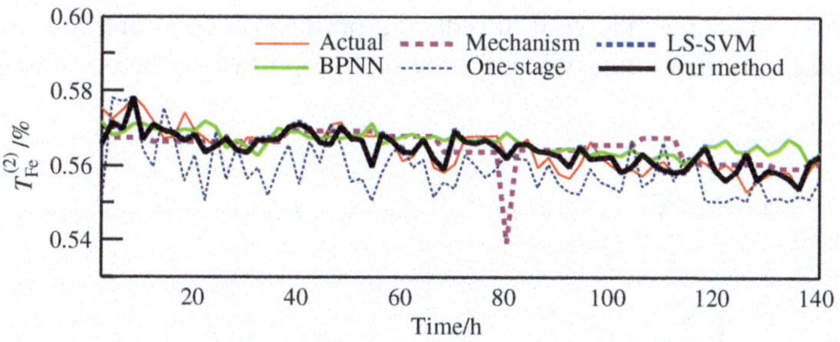

Fig. 2.15 Prediction results for iron grade

Table 2.2 Prediction results for iron grade for different models (OSIM: one-stage integrated model)

Model	e	σ	e_{max}	ρ ($\zeta = 0.005$)	P_r
Mechanism model	0.0046	0.0061	0.0233	0.6262	0.9892
LS-SVM	0.0027	0.0041	0.0138	0.8037	0.9270
BPNN	0.0047	0.0057	0.0140	0.5981	0.9898
OSIM	0.0069	0.0085	0.0248	0.3738	0.9849
CIQPM	0.0026	0.004	0.0125	0.8131	0.9929

prediction errors than the one-stage integrated model did. This shows that the state-parameter prediction part increases the prediction precision.

2.4.3 *Optimization Model of Proportioning*

This subsection explains the optimization of the first and second proportionings, and the LP and GA-PSO in the IIOS.

The objective is to find an optimal dosing scheme that minimizes production costs and SO_2 emissions. This optimization problem is nonlinear and has multiple constraints. Conventional optimization methods do not provide a satisfactory solution. In this subsection, we first formulated the design problem as a multi-objective optimization problem. Then, we devised an intelligent integrated optimization method that integrates the LP, GA, and PSO methods to solve it. The result is sent to a DCS as a reference input for quality control of the sinter.

Let n_1 be the number of different kinds of iron ore. For $i = 1, 2, \ldots, n$ and $k = T_{Fe}^{(2)}, SiO_2, CaO, MgO, Al_2O_3, S, P$, assume that $y_{ci}^{(1)}$ is the cost of the ith kind of iron ore and $y_{Si}^{(1)}$ is its S content; $x_i^{(1)}$ is the proportion of the ith kind of iron ore in the first proportioning phase; $x_{i\min}^{(1)}$ and $x_{i\max}^{(1)}$ are the lower and upper bounds on $x_i^{(1)}$, respectively; $y_{ki}^{(1)}$ is the percentage of the kth chemical component of the ith kind of iron ore; and $y_{k\min}^{(1)}$ and $y_{k\max}^{(1)}$ are the lower and upper bounds on $y_k^{(1)}$ of the iron ore mix. To achieve the goal of reducing production costs and SO_2 emissions, we define the optimization problem for the first proportioning phase to be

$$\min F^{(1)}, \tag{2.84}$$

$$F^{(1)} = \sum_{i=1}^{n} \left(y_{ci}^{(1)} + K_S^{(1)} y_{Si}^{(1)} \right) x_i^{(1)}, \quad K_S^{(1)} = \frac{1}{n} \sum_{i=1}^{n} \frac{y_{ci}^{(1)}}{y_{Si}^{(1)}}, \tag{2.85}$$

$$\text{s.t.} \begin{cases} \sum_{i=1}^{n} x_i^{(1)} = 1, \\ x_{i\min}^{(1)} \leqslant x_i^{(1)} \leqslant x_{i\max}^{(1)}, \\ y_{k\min}^{(1)} \leqslant \sum_{i=1}^{n} y_{ki}^{(1)} x_i^{(1)} \leqslant y_{k\max}^{(1)}. \end{cases} \tag{2.86}$$

In the same manner, we define the optimization problem for the second proportioning phase to be

$$\min F^{(2)}, \tag{2.87}$$

$$F^{(2)} = \sum_{j=1}^{m} \left(y_{cj}^{(2)} + K_S^{(2)} y_{Sj}^{(2)} \right) x_j^{(2)}, \quad K_S^{(2)} = \frac{1}{m} \sum_{j=1}^{m} \frac{y_{cj}^{(2)}}{y_{Sj}^{(2)}}, \tag{2.88}$$

$$\text{s.t.} \begin{cases} \sum_{j=1}^{m} x_j^{(2)} = 1, \\ 0 \leqslant x_{j\min}^{(2)} \leqslant x_j^{(2)} \leqslant x_{j\max}^{(2)} \leqslant 1, \\ y_{k\min}^{(2)} \leqslant y_k \leqslant y_{k\max}^{(2)}, \\ R_{\min} \leqslant R \leqslant R_{\max}. \end{cases} \tag{2.89}$$

The raw materials used in the second proportioning phase are returned sinter, iron ore mix, coke, limestone, and dolomite. We assume that there are m kinds of raw materials and use the order number j to indicate them ($j = 1, 2, \ldots, m$). In Eqs. (2.87) and (2.89), $x_j^{(2)}$ is the proportion of the jth raw material; $x_{j\min}^{(2)}$ and $x_{j\max}^{(2)}$ are the lower and upper bounds on $x_j^{(2)}$, respectively; $y_{cj}^{(2)}$ and $y_{Sj}^{(2)}$ are the cost and S content of the jth raw material; $y_{kj}^{(2)}$ is the percentage of the kth chemical component of the jth raw material; $y_{k\min}^{(2)}$ and $y_{k\max}^{(2)}$ are the lower and upper bounds on $y_k^{(2)}$ of the sinter; and $R_{\min}$ and $R_{\max}$ are the lower and upper bounds on R.

If we let $d_j^{(2)}$ be the evaporation rate of the jth raw material, then the predicted value of the kth chemical component, $\hat{y}_{k\text{Mech}}$, is given by

$$\hat{y}_{k\text{Mech}} = \frac{\sum_{j=1}^{m} x_j^{(2)} y_{kj}^{(2)}}{\sum_{j=1}^{m} \left(1 - d_j^{(2)}\right) x_j^{(2)}}, \quad k = SiO_2, CaO, MgO, Al_2O_3, \tag{2.90}$$

$$\hat{y}_{k\mathrm{Mech}} = \frac{\sum_{j=1}^{m}\left(1 - d_j^{(2)}\right) x_j^{(2)} y_{kj}^{(2)}}{\sum_{j=1}^{m}\left(1 - d_j^{(2)}\right) x_j^{(2)}}, \quad k = \mathrm{S}, \mathrm{P}. \tag{2.91}$$

And $\hat{T}_{\mathrm{FeMech}}^{(2)}$ and $\hat{R}_{\mathrm{Mech}}$ are given in Eqs. (2.66) and (2.67), respectively.

Although the constraints on the optimization problem for the second proportioning phase are nonlinear, an equivalent transformation converts the nonlinear constraints into linear ones. We take the S content and R as examples to illustrate the process. The original nonlinear constraint on $y_{\mathrm{S}j}^{(2)}$ is

$$y_{\mathrm{S}\min}^{(2)} \leqslant \frac{\sum_{j=1}^{m}\left(1 - d_j^{(2)}\right) x_j^{(2)} y_{\mathrm{S}j}^{(2)}}{\sum_{j=1}^{m}\left(1 - d_j^{(2)}\right) x_j^{(2)}} \leqslant y_{\mathrm{S}\max}^{(2)}. \tag{2.92}$$

It can be converted into the linear constraints

$$\begin{cases} \sum_{j=1}^{m}\left(1 - d_j^{(2)}\right) x_j^{(2)} \left(y_{\mathrm{S}j}^{(2)} - y_{\mathrm{S}\min}^{(2)}\right) \geqslant 0, \\ \sum_{j=1}^{m}\left(1 - d_j^{(2)}\right) x_j^{(2)} \left(y_{\mathrm{S}j}^{(2)} - y_{\mathrm{S}\max}^{(2)}\right) \leqslant 0. \end{cases} \tag{2.93}$$

In a similar manner, the constraints on $\hat{R}$:

$$R_{\min} \leqslant R = \frac{\sum_{j=1}^{m} y_{\mathrm{CaO}j}^{(2)} x_j^{(2)}}{\sum_{j=1}^{m} y_{\mathrm{SiO_2}j}^{(2)} x_j^{(2)}} \leqslant R_{\max}, \tag{2.94}$$

can be converted into

$$\begin{cases} \sum_{j=1}^{m} x_j^{(2)} \left(y_{\mathrm{CaO}j}^{(2)} - R_{\min} y_{\mathrm{SiO_2}j}^{(2)}\right) \geqslant 0, \\ \sum_{j=1}^{m} x_j^{(2)} \left(y_{\mathrm{CaO}j}^{(2)} - R_{\max} y_{\mathrm{SiO_2}j}^{(2)}\right) \leqslant 0. \end{cases} \tag{2.95}$$

The optimal dosing scheme is sent to the CIQPM (Fig. 2.4) to predict the quality of the sinter. If the predicted quality satisfies the requirements, then that scheme is used for the proportioning step. However, if the predicted quality does not meet the requirements, then we have to adjust the specifications in Eqs. (2.86) and (2.89) based on the experience of experts, and carry out the optimization again to find a feasible dosing scheme. For example, if $T_{\mathrm{Fe}}^{(2)}$ is out of the allowable range, adjusting $y^{(1)}_{T_{\mathrm{Fe}}\,\min}$ and $y^{(1)}_{T_{\mathrm{Fe}}\,\max}$ based on the differences $\Delta\underline{T}_{\mathrm{Fe}}^{(2)} = \hat{T}_{\mathrm{Fe}}^{(2)} - T^{(2)}_{\mathrm{Fe}\,\min}$ and $\Delta\overline{T}_{\mathrm{Fe}}^{(2)} = T^{(2)}_{\mathrm{Fe}\,\max} - \hat{T}_{\mathrm{Fe}}^{(2)}$ can regulate $T_{\mathrm{Fe}}^{(2)}$ over a large range. In addition, adjusting $y^{(2)}_{T_{\mathrm{Fe}}\,\min}$ and $y^{(2)}_{T_{\mathrm{Fe}}\,\max}$ based on $\Delta\underline{T}_{\mathrm{Fe}}^{(2)}$ and $\Delta\overline{T}_{\mathrm{Fe}}^{(2)}$ can regulate $T_{\mathrm{Fe}}^{(2)}$ over a small range. In the same manner, if R is out of the allowable range, adjusting $y^{(1)}_{\mathrm{CaO}\,\min}$, $y^{(1)}_{\mathrm{CaO}\,\max}$, $y^{(1)}_{\mathrm{SiO_2}\,\min}$, and $y^{(1)}_{\mathrm{SiO_2}\,\max}$ based on $\Delta\underline{R} = \hat{R} - R_{\min}$ and $\Delta\overline{R} = R_{\max} - \hat{R}$ can regulate R over a small range. And adjusting $y^{(2)}_{\mathrm{SiO_2}\,\min}$, $y^{(2)}_{\mathrm{SiO_2}\,\max}$, $y^{(2)}_{\mathrm{CaO}\,\min}$, and $y^{(2)}_{\mathrm{CaO}\,\max}$ can regulate R over a large range.

2.4.4 Optimization Method

The LP method is the simplest and most reliable algorithm for solving a linear optimization problem. However, it may not find a solution for two reasons:

(1) A computing error causes degradation in the simplex algorithm.

(2) Irrational constraints may make it difficult to find a solution.

These problems are solved by combining the LP with the GA-PSO methods. The first problem is solved using the GA-PSO method to carry out an additional search. And the second problem is solved using a GA-PSO fitness function. Even though the original problem may not have a feasible solution, we can find an infeasible solution that minimizes the fitness function. This solution is meaningful in actual production because it can be used as a guide in the production process.

We first use the LP method to optimize the proportioning step. If either of the above problems occurs, we turn to use the GA-PSO to find an optimal solution.

The selection of the fitness function and the initial particles in the swarm for the GA-PSO method are explained below.

The penalty-function method is used to convert the constraints into a fitness function with penalty factors. This transforms the original constrained optimization problem into an unconstrained optimization problem. So, the fitness function of the GA-PSO method has two parts: a performance index and a penalty function.

For $k = T_{\mathrm{Fe}}^{(2)}$, $\mathrm{SiO_2}$, CaO, MgO, $\mathrm{Al_2O_3}$, S, P, the fitness function for the first proportioning phase is

$$J^{(1)} = F^{(1)} + Q_0^{(1)} + Q_1^{(1)} + Q_2^{(1)}, \tag{2.96}$$

where

$$\begin{cases} Q_0^{(1)} = r_0^{(1)}\left(\sum_{i=1}^{n} x_i^{(1)} - 1\right)^2, \\ Q_1^{(1)} = \sum_{i=1}^{n} r_i^{(1)}\left(\max\left\{\max\left\{0, x_i^{(1)} - x_{i\max}^{(1)}\right\}, \max\left\{0, x_{i\min}^{(1)} - x_i^{(1)}\right\}\right\}\right)^2, \\ Q_2^{(1)} = \sum_{i=n+1}^{n+7} r_i^{(1)}\left(\max\left\{\max\left\{0, \sum_{i=1}^{n} y_{ki}^{(1)} x_i^{(1)} - y_{k\max}^{(1)}\right\},\right.\right. \\ \left.\left.\max\left\{0, y_{k\min}^{(1)} - \sum_{i=1}^{n} y_{ki}^{(1)} x_i^{(1)}\right\}\right\}\right)^2. \end{cases} \tag{2.97}$$

The first proportioning is given by

$$\min J^{(1)}. \tag{2.98}$$

The fitness function for the second proportioning phase is

$$J^{(2)} = F^{(2)} + Q_0^{(2)} + Q_1^{(2)} + Q_2^{(2)} + Q_3^{(2)}, \tag{2.99}$$

where

$$\begin{cases} Q_0^{(2)} = r_0^{(2)}\left(\sum_{j=1}^{m} x_j^{(2)} - 1\right)^2, \\ Q_1^{(2)} = \sum_{j=1}^{m} r_j^{(2)}\left(\max\left\{\max\left\{0, x_j^{(2)} - x_{j\max}^{(2)}\right\}, \max\left\{0, x_{j\min}^{(2)} - x_j^{(2)}\right\}\right\}\right)^2, \\ Q_2^{(2)} = \sum_{j=m+1}^{m+7} r_j^{(2)}\left(\max\left\{\max\left\{0, y_k^{(2)} - y_{k\max}^{(2)}\right\}, \max\left\{0, y_{k\min}^{(2)} - y_k^{(2)}\right\}\right\}\right)^2, \\ Q_3^{(2)} = r_R^{(2)}\left(\max\left\{\max\left\{0, R - R_{\max}\right\}, \max\left\{0, R_{\min} - R\right\}\right\}\right)^2. \end{cases} \tag{2.100}$$

The second proportioning is given by

$$\min J^{(2)}. \tag{2.101}$$

In the above equations, $F^{(1)}$ and $F^{(2)}$ are the performance indices given in Eqs. (2.85) and (2.88), respectively; $Q_0^{(1)}$ and $Q_0^{(2)}$ are penalty items for normalization; $Q_1^{(1)}$ and $Q_1^{(2)}$ are penalty items for the constraints in the proportioning phases; $Q_2^{(1)}$ and $Q_2^{(2)}$ are penalty items for the constraints on the chemical components; $Q_3^{(2)}$ is a penalty item for R in the second proportioning phase and $r_i^{(1)}$ $(i = 0, 1, \ldots, n+7)$ and $r_j^{(2)}$ $(j = 0, 1, \ldots, m+7)$ are penalty parameters.

We take the first proportioning phase as an example to explain how to select the penalty parameters. $r_0^{(1)}$ is a penalty parameter for the normalization constraint,

Table 2.3 Priorities and factors for penalty parameters for first proportioning phase

Penalty parameter	Priority	Factor
$r_0^{(1)}$	First	10^4
$r_i^{(1)}$ $(i = 1, 2, \ldots, n)$	Second	10^3
$r_i^{(1)}$ $(i = n+1, n+2, n+3, n+6)$	Third	10^2
$r_i^{(1)}$ $(i = n+4, n+5, n+7)$	Fourth	10

$r_i^{(1)}$ are those for the constraints on iron ore proportioning for $i = 1, 2, \ldots, n$, and are those for the constraints on the chemical components of the iron ore mix for $i = n+1, n+2, \ldots, n+7$. Priorities and correlated factors are assigned to these parameters (Table 2.3) on the following basis. The highest priority is assigned to $r_0^{(1)}$ because the proportioning has to be carried out at least under this normalization constraint. The penalty parameters for constraints on iron ore proportioning are the second because the plan can only be carried out if the raw materials is in stock. Those for T_{Fe}, SiO_2, CaO, and S are the third because these chemical components are directly related to the quality of sinter and the amount of SO_2 emissions. And those for MgO, Al_2O_3, and P are the fourth because they also affect the quality of sinter. These penalty parameters have to meet the requirements

$$
\begin{aligned}
r_0^{(1)} &> \max\left\{r_1^{(1)}, r_2^{(1)}, \ldots, r_n^{(1)}\right\} \\
&> \max\left\{r_{n+1}^{(1)}, r_{n+2}^{(1)}, r_{n+3}^{(1)}, r_{n+6}^{(1)}\right\} \\
&> \max\left\{r_{n+4}^{(1)}, r_{n+5}^{(1)}, r_{n+7}^{(1)}\right\}.
\end{aligned}
\tag{2.102}
$$

Regarding the selection of $r_0^{(1)}$, since its precision must be as high as 0.01%, we chose $r_0^{(1)}$ based on the factor assigned in Table 2.3 and the precision and selected a value around $10^4/0.01\% = 10^8$.

Since the constraints on the proportions of the first and second proportioning phases in the fitness functions are related to the stock of raw materials, they must be in proper ranges and the resulting proportions must meet them. As a result, $Q_0^{(1)}$, $Q_1^{(1)}$, $Q_0^{(2)}$, and $Q_1^{(2)}$ must be zero when we select the initial particles in the swarm.

Selection of initial particles in the first proportioning phase:

Step 1: Produce initial particles in the swarm as follows:

$$
\begin{cases}
X_l^{(1)} = \{x_{li}^{(1)}\}, \\
x_{li}^{(1)} = \alpha_{li}^{(1)} x_{i\,\min}^{(1)} + \left(1 - \alpha_{li}^{(1)}\right) x_{i\,\max}^{(1)}, \\
\qquad i = 1, 2, \ldots, n_{\mathrm{d}},\ l = 1, 2, \ldots, n_{\mathrm{p}},
\end{cases}
\tag{2.103}
$$

where l is the number of particles, i is the number of decision variables, $X_l^{(1)}$ is the coordinate of the lth particle, and $\alpha_{li}^{(1)}$ is a coefficient randomly selected in the range [0, 1].

Step 2: Calculate the one-norm of $X_l^{(1)}$ and its gain margin from 1, $G_l^{(1)}$:

$$\begin{cases} \left\| X_l^{(1)} \right\|_1 = \sum_{i=1}^{n_d} \left| x_{li}^{(1)} \right|, \\ G_l^{(1)} = 1 - \left\| X_l^{(1)} \right\|_1. \end{cases} \tag{2.104}$$

Step 3: Initialize the particle swarm positions, $P_l^{(1)}$:

$$\begin{cases} P_l^{(1)} = \left\{ p_{li}^{(1)} \right\}, \\ p_{li}^{(1)} = x_{li}^{(1)} + \Delta x_{li}^{(1)}, \\ \Delta x_{li}^{(1)} = \dfrac{\delta x_{li}^{(1)}}{\sum_{i=1}^{n_d} \delta x_{li}^{(1)}} G_l^{(1)}, \end{cases} \tag{2.105}$$

$$\delta x_{li}^{(1)} = \begin{cases} x_{i\max}^{(1)} - x_{li}^{(1)}, & G_l^{(1)} \geqslant 0, \\ x_{li}^{(1)} - x_{i\min}^{(1)}, & G_l^{(1)} < 0. \end{cases} \tag{2.106}$$

Step 4: Calculate the maximum initialized velocity of the lth particle, $v_{l\max}^{(1)}$:

$$v_{l\max}^{(1)} = \min \left\{ \left| \delta x_{li}^{(1)} \right| \right\}. \tag{2.107}$$

Step 5: Produce an initial velocity for the lth particle, $v_{li}^{(1)}$:

$$\begin{cases} v_{li}^{(1)} = \gamma_{li}^{(1)} v_{l\max}^{(1)}, \\ V_l^{(1)} = \left\{ v_{li}^{(1)} \right\}, \end{cases} \tag{2.108}$$

where

$$\gamma_{li}^{(1)} = \chi_{li}^{(1)} - \frac{1}{n_{\mathrm{d}}} \sum_{i=1}^{n_d} \beta_{li}^{(1)}, \tag{2.109}$$

and $\chi_{li}^{(1)}$ is a random number in the range [0, 1].

The above steps yield the initial positions $P_l^{(1)}$ and velocities $V_l^{(1)}$ that meet the requirements $Q_0^{(1)} = 0$ and $Q_1^{(1)} = 0$.

The PSO and GA have the advantages of easy implementation, high accuracy, and fast convergence. However, they also have the drawback of being easily trapped in a local minimum. To overcome it, we combine the two algorithms. That is, we first carry out a search using the PSO algorithm; and if it falls into a local minimum, we use the GA algorithm to carry out a mutation and deviation to jump out of the local minimum so as to continue searching. The search algorithm is given below. Note that, since this algorithm is used for both the first and second proportioning optimizations, we removed the superscripts (1) and (2).

GA-PSO search algorithm:

Step 1: Choose four numbers: the maximum number of iterations, $n_{I\,\max}$; the maximum number of iterations that avoids the stagnation of PSO, $n_{1\,\max}$; the maximum number of iterations that avoids the stagnation of the GA and PSO, $n_{2\,\max}$; and the critical number of iterations for switching from PSO to the GA, ε_{p}. Initialize the particle swarm, the number of iterations ($i = 0$), the counter for the stagnation of PSO ($n_1 = 0$), and the counter for overall stagnation ($n_2 = 0$).

Step 2: Calculate the fitness value of each particle, $pIbest_j(0)$, and find the optimal global value $pGbest(0)$.

Step 3: IF ($i > n_{I\,\max}$) OR ($n_2 > n_{2\,\max}$), go to Step 8. Otherwise, let $i = i + 1$, and renew the velocities and positions of the particle swarm:

$$\begin{cases} V_l(i+1) = wV_l(i) + C_1R_1[pIbest_l(i) - P(i)] + C_2R_2[pGbest(i) - P(i)], \\ P_l(i+1) = P_l(i) + V_l(i+1). \end{cases} \tag{2.110}$$

Step 4: Calculate the fitness function and renew $pIbest_j(i)$ and $pGbest(i)$.

Step 5: If $pGbest(i)$ is the same as for the last search, then let $n_1 = n_1 + 1$ and $n_2 = n_2 + 1$.

Step 6: If $n_1 \leqslant n_{1\,\max}$, go to Step 3. Otherwise, calculate the standard deviation of $P_l(k)$:

$$\begin{cases} \sigma_{\mathrm{p}} = \sqrt{\dfrac{1}{n_{\mathrm{p}}} \sum\limits_{l=1}^{n_{\mathrm{p}}} \left\| P_l(i) - P_{\mathrm{avg}}(i) \right\|_2}, \\ P_{\mathrm{avg}}(i) = \left\{ p_{\mathrm{avg}j}(i) \right\}, \quad p_{\mathrm{avg}j}(i) = \dfrac{1}{n_{\mathrm{p}}} \sum\limits_{l=1}^{n_{\mathrm{p}}} p_{lj}(i). \end{cases} \tag{2.111}$$

Step 7: If $\sigma_{\mathrm{p}} > \varepsilon_{\mathrm{p}}$, let $n_1 = n_1 - 1$ and go to Step 3. Otherwise, let $n_1 = 0$, switch from PSO to the GA, and renew the positions and velocities of the particle swarm. Choose the probability of crossover to be 0.3 and carry out crossover as follows:

$$\begin{cases} P_l(i+1) = \beta P_l(i) + (1-\beta) P_m(i), \\ P_m(i+1) = (1-\beta) P_l(i) + \beta P_m(i), \\ V_l(i+1) = \gamma V_l(i) + (1-\gamma) V_m(i), \\ V_m(i+1) = (1-\gamma) V_l(i) + \gamma V_m(i), \end{cases} \tag{2.112}$$

where β and γ are random numbers in the range [0, 1]. Choose the probability of mutation to be 0.05 and carry out a mutation (Fig. 2.16). In the figure, p_{ls} and p_{lt} are selected randomly. Then, go to Step 3.

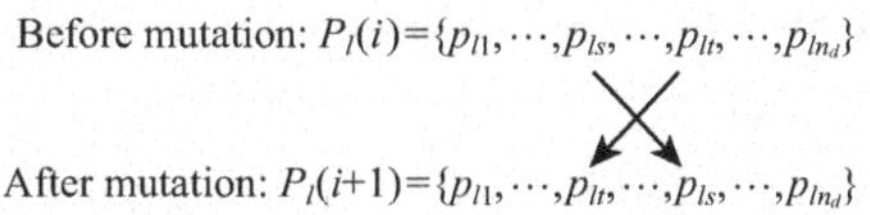

Fig. 2.16 Mutation

Step 8: Output the global optimal value $pGbest(k)$, and the corresponding particle positions, P_l.

Step 9: Stop.

2.4.5 *Verification of Optimization Algorithms*

We used data from actual runs of the sintering process in an iron and steel plant to verify our proportioning method. Note that the effect of reducing sulfur emissions and material costs depend strongly on the allowable ranges of the inputs for the first and second proportionings, and they are not always as large as expected because they are determined by the available raw materials.

Table 2.4 shows the chemical compositions and prices of iron ores. The allowable ranges of the proportions of the iron ores for the first proportioning phase are as follows:

(1) $x_{1\,\min}^{(1)} = x_{1\,\max}^{(1)} = 1\%$.
(2) $x_{2\,\min}^{(1)} = 15\%,\ x_{2\,\max}^{(1)} = 25\%$.
(3) $x_{3\,\min}^{(1)} = x_{3\,\max}^{(1)} = 6.5\%$.
(4) $x_{4\,\min}^{(1)} = 8.5\%,\ x_{4\,\max}^{(1)} = 8.5\%$.
(5) $x_{5\,\min}^{(1)} = 15\%,\ x_{5\,\max}^{(1)} = 25\%$.
(6) $x_{6\,\min}^{(1)} = 10\%,\ x_{6\,\max}^{(1)} = 20\%$.
(7) $x_{7\,\min}^{(1)} = x_{7\,\max}^{(1)} = 2\%$.
(8) $x_{8\,\min}^{(1)} = x_{8\,\max}^{(1)} = 5\%$.
(9) $x_{9\,\min}^{(1)} = 10\%,\ x_{9\,\max}^{(1)} = 20\%$.

And the specifications for the first proportioning phase are as follows:

(1) $T_{\mathrm{Fe}\,\min}^{(1)} = 60.75\%,\ T_{\mathrm{Fe}\,\max}^{(1)} = 100.00\%$.
(2) $y_{\mathrm{SiO_2}\,\min}^{(1)} = 4.00\%,\ y_{\mathrm{SiO_2}\,\max}^{(1)} = 6.00\%$.
(3) $y_{\mathrm{CaO}\,\min}^{(1)} = 0.00\%,\ y_{\mathrm{CaO}\,\max}^{(1)} = 1.00\%$.
(4) $y_{\mathrm{MgO}\,\min}^{(1)} = 0.00\%,\ y_{\mathrm{MgO}\,\max}^{(1)} = 1.00\%$.
(5) $y_{\mathrm{Al_2O_3}\,\min}^{(1)} = 0.00\%,\ y_{\mathrm{Al_2O_3}\,\max}^{(1)} = 2.10\%$.
(6) $y_{\mathrm{S}\,\min}^{(1)} = 0.00\%,\ y_{\mathrm{S}\,\max}^{(1)} = 1.00\%$.
(7) $y_{\mathrm{P}\,\min}^{(1)} = 0.00\%,\ y_{\mathrm{P}\,\max}^{(1)} = 1.00\%$.

Table 2.5 shows the optimal dosing schemes for the first proportioning phase obtained by the LP method and by the LP and GA-PSO method. It lists the proportions of the iron ores. Table 2.6 shows performance results for these dosing schemes. While the goal of the LP method is just the optimization of cost, the goal of the LP and

Table 2.4 Chemical compositions (%) and prices of iron ores

Iron ore	$T_{Fe}^{(1)}$	$y_{SiO_2}^{(1)}$	$y_{CaO}^{(1)}$	$y_{MgO}^{(1)}$	$y_{Al_2O_3}^{(1)}$	$y_S^{(1)}$	$y_P^{(1)}$	Price/(CNY/t)
Ore 1	64.87	4.55	0.15	0.10	1.78	0.02	0.08	1185.00
Ore 2	59.11	4.30	0.10	0.05	2.00	0.03	0.05	1011.00
Ore 3	60.74	7.70	0.39	0.10	1.70	0.01	0.04	1115.00
Ore 4	65.50	4.58	0.03	0.03	0.70	0.01	0.03	1175.00
Ore 5	61.40	4.40	0.10	0.11	1.90	0.04	0.07	1082.00
Ore 6	61.20	4.20	0.10	0.10	2.03	0.02	0.06	1082.00
Ore 7	64.90	3.87	0.03	0.03	0.80	0.01	0.05	1140.00
Ore 8	55.30	5.50	0.09	0.10	6.13	0.01	0.03	864.00
Ore 9	63.20	7.50	1.50	0.86	1.50	0.31	0.04	1075.00

Table 2.5 Optimal proportioning schemes for iron ores for first proportioning phase

Iron ore	LP (cost reduction) (%)	LP and GA-PSO (reduction of cost and S) (%)
Ore 1	1.00	1.00
Ore 2	25.00	25.00
Ore 3	6.50	6.50
Ore 4	8.50	8.50
Ore 5	18.51	22.00
Ore 6	13.49	20.00
Ore 7	2.00	2.00
Ore 8	5.00	5.00
Ore 9	20.00	10.00

GA-PSO method is the optimization of both cost and SO_2 emissions. It is clear from these tables that the LP method yielded the lowest unit cost of the raw materials (1064.20 CNY/t). Compared to the LP method, the LP and GA-PSO method uses about 34% less S but at a unit cost that is 0.7 CNY/t higher. This shows that the LP and GA-PSO method yields a balanced dosing scheme that optimizes both cost and SO_2 emissions.

Then, we optimized the second proportioning based on the results for the first proportioning phase. The chemical compositions and prices of the raw materials for this phase are shown in Table 2.7.

The rate of burning loss, d_i $(i = 1, 2, \ldots, 7)$, and the specifications for the second proportioning phase are as follows:

(1) $x_{1\min}^{(2)} = 50\%$, $x_{1\max}^{(2)} = 80\%$, $d_1 = 3.71\%$.
(2) $x_{2\min}^{(2)} = 3.36\%$, $x_{2\max}^{(2)} = 3.90\%$, $d_2 = 84.00\%$.
(3) $x_{3\min}^{(2)} = x_{3\max}^{(2)} = 3.17\%$, $d_3 = 8.70\%$.
(4) $x_{4\min}^{(2)} = x_{4\max}^{(2)} = 3.83\%$, $d_4 = 43.00\%$.

Table 2.6 Results of two optimization methods for first proportioning phase

Performance	LP (%)	LP and GA-PSO (%)
$T_{\mathrm{Fe}}^{(1)}$	61.27	61.07
$y_{\mathrm{SiO_2}}^{(1)}$	5.244	4.921
$y_{\mathrm{CaO}}^{(1)}$	0.392	0.252
$y_{\mathrm{MgO}}^{(1)}$	0.234	0.158
$y_{\mathrm{Al_2}O_3}^{(1)}$	1.936	1.984
$y_{\mathrm{S}}^{(1)}$	0.082	0.054
$y_{\mathrm{P}}^{(1)}$	0.050	0.052
Unit cost	1064.20 CNY/t	1064.90 CNY/t

Table 2.7 Chemical compositions (%) and prices of raw materials for second proportioning phase

Raw material	$T_{\mathrm{Fe}}^{(2)}$	$y_{\mathrm{SiO_2}}^{(2)}$	$y_{\mathrm{CaO}}^{(2)}$	$y_{\mathrm{MgO}}^{(2)}$	$y_{\mathrm{Al_2O_3}}^{(2)}$	$y_{\mathrm{S}}^{(2)}$	$y_{\mathrm{P}}^{(2)}$	Price/(CNY/t)
Material 2	0.00	10.00	0.92	0.60	5.50	0.40	0.00	965.00
Material 3	0.00	4.50	83.00	3.30	1.50	0.00	0.00	219.50
Material 4	0.00	1.30	53.00	2.50	0.80	0.00	0.00	78.45
Material 5	0.00	1.00	31.00	20.00	0.90	0.00	0.00	53.30
Material 6	55.00	5.80	10.00	1.85	2.66	0.04	0.07	0.00
Material 7	55.00	5.80	10.00	1.85	2.66	0.04	0.07	0.00

(5) $x_{5\min}^{(2)} = 0.00\%$, $x_{5\max}^{(2)} = 20.00\%$, $d_5 = 44.50\%$.
(6) $x_{6\min}^{(2)} = 10\%$, $x_{6\max}^{(2)} = 35\%$, $d_6 = 0\%$.
(7) $x_{7\min}^{(2)} = x_{7\max}^{(2)} = 1.26\%$, $d_7 = 0\%$.
(8) $T_{\mathrm{Fe}\min}^{(2)} = 55.00\%$, $T_{\mathrm{Fe}\max}^{(2)} = 100.00\%$.
(9) $y_{\mathrm{SiO_2}\min}^{(2)} = 4.00\%$, $y_{\mathrm{SiO_2}\max}^{(2)} = 6.00\%$.
(10) $y_{\mathrm{CaO}\min}^{(2)} = 9.00\%$, $y_{\mathrm{CaO}\max}^{(2)} = 11.00\%$.
(11) $y_{\mathrm{MgO}\min}^{(2)} = 1.00\%$, $y_{\mathrm{MgO}\max}^{(2)} = 2.00\%$.
(12) $y_{\mathrm{Al_2O_3}\min}^{(2)} = 1.00\%$, $y_{\mathrm{Al_2O_3}\max}^{(2)} = 2.00\%$.
(13) $y_{\mathrm{S}\min}^{(2)} = 0.00\%$, $y_{\mathrm{S}\max}^{(2)} = 1.00\%$.
(14) $y_{\mathrm{P}\min}^{(2)} = 0.00\%$, $y_{\mathrm{P}\max}^{(2)} = 1.00\%$.
(15) $R_{\min} = 1.78$, $R_{\max} = 1.82$.

Table 2.8 shows the optimal dosing schemes for the second proportioning phase produced by the LP and the LP and GA-PSO methods. It lists the proportions of the raw materials for this phase.

Table 2.8 Optimal proportioning schemes for raw materials for second proportioning phase produced by LP and LP and GA-PSO

Raw material	LP (Cost reduction) (%)	LP and GA-PSO (Reduction of cost and S) (%)
Material 1	50.00	50.43
Material 2	3.36	3.36
Material 3	3.17	3.17
Material 4	3.83	3.83
Material 5	3.38	2.95
Material 6	35.00	35.00
Material 7	1.26	1.26

Table 2.9 Predictions of two methods for second proportioning phase

Performance	LP (%)	LP and GA-PSO (%)
$T_{Fe}^{(2)}$	55.038	55.116
y_{SiO_2}	5.754	5.586
y_{CaO}	10.405	10.166
y_{MgO}	1.833	1.696
$y_{Al_2O_3}$	2.422	2.449
y_S	0.061	0.047
y_P	0.054	0.055
Unit cost/(CNY/t)	576.255	580.954
S_e	1.27	1.26

Table 2.9 shows the predictions produced by the two optimization methods. The results for the second proportioning phase show that the S content of the sinter was 0.047% for the LP and GA-PSO method and 0.061% for the LP method. So, the LP and GA-PSO method reduced the S content by 23% but increased the unit cost of the raw materials by 4.7 CNY/t.

The amount of SO_2 emissions is given by

$$S_e = \left(\sum_{j=1}^{m} d_j^{(2)} x_j^{(2)} y_{Sj}^{(2)} \right) \times 100\%, \tag{2.113}$$

where S_e is the ratio of the S content of discharged SO_2 to the total S content of the raw mix, and m is the number of kinds of raw materials in the second proportioning phase. Table 2.9 shows that the amount of SO_2 emissions is smaller for the LP and GA-PSO method than for the LP method.

2.5 System Implementation for Proportioning of Iron Ore in Sintering Process

The designed IIOS was implemented in an iron and steel plant to optimize the proportioning of iron ores, and the validity of the LP and GA-PSO method was demonstrated through the results of actual runs.

2.5.1 System Configuration and Implementation

Figure 2.17 shows the hardware architecture of the IIOS and production management, and Fig. 2.18 shows the flow of information and data in the system.

The IIOS mainly performs proportioning optimization, quality prediction, and data management. It was implemented especially to have the features of easy, stable operation; easy development and debugging; and simple maintenance. It consists of two levels: the basic automation level (L1) and the intelligent integrated optimization level (L2). There are also a production management level (L3) and a business planning level (L4) at the top of the whole control system. The system employs the client/server model.

The L4 layer contains a business planning client computer, which plans and monitors the whole business, and a database, ERP-DB, which stores the prices and chemical compositions of the raw materials.

The L3 layer contains three client computers: the Sin-client is in the sintering control room, the Mat-client is in the material field management room, and the Mill-client is in the proportioning control room. It also has a database, Sys-DB, that uses the Oracle 9i database program, which manages all the information involved in the sintering process and the information from the ERP-DB so that the clients can manage and monitor the process of sinter production.

The L2 layer has two engineering stations and two operator stations, all of which are running WinCC (Windows control center) HMI (human–machine interface). There are two databases in the IIOS: the Mat-DB stores information about the material field and the Sin-DB stores information about the sintering process. Operator Station 1 and Engineering Station 1 are used for the material field, and Operator Station 2 and Engineering Station 2 are used for the sintering process. An OPC (OLE for process control) data acquisition server connects the engineering stations and programmable logic controllers (PLCs) so that data from the PLCs can be collected. The IIOC in Engineering Station 2 gathers real-time data from the Mat-DB and the Sin-DB, calculates an optimal dosing scheme, sends it to the OPC data acquisition server through the OPC client, and then forwards it to the PLCs.

The layer 1 consists of 6 PLCs in charge of data acquisition and executing control commands. PLCs 1 and 2 are in the material field, and PLCs 3, 4, 5, and 6 are located in different parts of the sintering machine.

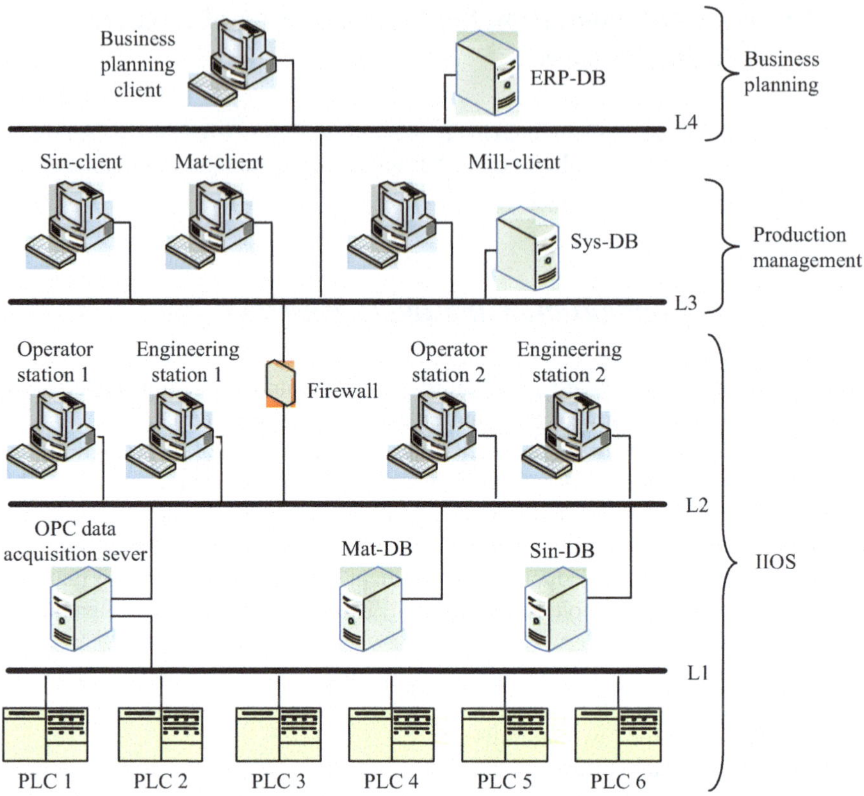

Fig. 2.17 Hardware architecture of the IIOS and production management

Fig. 2.18 Data flow in the IIOS

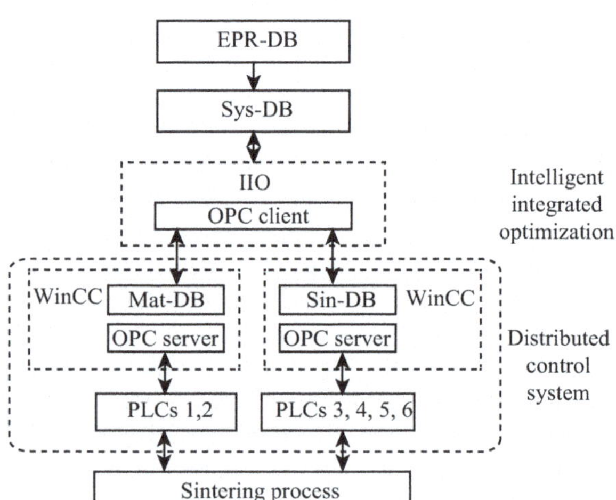

Table 2.10 Comparison of *M* before and after the IIOS was put into operation

Data set (w/o IIOS)	Unit cost/(CNY/t)	Data set (w/o IIOS)	Unit cost/(CNY/t)
A-1	3.696	B-1	−9.490
A-2	6.948	B-2	3.545
A-3	19.174	B-3	−42.09
A-4	17.456	B-4	−41.09
A-5	10.072	B-5	−30.04
A-6	18.655	B-6	−32.37
A-7	25.014	B-7	−54.89
A-8	−7.230	B-8	−50.85
A-9	−18.636	B-9	−52.49
A-10	−8.558	B-10	−53.78
Average	6.659	Average	−36.355

The system was developed under Windows XP operating system and was written in Visual C++ 6.0. This environment allows us to build easy access to a database and compile stable application programs by using the MFC (Microsoft Foundation Classes) series of class libraries. More specifically, activeX data object (ADO) technology was used for database access. Object-oriented OLE DB (object linking and embedding database) technology ensures that the database provides easy use, fast access, and low memory usage and disk space.

In order to simplify the debugging and maintenance of the system, we used modular-design technology and divided the system into two levels for testing: a module level and a system-integration level.

An important measure of the effectiveness of our system is the difference between the expected and actual unit cost of production, M.

2.5.2 Results of Actual Runs in Sintering Process

First, the system optimizes the proportions of materials. Then, experts check the results and approve them. To show how effective the method is, we randomly chose 10 sets of data from actual runs before and after the IIOS was put into operation. Table 2.10 shows that M was generally positive before the IIOS was put into operation. That means that the actual unit cost was higher than expected in most cases. However, the IIOS generally produces a negative M, which means that the actual unit cost of production was usually lower than expected. The IIOS reduced the average value of M from 6.659 CNY/t to −36.355 CNY/t (Table 2.10). This proves that the system has a large economic impact.

SO_2 cannot be measured online, but it can be converted into the S content of the raw materials. We used Eq. (2.113) to estimate the amount of SO_2 emissions.

Table 2.11 S_e **before and after the IIOS was put into operation**

Data set (w/o IIOS)	S_e/%	Data set (w/o IIOS)	S_e/%
A-1	0.0137	B-1	0.0121
A-2	0.0147	B-2	0.0123
A-3	0.0131	B-3	0.0124
A-4	0.0134	B-4	0.0127
A-5	0.0131	B-5	0.0128
A-6	0.0131	B-6	0.0122
A-7	0.0141	B-7	0.0123
A-8	0.0125	B-8	0.0116
A-9	0.0123	B-9	0.0123
A-10	0.0125	B-10	0.0125
Average	0.0133	Average	0.0123

Table 2.11 contains S_e for the data sets in Table 2.10. It shows that the IIOS reduced SO_2 emissions from 13.3 to 12.3 kg/t. This demonstrates that the presented method has significant environmental benefits.

2.6 Conclusion

An expert control strategy based on BPNNs and mathematical models was presented for the control of the coal blending process in an iron and steel plant. It involved the determination and tracking of the target percentage of each type of coal to be blended. The determination was implemented through two iterative algorithms that were based on the combination of BPNNs, mathematical models, and rule models, and used forward chaining and model-based reasoning. The prediction of the coal blend and coke quality was carried out repetitively based on the BPNNs and mathematical models during the determination of the target percentages. The BPNNs, mathematical models, and rule models were constructed based on statistical data and empirical knowledge. The tracking of the target percentages was achieved by a distributed control technique employing the PI control algorithm. The run results show that the presented strategy is an effective way to control the coal blending process.

The proportioning step in the iron ore sintering process is complex and strongly nonlinear, and has a large time delay. Moreover, the optimal proportions change depending on the ore sources. These characteristics make it difficult to find an optimal dosing scheme for the first and second proportioning phases. An IIOS with two levels (intelligent integrated optimization, basic automation) was built. In the intelligent integrated optimization level, an intelligent integrated optimizer produced an optimal dosing scheme. The intelligent integrated optimizer has three parts: a cascade integrated quality-prediction model, the optimization of the first proportioning, and the optimization of the second proportioning. Computational intelligence methods predicted the quality of sinter. Then, the predicted quality indices were fed back to the optimizations of the first and second proportionings to find feasible optimal dosing schemes. The IIOS was implemented in an iron and steel plant. Actual runs show that the system reduced production costs by 43.014 CNY/t and SO_2 emissions by 0.001% on average.

References

1. Hashimoto S (1989) Coke in the blast furnace: property requirements and how to realize them. In: Proceeding of the symposium on coke properties required by the blast furnace for stable operation, pp 1–22
2. Miura Y, Okuhara T, Nishi T, Yamaguchi T, Haraguchi H (1979) Study on coal blending theory. Seitetsu Kenkyu 299:13060–13074
3. Shi MY (1989) A practical method for predicting coke quality. Fuel Chem Eng 20(3):114–117
4. Wen ZM, Wu M, Shen DY (1994) Design and implementation of a computer control system for the coal blending process. Autom: Theory, Tech Appl 2:610–613
5. Yang G, Fan X, Chen X, Huang X, Li Z (2016) Intelligent control of grate-kiln-cooler process of iron ore pellets using a combination of expert system approach and Takagi-Sugeno fuzzy model. J Iron Steel Res, Int 23(5):434–441
6. Fan X, Wang Y, Chen X (2012) Mathematical models and expert system for grate-kiln process of iron ore oxide pellet production. Part II: rotary kiln process control. J Cent South Univ 19(6):1724–1727
7. Åström KJ, Anton JJ, Årzen KE (1986) Expert control. Automatica 22(3):277–286
8. Efstathiou J (1989) Expert systems in process control. Longman, Essex
9. Ishizuka M, Kobayashi S (1991) Expert systems. Maruzen, Tokyo
10. Liebowitz J (1955) Expert systems: a short introduction. Eng Fract Mech 50:601–607
11. Garg H (2016) A new generalized improved score function of interval-valued intuitionistic fuzzy sets and applications in expert systems. Appl Soft Comput 38(C):988–999
12. Narendra KS, Parthasarathy K (1990) Identification and control of dynamic system using neural networks. IEEE Trans Neural Netw 1(1):4–27
13. Su HT, McAvoy TJ (1997) Artificial neural networks for nonlinear process identification and control. Henson MA, Seborg DE. Nonlinear process control. Upper Saddle River: Prentice Hall PTR
14. Rumelhart DE, Hinton GE, Williams RJ (1986) Learning internal representations by error propagation. Rumelhart DE, Mecclellanl JL, Parallel data processing. MIT Press, Cambridge
15. Hornik K, Stinchcombe M, White H (1989) Multilayer feedforward networks are universal approximators. Neural Netw 2(5):359–366
16. Hagan MT, Demuth HB, Beale MH (1996) Neural network design. PWS Publishing, Boston
17. Ghosh A, Chatterjee A (2008) Ironmaking and steelmaking: theory and practice. Rajkamal Electric Press, New Delhi
18. Frolov YA, Polotskii LI (2015) Three-dimensional mathematical (dynamic) model of the sintering process. Part I. Metallurgist 58(11–12):1071–1079
19. Frolov YA, Polotskii LI (2015) Three-dimensional mathematical (dynamic) model of the sintering process. Part II. Metallurgist 59(1–2):9–15
20. Wu M, Chen XX, Cao WH, She JH, Wang CS (2014) An intelligent integrated optimization system for the proportioning of iron ore in a sintering process. J Process Control 24(1):182–202
21. Zhao JP, Loo CE, Dukino RD (2015) Modelling fuel combustion in iron ore sintering. Combust Flame 162(4):1019–1034

22. Kawaguchi T, Yoshinaga M (1987) Development and application of an integrated simulation model for iron ore sintering. Iron Mak Proc 46:99–106
23. Song Q, Wang AM (2009) Simulation and prediction of alkalinity in sintering process based on grey least squares support vector machine. Int J Iron Steel Res 16(5):1–6
24. Wang R, Wang AM, Song Q (2012) Research on the alkalinity of sintering process based on LS-SVM algorithms. Adv Intell Soft Comput 168:449–454
25. Han HG, Qiao JF (2012) Prediction of activated sludge bulking based on a self-organizing RBF neural network. J Process Control 22(6):1103–1112
26. Wang JS, Wang W (2006) A predictive model of sinter chemical composition and its application. In: Proceedings of the 6th world congress on intelligent control and automation, pp 4856–4860
27. Wang B, Yang B, Sheng J, Chen M, He G (2009) An improved neural network algorithm and its application in sinter cost prediction. In: Second international workshop on knowledge discovery and data mining, pp 112–115
28. Chen XX, Chen X, She JH, Wu M (2017) A hybrid just-in-time soft sensor for carbon efficiency of iron ore sintering process based on feature extraction of cross-sectional frames at discharge end. J Process Control 54:14–24
29. Chen XX, Chen X, She JH, Wu M (2017) A hybrid time series prediction model based on recurrent neural network and double extreme learning machine for carbon efficiency prediction in iron ore sintering process. Neurocomputing 292:128–139
30. Chen XX, Chen X, She JH, Wu M (2017) Hybrid multistep modeling for calculation of carbon efficiency of iron ore sintering process based on yield prediction. Neural Comput Appl 28(6):1193–1207
31. Zheng DL, Zhao ZL (2000) Application of gray linear programming in sintering mixing calculation. J Univ Sci Technol Beijing 7(4):273–276 (In Chinese)
32. Zhang JH, Xie AG, Shen FM (2007) Multi-objective optimization and analysis model of sintering process based on BP neural network. Int J Iron Steel Res 14(2):1–5
33. Chen X, Chen XX, Wu M (2016) Modeling and optimization method featuring multiple operating modes for improving carbon efficiency of iron ore sintering process. Control Eng Pract 54:117–128
34. Giri BK, Roy GG (2012) Mathematical modeling of iron ore sintering process using genetic algorithm. Ironmak Steelmak 39(1):59–66
35. Zhang H, Zhu YL, Zou WP, Yan XH (2012) A hybrid multi-objective artificial bee colony algorithm for burdening optimization of copper strip production. Appl Math Model 36(6):2578–2591
36. Liu C, Xie Z, Sun F, Chen L (2016) Optimization for sintering proportioning based on energy value. Appl Therm Eng 103:1087–1094
37. Yu Z, Xiao L, Li H, Zhu X, Huai R (2017) Model parameter identification for lithium batteries using the coevolutionary particle swarm optimization method. IEEE Trans Ind Electron 64(7):5690–5700
38. Lu XJ, Li HX, Yuan X (2010) PSO-based intelligent integration of design and control for one kind of curing process. J Process Control 20(10):1116–1125
39. Yu G, Chai TY, Luo XC (2011) Multiobjective production planning optimization using hybrid evolutionary algorithms for mineral processing. IEEE Trans Evol Comput 15(4):487–514
40. Yildiz AR, Solanki KN (2012) Multi-objective optimization of vehicle crashworthiness using a new particle swarm based approach. Int J Adv Manuf Technol 59(1–4):367–376

Chapter 3
Intelligent Optimization and Control of Coking Process

Coke, the product of a coking process, is an important material in the metallurgical industry. In a blast furnace that produces iron, it functions as the main supplier of heat, a reducing reagent, and a support framework for other materials. Its quality directly influences the metallurgical process that occurs in a blast furnace. Coke-oven temperature (COT) is a key parameter that reflects the thermal state of the whole oven. It directly influences both the quality of coke and the lifetime of an oven [1].

Coking is a complex process. It features a large time constant, strong nonlinearities, and time-varying characteristics. To achieve good control performance, it is important to obtain a precise COT in a real-time fashion; but this is difficult owing to restrictions imposed by the structure of a coke-oven and operating costs. The COT is generally measured manually, but that gives rise to two problems:

(1) The sampling period (usually 4 h) is too long for real-time control.
(2) Since the measurements rely to some extent on subjective judgments, different workers produce different values when measuring exactly the same thing.

A straightforward method of obtaining the COT in a real-time fashion is to install a thermocouple in each combustion chamber to measure the flue temperature and to use those values to calculate the COT [2]. However, the structure of a coke-oven makes it very difficult to install thermocouples in all the combustion chambers. And even if it were possible, the high flue temperature (about 1300°C) would result in a very high cost and difficulty in maintenance. The solution is the online estimation of the COT and one way of doing that is to use soft-sensing technology. The idea of soft sensing first arose in inferential control [3]. Since it can overcome the drawbacks of the large time delay involved in offline analysis and a long sampling period [4, 5], it has seen rapid development and is now an important tool in practical process control [6].

An analysis of the mathematical relationships between easily measured variables and process variables that are difficult to measure directly enables the use of various

M. Wu et al., *Intelligent Optimization and Control of Complex Metallurgical Processes*, Engineering Applications of Computational Methods 3,
https://doi.org/10.1007/978-981-15-1145-5_3

algorithms for the estimation of those process variables. This measurement method is called soft sensing. Soft sensing can be used for process modeling, of which there are three main types: mechanism modeling [7]; data-driven modeling [8, 9]; and hybrid modeling, which combines the first two [10].

Mechanism modeling relies on an analysis of the physical and/or chemical mechanism of a process. The parameters have explicit physical meanings and the resulting model is relatively simple. However, if a process is complex, it is usually difficult to determine the values of the parameters with any high precision. Many important parameters of the coking process are difficult to obtain due to the high temperature and extreme conditions. So, physically modeling the COT is very difficult and time-consuming.

Data-driven modeling methods include regression analysis [11, 12] and AI approaches [13]. Over the last couple of decades, AI-based modeling methods have been used for many industrial processes [14, 15]. Neural network (NN)-based modeling is the most common type. It has many advantages, such as large-scale parallel processing, the distributed storage of knowledge, and a strong self-learning capability. Since the NN technique does not require any prior knowledge of a model of a process, it is widely used in industry to estimate process variables in a real-time fashion for monitoring and control [16] and for online sensing in a process with a large time delay [17].

Data-driven modeling based on soft sensing can be used to estimate the COT by employing process parameters that are strongly related to the COT. The collection of a large amount of historical monitoring data enables a model of the process to be established based on the NN technique [18, 19]. For instance, the temperature at the top of a regenerating chamber (TTR) can be used to establish a dual-parabola model or an linear regression (LR) model to indirectly measure the COT [20, 21].

In the early stages of their development, soft-sensing models employed only a single modeling method; but now they often integrate multiple methods [22, 23]. Hybrid modeling has been shown to produce more precise estimates than a single modeling method does [24, 25].

On the other hand, combustion in a coke-oven is a complex physical and chemical process. For given proportions of coals, COT is the most important parameter in the combustion process and is thus the key to obtaining good-quality coke [26, 27]. Optimization and control of the combustion process of a coke-oven have been demonstrated to be of practical importance [28, 29]. However, the time delay, in particular, makes control of the oven temperature difficult. A Smith predictor is an effective way to handle a time delay, but it strongly depends on the exact mathematical model of the plant. Predictive control is also a practical method for process control, but it is sensitive to the precision of the plant model, as well. Although many studies on the Smith predictor and predictive control have been reported [30], few are directly applicable to the combustion process of a coke-oven because no exact mathematical model of the process is available. On the other hand, since fuzzy control does not require an exact mathematical model of a plant and is very robust, it is widely used for process control in industry [31]. One of its advantages is that, if we take the trend in the change in oven temperature into consideration, it can suppress the effects of the time delay to some extent. In actual practice, fuzzy control has already been shown to be effective in controlling the temperature of a coke-oven [32]. Industrial processes are generally becoming larger, more precise, and more complicated; and it is difficult to obtain satisfactory results with a single control method. One way to solve this problem is to combine intelligent methods, and in fact, this constitutes a new way to control the temperature of a coke-oven. Simulation studies have demonstrated the validity of the method [33]. However, there are usually two problems with the methods of controlling the combustion process of a coke-oven:

(1) Only tracking error information is utilized in the control of the oven temperature, and no consideration is given to the influence of discrete events, such as those arising from a production plan and coking operations; even if they strongly affect process conditions.

(2) The parameters of a fuzzy controller are selected based on operators' experience and are not optimized.

The operating state of a coke-oven is defined to be a set of parameters that represents the process conditions in the oven. Determination of the operating state is important for good decision-making and proper control of the combustion process. It is also important for improving the intelligence of the control system and increasing production efficiency. A simple determination of the operating state can be made just by checking whether or not the oven is in the maintenance period, but this method does not permit a response to changes in the operating state arising from other situations.

A hybrid system is a dynamic system that contains continuous systems as well as discrete events [34]. It places great emphasis on the combined effect of time- and event-driven processes during system evolution. The combustion process can be regarded as a hybrid system because it can be divided into two layers: a physical layer described by the continuous variables of a dynamic evolution mechanism, and a layer of high-level coordination that features symbol manipulation and discrete monitoring and decisions.

This chapter first describes the coking process and the configuration of a control system for the combustion process. Next, it presents an integrated method combining LR and NN-based modeling to avoid the drawbacks of mechanism modeling and to create a precise model for estimating the COT based on actual data. Then, it explains a hierarchical intelligent integrated control structure for the combustion process based on the determination of the operating state. Then, it concerns the operating plan and optimal scheduling of the coking operations. Finally, it discusses results of actual runs.

3.1 Characteristic Analysis and System Configuration

This section first describes soft sensing of the COT and control of the combustion process of a coke-oven. Then, it focuses on the analysis of control requirements. Finally, it explains the system configuration of the coking process.

3.1.1 Process Description

This study concerned a JN-60 type coke-oven, which has 6-m-high combustion and coking chambers and usually 50–100 heating units. It consists of a line of coking chambers, each of which is sandwiched between two combustion chambers to ensure uniform heating, show as Fig. 3.1. In the combustion process, gas and air are fed to the combustion chambers, where they burn. The target coke-oven has 60 coking chambers and 61 combustion chambers arranged in a line. The coal in the coking chamber is carbonized by the heat supplied by combustion chambers from two directions to become coke.

There is one regenerating chamber below each coking chamber, and there is an additional one at each end of the line, for a total of 62. Oblique conduits connect the tops of the regenerating chambers to vents in the bottoms of the combustion chambers. Each combustion chamber has 16 pairs of vents along the bottom. Half connect to the regenerating chamber on one side, and the other half connect to the regenerating chamber on the other side (Fig. 3.2). Half the vents feed in gas to be burned, and the other half draw off the exhaust gas.

The coking chambers are numbered from 1 to 60; the combustion chambers, from 0 to 60; the regenerating chambers, from 0 to 61; and the pairs of vents, from 1 to 16 (see Fig. 3.2 for details).

The lateral direction is defined to be along the line of coking chambers, and the longitudinal direction to be perpendicular to that, along the direction in which the coke moves. A coke-oven is divided longitudinally into two parts: the coke side is the side on which finished coke exits (roughly Vents 9 to 16 in Fig. 3.2), and the

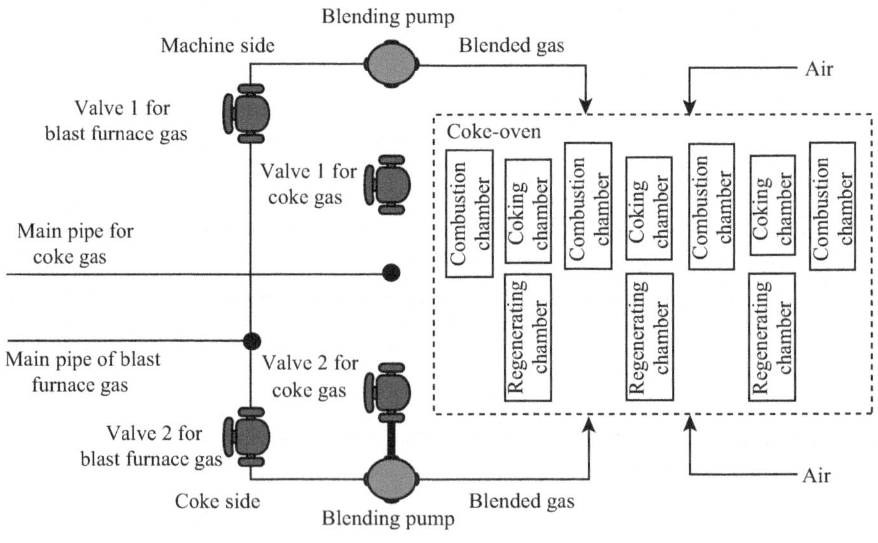

Fig. 3.1 Coke-oven system

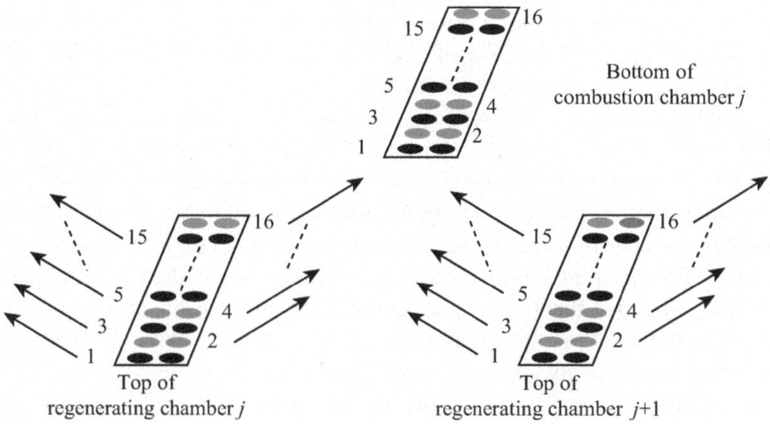

Fig. 3.2 Vents connection

machine side is the other one (roughly Vents 1 to 8 in Fig. 3.2). The coking chambers are wedge-shaped, with the coke side being wider, to allow coke to be pushed out easily. The temperature of a combustion chamber is higher on the coke side than on the machine side.

In the combustion process, the coal in the coking chambers is carbonized in a hermetic environment and becomes coke. The exhaust gas is discharged into a neighboring regenerating chamber.

Two vents in each combustion chamber were selected for manual temperature measurements. Both vents handle the outflow from the chamber and are located close to the midpoint between the machine and coke sides. The temperature was measured every four hours with an infrared thermometer. The measurement result for one of the vents is called the longitudinal temperature on the machine (LTM) side, and the other is called the longitudinal temperature on the coke (LTC) side. The two values are not usually the same. Since the COT is the average longitudinal temperature of the combustion chambers, the COT was divided into two components to better describe it: the COTC, which is the COT on the coke side, and the COTM, which is the COT on the machine side. That is, the COTC is the average temperature of the outflow of all the combustion chambers on the coke side, and the COTM is the average on the machine side. Note that the COTC is higher than the COTM. While the COTC and COTM can only be measured manually, the TTR can be measured automatically. The TTR is also divided into two components: the TTRM, which is the TTR on the machine side, and the TTRC, which is the TTR on the coke side. In this chapter, the TTRM and TTRC were used to establish real-time models for estimating the COTM and COTC. In addition, the manually measured COTM and COTC, along with the outputs of the soft-sensing models, were used to train and verify the models.

The COT is a key factor determining coke quality, and it has to be maintained in a proper range by regulating the gas flow rate and the amount of air. The gas used in the combustion chambers can be coke-oven gas, blast furnace gas, or a mixture of the two. This chapter kept the flow of coke-oven gas constant and adjusted the flow of blast furnace gas by means of two valves.

The supply of coal and the discharge of coke are not performed at the same time in all the coking chambers; these operations follow an operating plan. The coking process in a coking chamber is a series of changes from blended coal to coke, and the length of time from coaling to pushing for one coking chamber is called the coking time. During the coking process, the material in a coking chamber absorbs the heat differently in the different stages of coking: In the first stage, the fresh coal supplied to the coking chamber absorbs a great deal of heat; but near the end of the process, the coke does not need much heat and even releases heat. When coke is pushed out of the oven, it carries away a large amount of heat.

The expected coking time (normally in the range 18–22 h) is specified in the operating schedule, which is made before the process begins. If the combustion process closely follows the schedule, the timewise variation in heat consumption for all the chambers together is small. However, a delay in pushing may cause a difference between the actual and expected coking times for a chamber. And a large difference changes the heat consumption of the coke-oven and the actual average coking time of all the coking chambers, which is called the actual coking time in this chapter.

Moreover, the behavior of the COT is closely related to the current calorific value of the gas, which usually cannot be measured in an online manner.

The combustion process has two operating phases: heating-up and production. Heating an oven up to a target COT usually takes 50–60 days. Once the phase is finished, the process enters into the production phase, which is the main stage of coke production. In this phase, the gas flow rate is regulated to maintain the COTC and COTM at the target values. This chapter considers the COT control in the production phase.

3.1.2 Analysis of Characteristics

The two main factors affecting the COT in coking are the change in the direction of gas flow and the coke-pushing series.

1. Changing Direction of Gas Flow

Changing the direction of gas flow is an important operation in coking. It is necessitated by the structure of the coke-oven and the method of heating. The rule for changing the direction is as follows. Regenerating Chamber j is connected to Combustion Chamber $j-1$ through odd-numbered vents, and to Combustion Chamber j through even-numbered vents. If gas flows into a combustion chamber through odd-numbered vents, then gas is discharged through even-numbered vents, and vice versa. The direction of flow changes every 30 min. Take Combustion Chamber j as an example. First, gas flows into Regenerating Chamber $j+1$ and is preheated.

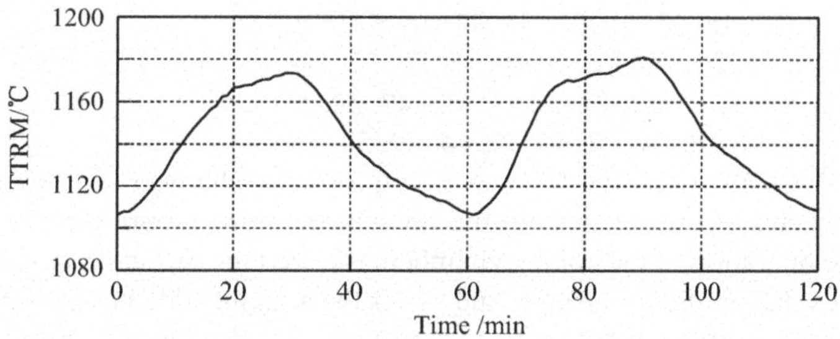

Fig. 3.3 Typical curve of TTRM for Regenerating Chamber 20 (two periods)

Next, it flows into Combustion Chamber j through odd-numbered vents and burns. The exhaust gas flows out through even-numbered vents into Regenerating Chamber j, where the heat is transferred to the walls. Finally, the exhaust gas is discharged. Changing the direction of flow results in that gas flows from Regenerating Chamber j to Combustion Chamber j through even-numbered vents, and exhaust gas flows from Combustion Chamber j to Regenerating Chamber $j + 1$ through odd-numbered vents.

As the above explanation suggested, the regenerating chambers function as heat exchangers. Exhaust gas is quickly drawn from the combustion chambers into the regenerating chambers. So, the COTM and COTC are related to the temperature of the exhaust gas. Furthermore, since the regenerating chambers are located along the bottom of the coke-oven, the ambient conditions of the oven do not affect the exhaust gas temperature. This guarantees that the TTRMs and TTRCs are closely related to the COTM and COTC, respectively. So, the TTRMs and TTRCs can be considered to be auxiliary variables for the soft-sensing models of the COTM and COTC, respectively.

Due to the change in the direction of gas flow, the TTRMs and TTRCs vary periodically and exhibit the same behavior. For example, Fig. 3.3 shows a typical curve of the TTRM measured in Regenerating Chamber 20 for two periods. The curve in each period has two parts: The TTRM increases when exhaust gas enters the chamber because the gas is hot and radiates heat, and the TTRM decreases when new gas for combustion enters the regenerating chamber and is sent to a combustion chamber because the gas is cold.

Since the gas preheated in a regenerating chamber has not yet burned, the TTR at this moment has little connection with the COT. On the other hand, the exhaust gas flowing into a regenerating chamber has been produced by a few seconds of combustion and is closely related to the COT.

2. Coke-Pushing Series

Coal is not supplied to, and coke is not discharged from, all 60 coking chambers at the same time. The chambers are divided into 5 groups to facilitate the making of a plan of operations. The five groups are as follows:

First: 1, 6, 11, **16**, 21, 26, 31, **36**, 41, 46, 51, 56;

Third: 3, 8, **13**, 18, 23, 28, **33**, 38, 43, 48, **53**, 58;
Fifth: 5, 10, 15, **20**, 25, 30, 35, **40**, 45, 50, 55, 60;
Second: 2, **7**, 12, 17, 22, **27**, 32, 37, 42, **47**, 52, 57;
Fourth: 4, 9, 14, **19**, 24, 29, 34, 39, **44**, 49, 54, 59.

Coke is discharged in order from each of the coking chambers in the first group; next, it is discharged in order from the ones in the third group; etc. This way of ordering the operation of the coking chambers is called the 5–2 coke-pushing series. "5" indicates the number of groups; and "2" is the amount added to a group number to get the next group number (that is, first group + 2 = third group; third group + 2 = fifth group; fifth group + 2 (modulo 5) = second group; and second group + 2 = fourth group). It ensures that coking proceeds smoothly, and it keeps the temperature variation among the combustion chambers small.

During the coking process, the material in the coking chambers absorbs the heat produced in the combustion chambers at different rates depending on the stage of the process. At the first stage, when the coking chamber is loaded with fresh coal, the coal absorbs a great deal of heat. This makes the flue temperature of the combustion chambers that sandwich the coking chamber low. On the other hand, near the end of the coking process, the coke does not need much heat and the flue temperatures of the combustion chambers that sandwich the coking chamber are high. So, the numbers of coking chambers at the different stages of the coking process should be about equal to keep the change in the COT small and to suppress temperature variations during the different stages.

3.1.3 Control Requirements

The most basic control requirement is to maintain the COT in a proper range by regulating the gas flow rate and the amount of air.

The control of the coke-oven combustion process is to adjust heat supply based on the change of the COT and to maintain a stable target COT. The target COT contains the target COTM and the target COTC. It is a main technical specification that ensures coke maturity within a specified coke time. Note that a very high target COT not only greatly increases the energy consumption per unit of output, but also may cause the phenomenon of grilled coke and degrade the quality of coke. On the other hand, a very low target COT leads to non-uniform heating and prolongs the time of dry distillation in the coking chamber. This affects the quality and production of coke. So, the setting of the COT is important to ensure the quality and production of coke and to reduce energy consumption.

Iron and steel enterprises try to maximize the coke production and minimize the energy consumption under the condition that the quality of coke meets prescribed requirements. The indexes related to the quality and production of coke, and the energy consumption are explained as follows.

The quality of coke is an important production indicator for the optimization of the coking process. The related main indices are moisture content (M_{ad}), ash content

(A_{d}), volatile matter content (V_{daf}), sulfur content ($S_{\mathrm{t,d}}$), crushing strength M_{40}, the abrasion resistance (M_{10}), the coke reactivity index (CRI), and coke strength after reaction (CSR).

The quality of the first-class coke has to meet

$$\begin{cases} 3 < M_{\mathrm{ad}} < 6, \quad 0 < V_{\mathrm{daf}} < 1.9, \\ 0 < A_{\mathrm{d}} \leqslant 12, \quad 0 < S_{\mathrm{t,d}} \leqslant 0.6, \\ 92 < M_{40} < 100, \quad 0 < M_{10} \leqslant 7. \end{cases} \tag{3.1}$$

The quality of the second-class coke has to meet

$$\begin{cases} 3 < M_{\mathrm{ad}} < 6, \quad 0 < V_{\mathrm{daf}} < 1.9, \\ 12 < A_{\mathrm{d}} \leqslant 13.5, \quad 0.6 < S_{\mathrm{t,d}} \leqslant 0.8, \\ 88 < M_{40} < 92, \quad 7 < M_{10} \leqslant 8.5. \end{cases} \tag{3.2}$$

And the quality of the third-class coke has to meet

$$\begin{cases} 3 < M_{\mathrm{ad}} < 6, \quad 0 < V_{\mathrm{daf}} < 1.9, \\ 13.5 < A_{\mathrm{d}} \leqslant 15, \quad 0.8 < S_{\mathrm{t,d}} \leqslant 1, \\ 83 < M_{40} < 88, \quad 8.5 < M_{10} \leqslant 10.5, \end{cases} \tag{3.3}$$

where CRI is in the range [22, 25%], and CSR is in the range [65, 70%].

3.1.4 System Configuration

The coke furnace production process includes three main operations, that is, the coal blending, the combustion, and the operation planning and optimal scheduling. Due to the complexity of a coke-oven during the production process with heat transfer, and physical and chemical changes, the optimization and control of the coking process is generally based on the three operations independently. But these operations are mutually restricted and influence each other. It is difficult to meet the requirements of coke quality and energy consumption at the same time. Therefore, it is necessary to establish an optimization and control system to coordinate the three operations of the coking process to achieve the production target of the steel enterprises. The configuration of this system is shown in Fig. 3.4.

The optimization and control model for the coking process takes the coking energy consumption as optimization target; the coal blend moisture, the target temperature, and the coking time as decision variables; the coke quality and coking production process as constraint conditions. Then the regional optimization set values of the three operations, that is, coal blending, combustion, and operation planning and optimal scheduling, are calculated. The three operations carry out their own regional optimization and control for the proportioning, valve opening, and scheduling scheme of the basic automation layer, thus achieving the guidance to the coking process.

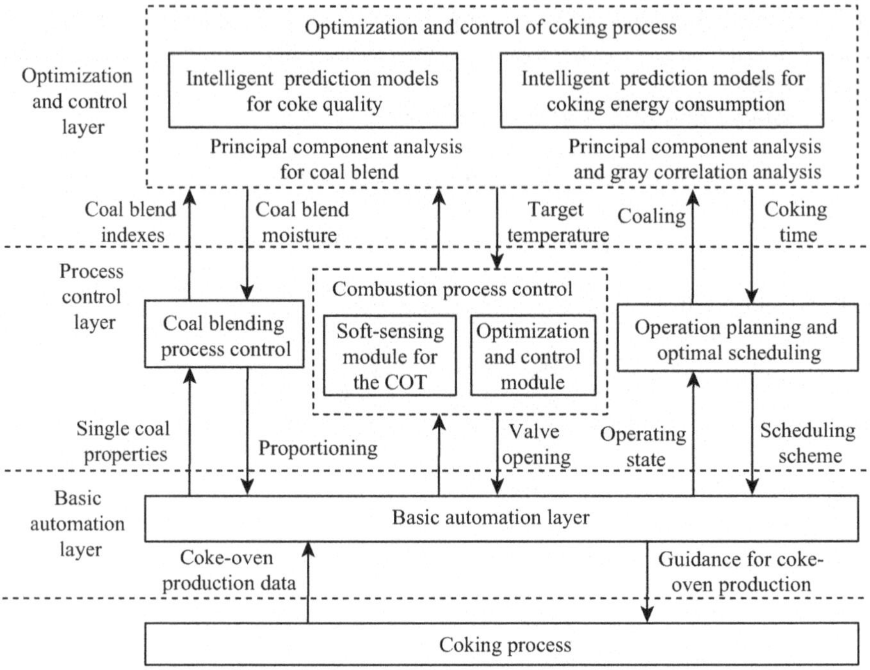

Fig. 3.4 Configuration of the intelligent optimization and control system for the coking process

For the optimization and control layer, the intelligent prediction models for coke quality are established and analyzed in Sect. 3.2. And the intelligent prediction models for coking energy consumption are established integrating the principal component regression model and distributed complex NN models. The models adjust the structure in a real-time fashion according to the change of operating states so as to ensure the prediction accuracy.

On the process control layer, the coal blending process employs an expert strategy to accurately determine the target percentage of each type of coal to be blended and to track it. This process is explained in Sect. 3.3. The control of the combustion process contains two modules: the soft-sensing module for the COT and the optimization and control module for the combustion process. The soft-sensing module employs an integrated model to estimate the COT, and the optimization and control module uses a hierarchical intelligent integrated control structure based on the determination of operating states to maintain the COT in a proper range. The operation planning and optimal scheduling control the timing sequence of the coking process. The optimal scheduling models are established under normal and abnormal operating states.

The basic automation layer mainly guides the coke-oven production based on the proportioning, valve opening, and scheduling scheme.

3.2 Integrated Soft Sensing of Coke-Oven Temperature

This section first explains the choice of auxiliary variables and measurement points, then presents a soft-sensing model of the COT.

3.2.1 Choice of Auxiliary Variables and Measurement Points

There are four steps in the construction of a soft-sensing model: the selection of auxiliary variables, data acquisition and processing, modeling, and online adjustment. The selection of the auxiliary variables is a key step because auxiliary variables not only determine the input information, but also greatly influence the structure and outputs of soft sensing. The TTRs were chosen to be auxiliary variables because they reflect the temperatures of the corresponding combustion chambers. They were measured on both the coke and machine sides (TTRCs and TTRMs) 15 min after each change in the direction of gas flow.

It is important to choose suitable regenerating chambers where thermocouples are to be installed. The twelve chambers selected in this chapter are

$$S = \{16,\ 36,\ 13,\ 33,\ 53,\ 20,\ 40,\ 7,\ 27,\ 47,\ 19,\ 44\}. \tag{3.4}$$

The rules for selection are based on practical experience and an analysis of the combustion mechanism:

(1) Thermocouples should not be installed in the regenerating chambers at the two ends of the coke-oven.

(2) Thermocouples should be installed in vents connected to regenerating chambers where the combustion process is normal.

(3) The interval between the coke discharge times for the coking chambers directly above two neighboring regenerating chambers in the series should be about the same. In this chapter, the interval is 40–60 min. For example, if coke is discharged from Chamber 13 at 1 p.m. and from Chamber 33 at 1:40 p.m., then it should be discharged from Chamber 53 at 2:20 p.m., etc.

(4) The number of regenerating chambers in which thermocouples are installed in even-numbered vents should be the same as the number of regenerating chambers in which thermocouples are installed in odd-numbered vents.

3.2.2 Structure of Soft-Sensing Model for Coke-Oven Temperature

Due to the complexity of the coking process and the relationship between the COT and the TTRs, a soft-sensing model (Fig. 3.5) for the COT was built that combines a

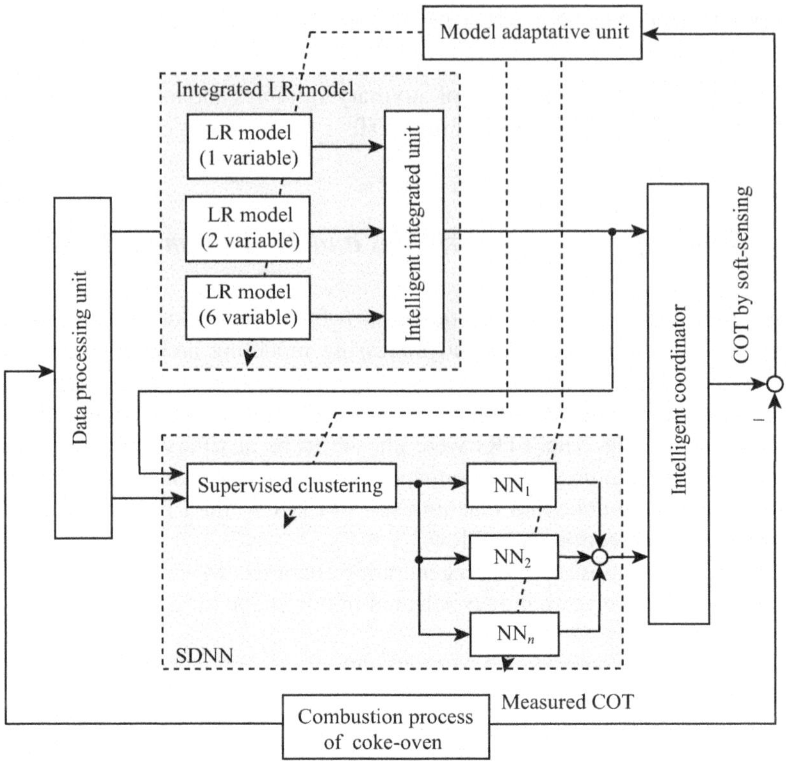

Fig. 3.5 Integrated soft-sensing model of COT

data processing unit, an integrated LR model, a supervised distributed NN (SDNN), an intelligent coordinator,, and a model adaptive unit.

The method of moving averages is employed to suppress measurement noise. An analysis of operating data revealed that the relationship between the COT and the TTRs is largely linear. However, due to the unique structure of the coke-oven, one LR model cannot handle both the normal and abnormal conditions. In order to balance the stability and accuracy of the model, different LR models are established to deal with various conditions, and they are integrated into an integrated LR model based on expert rules.

Although an LR model can reflect the linear relationships between the COT and the TTR and implement soft sensing of the COT, it fails to describe the nonlinear relationship between them. To improve the precision of the soft sensing and generalization ability of the model, the supervised clustering method is first used to cluster samples. Then, an NN model is established for each subspace of the samples. Finally, the NN models are integrated into an SDNN using the degree of fuzzy membership.

An intelligent coordinator is employed to coordinate the outputs of the two models to take full advantage of the integrated LR model and the SDNN model. Furthermore, to guarantee the accuracy of the soft sensing, a model adaptive unit updates the parameters of the integrated LR model and the SDNN based on the manually measured COT and automatically collected TTRs in a real-time fashion.

3.2.3 Integrated Linear Regression Model

An analysis of the process mechanism and operational data shows that the COT and the TTRs are closely related. An integrated LR model combining six-, bi-, and single-variable LR models is established to obtain the COT.

1. Six-Variable LR Model

A six-variable LR model based on measured TTRs uses a linear relationship to describe the dependence of the COT on the TTRs. Let y be the estimated COT and $x_1, x_2, x_3, x_4, x_5,$ and x_6 be the TTRs of regenerating chambers. Then, the relationship between the COT and the TTRs is

$$\begin{cases} y_1^{(6)} = \beta_0^{(6)} + \beta_1^{(6)} x_{11}^{(6)} + \beta_2^{(6)} x_{12}^{(6)} + \cdots + \beta_6^{(6)} x_{16}^{(6)} + \varepsilon_1^{(6)}, \\ y_2^{(6)} = \beta_0^{(6)} + \beta_1^{(6)} x_{21}^{(6)} + \beta_2^{(6)} x_{22}^{(6)} + \cdots + \beta_6^{(6)} x_{26}^{(6)} + \varepsilon_2^{(6)}, \\ \vdots \\ y_n^{(6)} = \beta_0^{(6)} + \beta_1^{(6)} x_{n1}^{(6)} + \beta_2^{(6)} x_{n2}^{(6)} + \cdots + \beta_6^{(6)} x_{n6}^{(6)} + \varepsilon_n^{(6)}, \end{cases} \tag{3.5}$$

where $\varepsilon_i^{(6)}$ $(i = 1, 2, \ldots, n)$ is a random error in the ith experiment, which is independent of each other and share the same distribution $\varepsilon^{(6)}$, and

$$Y^{(6)} = \begin{bmatrix} y_1^{(6)} \\ y_2^{(6)} \\ \vdots \\ y_n^{(6)} \end{bmatrix}, \quad \beta^{(6)} = \begin{bmatrix} \beta_0^{(6)} \\ \beta_1^{(6)} \\ \vdots \\ \beta_6^{(6)} \end{bmatrix}, \quad u^{(6)} = \begin{bmatrix} \varepsilon_1^{(6)} \\ \varepsilon_2^{(6)} \\ \vdots \\ \varepsilon_n^{(6)} \end{bmatrix},$$

$$X^{(6)} = \begin{bmatrix} 1 & x_{11}^{(6)} & \ldots & x_{16}^{(6)} \\ 1 & x_{21}^{(6)} & \ldots & x_{26}^{(6)} \\ \vdots & \vdots & & \vdots \\ 1 & x_{n1}^{(6)} & \ldots & x_{n6}^{(6)} \end{bmatrix}.$$

This yields

$$\begin{cases} Y^{(6)} = \beta^{(6)} X^{(6)} + u^{(6)}, \\ u^{(6)} \sim N_n\left(0, \sigma^2 I_n\right). \end{cases} \tag{3.6}$$

A least-squares solution for $\hat{\beta}$ is

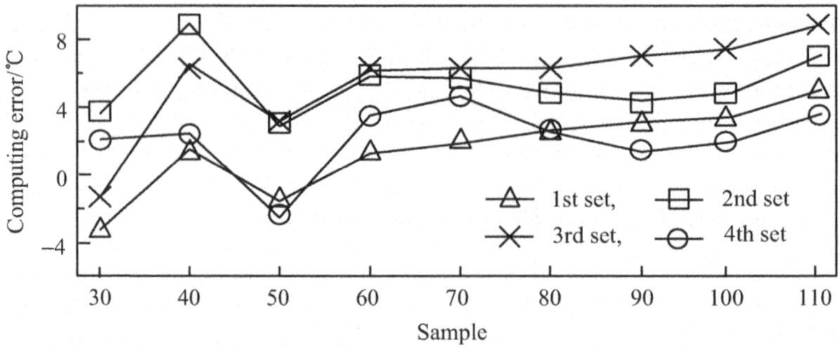

Fig. 3.6 Relationship between absolute estimation error of LR model and number of samples, n, for four sample sets ($n = 12$)

$$\hat{\beta}^{(6)} = \left(\left(X^{(6)}\right)^{\mathrm{T}} X^{(6)}\right)^{-1} \left(X^{(6)}\right)^{\mathrm{T}} Y^{(6)}. \tag{3.7}$$

The output of model (3.6) is $\hat{Y}^{(6)}$. A suitable number of samples, n, was chosen by first examining how n affects the estimation error. The absolute estimation error is defined to be the absolute value of the error between the COT calculated from Eq. (3.6) and a measured value. The relationship between the absolute estimation error and n (Fig. 3.6) shows that the error is very sensitive to n when n is small, but it does not change very much when $n \geqslant 90$. Thus, the value was chosen to be $n = 90$ in Eq. (3.6).

2. Bi-Variable LR Model

According to the production mechanism and process requirements of the coke-oven, the difference in the sampling time of the TTRs for the odd and the even numbers is half an hour. The time difference and the difference in the direction of the airflow cause the difference in the properties of the TTRs for odd and even numbers. To describe this difference, a bi-variable LR model is established by taking the average values of the TTRs of odd and even numbers as two inputs. It is given by

$$Y_i^{(2)} = \beta_0^{(2)} + \beta_1^{(2)} x_{i1}^{(2)} + \beta_2^{(2)} x_{i2}^{(2)} + \varepsilon_i^{(2)}, \quad i = 1, 2, \ldots, n, \tag{3.8}$$

where $x_{i1}^{(2)}$ and $x_{i2}^{(2)}$ are the average TTRs of the odd and even numbers, respectively. The output of model Eq. (3.8) is $\hat{Y}^{(2)}$.

3. Single-Variable LR Model

The single-variable LR model is built to handle abnormal conditions. Operational data shows that the coking process takes different amounts of time in different coking chambers. However, the changes in the mean value of the normal TTRs, $x_i^{(1)}$, and the mean value of COT, $y_i^{(1)}$, are relatively small; and the trends are similar. So, the relationship between $y_i^{(1)}$ and $x_i^{(1)}$ is

$$y_i^{(1)} = \beta_0^{(1)} + \beta_1^{(1)} x_i^{(1)} + \varepsilon_i^{(1)}, \quad i = 1, 2, \ldots, n, \tag{3.9}$$

where $\varepsilon_i^{(1)}$ is a random error in the ith experiment, and n is the sample size.

The output of model Eq. (3.9) is $\hat{Y}^{(1)}$. Note that the measured accuracy of this model is lower than that of the model Eq. (3.6).

4. Integrated Linear Regression Model

The integrated LR model is constructed by integrating the outputs of the three models using a rule model. It produces the final output that is close to the true value of the COT. So, the soft-sensing value of the integrated LR model is

$$\hat{T}_{\text{reg}} = \alpha_1 \hat{Y}^{(6)} + \alpha_2 \hat{Y}^{(2)} + \alpha_3 \hat{Y}^{(1)}, \tag{3.10}$$

where $\alpha_1 + \alpha_2 + \alpha_3 = 1$, and $0 \leqslant \alpha_1,\ \alpha_2,\ \alpha_3 \leqslant 1$. The initial values of α_1, α_2, and α_3 are set to 1/3. Optimal values of those coefficients are determined by optimizing the output error E, that is,

$$\frac{\partial E}{\partial \alpha_n} = 0, \quad n = 1, 2, 3. \tag{3.11}$$

Taking the measurement error of $\hat{Y}_1$ of the single-variable LR model as an example, we show some adjusting rules for the coefficients:

R_1: **IF** $|(\hat{Y}_1 - Y_{\text{real}})| \leqslant 1\,^{\circ}\text{C}$, **THEN** $(\alpha_3(k) = 1)$ AND $(\alpha_1(k) = \alpha_2(k) = 0)$.
R_2: **IF** $|(\hat{Y}_1 - Y_{\text{real}})|$ is minimum, **THEN** $\alpha_3(k) = 1.5\alpha_3(k-1)$.
R_3: **IF** $|(\hat{Y}_1 - Y_{\text{real}})|$ is maximum, **THEN** $\alpha_3(k) = 0.5\alpha_3(k-1)$.

3.2.4 Supervised Distributed Neural Network Model

As mentioned above, the nonlinearity in the relationship between the COT and the TTRs needs to be taken into account to improve the precision of soft sensing. This is performed by using the NN technology to fine-tune the model.

Since a huge amount of operational data has been collected, not all of which is consistent, using a single NN to model the nonlinear relationship results in both a large size and slow convergence. So, a nonlinear model is built that employs an SDNN, which is a kind of multi-NN modeling method. First, sample data is classified into various subspaces; and an NN is built for each. The SDNN outputs a weighted sum of the outputs of the NNs.

The partitioning of the sample space is a key to building an SDNN. Commonly used methods are the K-means clustering, the subtractive clustering, and the fuzzy C-means (FCM) clustering. In the FCM, the degree of membership of a given data point in a cluster is specified by a membership grade. Since FCM easily adapts to changing conditions, it is widely employed in fields such as pattern recognition and data classification [35, 36]. It is also employed in this subsection.

Note that the clustering methods mentioned above only consider the influence of the input space [37, 38]. If two data are both very close to the center of a cluster, then they are assigned to the same subspace. That means that data for which the TTRs are close to each other is classified into the same group. However, the operational data shows that COTs corresponding to TTRs with close values might be quite different. Conversely, COTs that are very close might correspond to very different TTRs. So, the output space (that is, the COTs) also influences the partitioning of the sample space. Thus, both the TTRs and the COTs must be considered in sample clustering. This study employed a supervised clustering method [39] and used information on the COTs to divide the sample space.

The input vector is augmented to

$$\tilde{X}_k = [X_k, \beta y_k], \quad k = 1, 2, \ldots, \tag{3.12}$$

where $\tilde{X}_k$ is the augmented input, X_k is the TTR, y_k is the measured COT, and β is the parameter that controls the effect of y_k on the clustering results. Reducing β decreases the effect of y_k on the clustering results. The clustered sample set is expressed to be

$$\left\{X_k,\ \beta y_k | X_k \in \mathbb{R}^6, y_k \in \mathbb{R}^1,\ k = 1, 2, \ldots\right\}. \tag{3.13}$$

The object function employed here is

$$Q = \sum_{j=1}^{c} \sum_{k=1}^{N} u_{jk}^2 \left\| \tilde{X}_k - V_j \right\|^2, \tag{3.14}$$

where u_{jk} indicates the degree of membership of the sample $\tilde{X}_k$ in V_j, $V = [V_1, V_2, \ldots, V_c]$ is a matrix of cluster centers consisting of c cluster center vectors.

The input space is partitioned after the characteristics of the augmented input Eq. (3.12) are examined. Too many cluster centers make the update process too slow when there are a large number of training samples. So, the number of centers is restricted to a relatively small value. This makes the SDNN simple and easy to implement, and provides a high learning efficiency.

Note that an FCM clustering algorithm needs to know the number of cluster centers when it starts, and both the possibility of convergence and the rate of convergence of the algorithm depend greatly on the initial number. If the initial number is suitable, the estimation precision will be high; but if not, the algorithm may easily fall into a local minimum. On the other hand, the subtractive clustering is widely used to estimate the number and position of cluster centers. It assumes that each data point is a potential cluster center and calculates the likelihood that each data point will become a cluster center based on the density of surrounding data points. Clustering efficiency is improved by combining the subtractive and the FCM clustering into the following clustering algorithm:

Step 1: Calculate the density index [40] for all data points and select the data point with the highest index to be the first cluster center.

Step 2: Remove all data points within a radius R_p from the first cluster center and determine the next subspace of data and its cluster center.

Step 3: Repeat Steps 1 and 2 until all the data points are within the prescribed radius of a cluster center.

According to the cluster centers and radius of each subspace, the degree of the corresponding subnet membership is determined by

$$\mu_j = \begin{cases} 1 - \dfrac{\left\|\tilde{X} - V_j\right\|}{r_j}, & \left\|\tilde{X} - V_j\right\| \leqslant r_j, \\ 0, & \left\|\tilde{X} - V_j\right\| > r_j. \end{cases} \tag{3.15}$$

Let the output of the jth BPNN be $\hat{Y}_{\mathrm{NNj}}$. Then, the output of the SDNN model is

$$\hat{Y}_{\mathrm{NN}} = \frac{\sum_{j=1}^{c} \mu_j \hat{Y}_{\mathrm{NN}_j}}{\sum_{j=1}^{c} \mu_j}, \quad \sum_{j=1}^{c} \mu_j = 1. \tag{3.16}$$

The basic soft-sensing values, T_{reg}, of the COTs are obtained by the LR integrated model. And the augmented input is constructed by T_{reg} and X_k:

$$\tilde{X}_k = [X_k, \ \beta \hat{T}_{\mathrm{reg}}], \quad k = 1, 2, \ldots. \tag{3.17}$$

The supervised clustering method is used to classify the sample data for the input Eq. (3.17).

The BPNN, using the basic three-layer structure, is applied to describe the nonlinearity between the COT and the TTRs. In this structure, X_i $(i = 1, 2, \ldots, 6)$ represents the input of six TTRs, $\hat{T}_{\mathrm{net}\ j}$ $(j = 1, 2)$ is the soft-sensing value of the output of a subnet. So, the input layer of the subnet has six neurons, and the output layer has one neuron. The number of the neurons of the hidden layer is set to be 18 according to the empirical formula and experiments. The hidden layer employs sigmoid function, and the output layer is simply $y = x$. The relationship of the input and output for a subnet is

$$\hat{T}_{\mathrm{net}\ j} = \sum_{m=1}^{18} w_m^{\mathrm{O}} \mathrm{tansig}\left(\sum_{i=1}^{6} w_{i,m}^{\mathrm{H}} \tilde{X}_i + b_m^{\mathrm{H}} \right) + b^{\mathrm{O}}, \tag{3.18}$$

where $w_{i,m}^{\mathrm{H}}$ is the weight of the ith neuron of the input layer to the mth neuron of the hidden layer, b_m^{H} is the threshold of the mth neuron of the hidden layer, w_m^{O} is the weight of the mth neuron of the hidden layer to the neuron of the output layer, b^{O} is the threshold of the neuron of the output layer, and tansig(·) represents the sigmoid function.

An SDNN has high accuracy in most cases. However, among the large number of samples, some of them may have conflicted with each other. This may result in a large prediction error. Since the integrated LR model takes full advantage of the three LR models, its predicting result is robust. For this reason, the prediction result of the integrated LR model is adopted when the deviation of the COT obtained by SDNN is large. The final result of the soft sensing is obtained by a series of expert rules according to the analysis of field data.

In the production process, the COT is measured every four hours. The calculated COT and the corresponding TTR construct a new sample. The number of samples in the sample library used to train the integrated LR model and the SDNN model is fixed, and these samples are continually updated. If a newly obtained sample is qualified, it will be added to the sample library, replacing the oldest one in the sample library, and written into the log file. And if the sample is not qualified, it will be discarded and recorded in a different document. When the new data in the sample library is accumulated to a certain amount, it is necessary to start the learning program to correct the established integrated soft-sensing model.

3.2.5 Model Adaptation

The model adaptive unit (Fig. 3.5) ensures that the integrated soft-sensing model of the COT adapts to changes in operating and environmental conditions by updating the following parameters:

(1) The data used for training the LR and SDNN models, including the TTRs and the COT.

(2) The structure and parameters of the soft-sensing model, such as the number and positions of the cluster centers.

The COT is measured once every 4 h. It is used to construct a new sample for training the LR and SDNN models. Before a new sample is added into the database, a check is made to determine whether or not it is normal. If it passes the test, it is added to the database and the oldest sample is deleted. The criteria for the check are as follows:

(1) Whether or not the value is within a specific range.

(2) Whether or not the difference between the current data and the last measured data is in the range [−30°C, +30°C].

(3) Whether or not the data upholds the linear relationship of the data in the database.

A new qualified sample is put into the category for which the sample has the largest degree of membership.

The adaptation program runs when the number of new samples reaches a particular value. That value is critical. A small value leads to frequent adaptations with consequent large variations in the output of the soft-sensing model. This increases the training frequency and results in a cumulatively long training time. On the other hand, a large value results in infrequent adaptations, which means that the model does not reflect changes in the operating conditions in a timely manner.

Since the LR and SDNN models have different structures, they have different frequencies of adaptation. The number of new samples for the LR model was chosen to be 6, but that number makes the adaptation period of the SDNN model very long.

Two aspects of the SDNN model need to adapt: the structure and the parameters. The partitioning of the sample space strongly influences the model precision. For structural adaptation, the space is partitioned adaptively according to the current operating conditions. This involves adding or removing samples, modifying cluster centers, and adjusting the number and parameters of the NNs. That usually takes a long time. To shorten the time, the soft-sensing model was built using a massive amount of historical data collected over a 2-year period. The samples cover an extensive range of conditions, and the partitioning of the sample space is nearly ideal. This enables a new sample to be readily classified into an appropriate subspace, thereby making frequent modification of the structure of the model unnecessary. Structural adaptation is carried out every two months and involved two steps:

(1) Choose new samples.

(2) Rebuild the SDNN model according to the procedure in this subsection.

Since the adaptation program does not need to modify the structure of the SDNN model, it only needs to modify the parameters of the soft-sensing model. For each category, when the number of new samples reaches 20, the oldest 20 samples are deleted, and the NN model for that category is trained again.

3.3 Intelligent Optimization and Control of Coke-Oven Combustion Process

This section describes a hierarchical intelligent integrated control method that employs the determination of the operating state of a coke-oven based on an analysis of the characteristics of the combustion process in the oven.

3.3.1 Configuration of Hybrid Hierarchical Control System

The following strategies are employed to control the combustion process:

(1) A hybrid control system, which combines continuous control and the manipulation of discrete symbols, is constructed to control this complex process.

(2) Due to the complexity and a large time delay of the combustion process, it is impossible to build a temperature control system with a single controller. To obtain

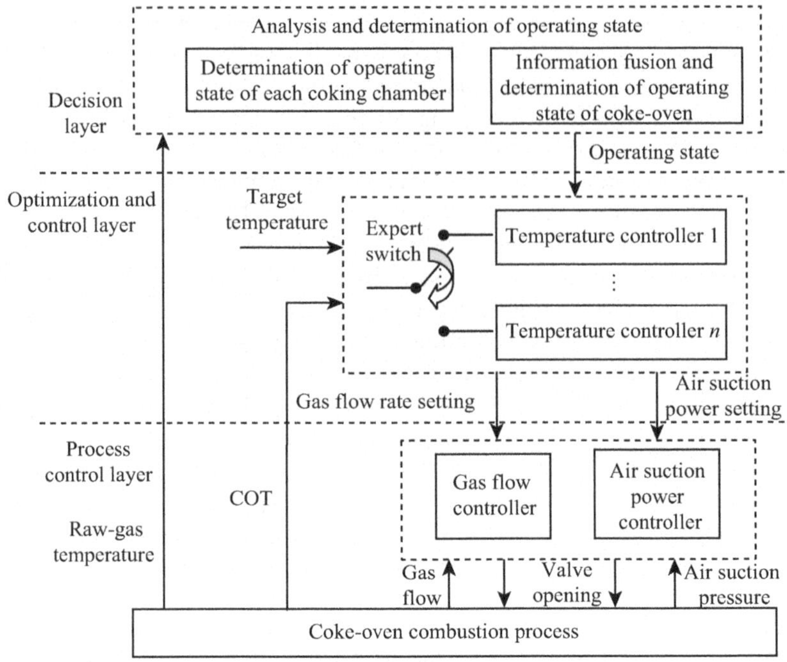

Fig. 3.7 Configuration of hybrid hierarchical control system for coke-oven combustion process

good control performance, the operating states are classified into several types, and a controller is designed for each one.

(3) A two-stage decision method based on information fusion is used to determine the current operating state.

(4) A self-adaptive GA optimizes the parameters of the fuzzy controllers.

The configuration of the hybrid hierarchical control system for the combustion process (Fig. 3.7) contains three layers: a decision layer, an optimization and control layer, and a process control layer.

The main function of the decision layer is to determine the operating state in a real-time fashion. In each ascending pipe, a thermocouple measures the temperature of the raw gas, which is an indicator of the maturity of the coke in the coking chambers. The total amount of required heat depends on the maturity of the coke in each of the coking chambers and is determined by the combined needs of the coking processes in all the chambers. The measured temperatures of the raw gas in the ascending pipes are used by a two-stage decision method based on information fusion to determine the operating state of the oven.

The optimization and control layer produces a set of optimal temperature controllers for various operating states. The parameters of the controllers are the scaling and proportion factors of fuzzy controllers. Adaptive GAs optimize them in a offline fashion. Expert rules are employed to switch the controllers online to provide quick adaptation to changes in the operating state. The control system is constructed using a two-loop control strategy: the outer loop controls the oven temperature, and the inner loop adjusts the gas flow rate and air suction power to reference values produced by the temperature controller. The outer loop is implemented on the optimization and control layer, and the inner loop is implemented on the process control layer.

3.3.2 Determination of Operating State

Since the thermal absorptivity of coal is different in the beginning, middle, and terminal stages of the coking process, even if the oven temperature is the same, the operating state is different at different stages. So, different controllers are used to handle this situation.

The amount of heat required for coking is an important factor in the selection of a temperature controller. It is basically determined by discrete events in coking operations. This section shows how the operating state is determined from an analysis of the maturity of the coke, which is obtained from the measured temperature of the raw gas in an ascending pipe, and how critical points in the coking process are extracted.

1. Classification of Operating States

When coke is pushed out of the oven, it carries away a large amount of heat. So, a certain rhythm in pushing operations is needed to enable good control of the oven temperature. That is, when coke is pushed strictly according to the production schedule, the heat balance in the oven is maintained; but a delay in pushing causes a change in the heating level of the whole oven, resulting in a change in the oven temperature.

Pushing operations are halted for two reasons. One is maintenance. There are about 6 h of maintenance per day. This time is equally divided among the operating periods and is added at the end of each period. So, maintenance has little influence on oven temperature. The other is any kind of equipment failure that prevents mature coke from being pushed out on time. The longer it takes to fix the equipment, the larger the number of coking chambers containing mature coke there are. This changes the operating state and leads to a rapid increase in oven temperature. Furthermore, after the problem is solved and the mature coke is pushed out, the coking chambers are quickly charged with coal. This creates a situation in which a large number of coking chambers are in the beginning stage of the coking process, which changes the operating state and dramatically reduces the oven temperature.

Based on the above analysis, the operating states are classified into three types:

S_1 : Normal operating state. In this state, operations (coaling, pushing, maintenance, etc.) are carried out according to the production schedule. The heat balance in the coking process is taken into consideration in the production schedule.

S_2 : No pushing. This state occurs when equipment failure has caused a relatively long halt in the coking process. A large number of coking chambers contain mature coke, and the oven temperature tends to be higher than usual.

S_3 : Quick pushing. After the equipment is fixed, the coking chambers must be cleared rapidly of mature coke and charged with new coal to return production to the normal operating state. Thus, a large number of coking chambers are in the beginning stage and absorb a great deal of heat. So, the oven temperature tends to be lower than usual.

2. Determination of Operating State

It is difficult to determine the operating state in a real-time fashion because of the structure of the oven and limitations on production conditions. Conventionally, the maturity of coke in a coking chamber is assessed from the color of the raw gas in the ascending pipe, which changes during the coking process. This method is very labor-intensive for the operators; and it is also very subjective, with the results strongly depending on the experience of an operator. To solve these problems and obtain an accurate determination of the operating state, a thermocouple was installed in each ascending pipe to measure the temperature of the raw gas, and a two-stage decision method based on information fusion was used to determine the operating state (Fig. 3.8).

In the first stage of the decision process, the temperatures of the raw gas are measured by the thermocouples in the ascending pipes, and the data is processed to extract critical points. Then, expert rules yield the operating state of each coking chamber ($p_1, p_2, \ldots, p_{60}$). In the second stage, the operating state of the whole coke-oven is obtained by the fusion of the first-stage outputs.

In the first stage, the temperature data is processed by the method of moving averages, which smoothes the data and filters out noise. Then, critical points are extracted from the processed data. Generally, the temperature of the raw gas in an ascending pipe gradually increases after the start of the coking process and drops sharply after coking finishes. The coking process is characterized by the point at which coking finishes and the coke becomes mature. On the other hand, the coking period, T, which ranges from 18 to 24 h, is an important parameter provided by the production scheduler. We define t_i to be the coking time of the ith coking chamber (that is, the time since the chamber was charged with coal) and t_C be the time when coking finishes. Then, the coking index, C_i, given by

$$C_i = \frac{t_i}{t_C}, \quad t_i \geqslant t_C, \tag{3.19}$$

indicates the maturity of the coke.

So, the operating state, p_i, of the ith coking chamber can be represented by the triplet (δ, a_i, C_i), where a_i is a flag indicating whether or not coking is finished ($a_i = 0$ when $t_i < t_C$ and $a_i = 1$ when $t_i \geqslant t_C$). The normal operating state is further classified as early, middle, or terminal stage; and the state of the coke in a coking chamber is classified as immature, mature, undercoked, and overcoked. Table 3.1 shows the rules for determining the operating state of the ith coking chamber.

In the first stage of the decision process, an output space consisting of the operating states of all the coking chambers is constructed. It is then categorized to make an input

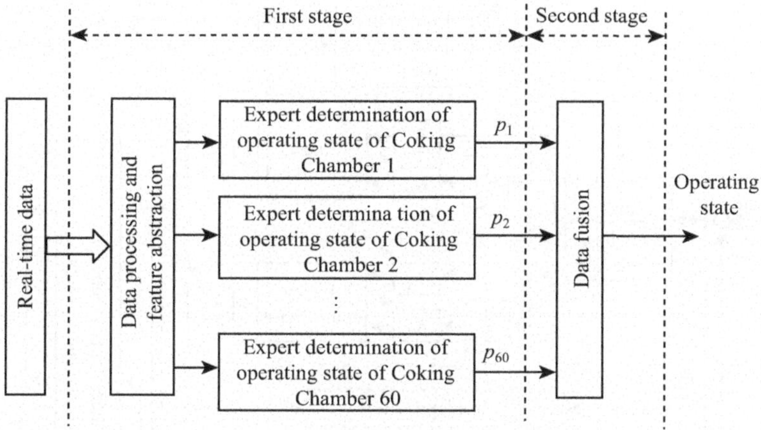

Fig. 3.8 Determination of operating state using two-stage decision method

Table 3.1 Rules for determining operating state of ith coking chamber

Rule	Parameter			Status of coke	
	δ	a_i	C_i	p_i	M_i
1	0–1/3	0	–	Early stage	Immature
2	1/3–2/3	0	–	Middle stage	Immature
3	2/3–1	1	1–1.2	Terminal stage	Undercoked
4	2/3–1	1	1.2–1.3	Terminal stage	Mature
5	2/3–1	1	>1.3	Terminal stage	Overcoked

space for information fusion in the second stage. Statistics on the coking chambers in different operating states and different coking states are collected: cp_{e}, cp_{m}, and cp_{t} are the numbers of coking chambers in the early, middle, and terminal stages, respectively; and cm_{uc}, cp_{m}, and cp_{oc} are the numbers of coking chambers in the terminal stage for which the coke is undercoked, mature, and overcoked. Some typical rules are listed below as examples, where $S(k)$ indicates the present (kth sampling time) operating state of the ith coking chamber.

R_{OS1}: **IF** ($cp_{\mathrm{oc}} > 5$) AND ($cp_{\mathrm{t}} \geqslant 25$), **THEN** $S(k) = S_2$.
R_{OS2}: **IF** ($S(k-1) =$ Stop) AND ($cp_{\mathrm{e}} \geqslant 20$), **THEN** $S(k) = S_3$.

3.3.3 *Design of Coke-Oven Temperature Controller*

The target of the intelligent integrated optimization and control of the combustion process is to stabilize the oven temperature in spite of disturbances by making a

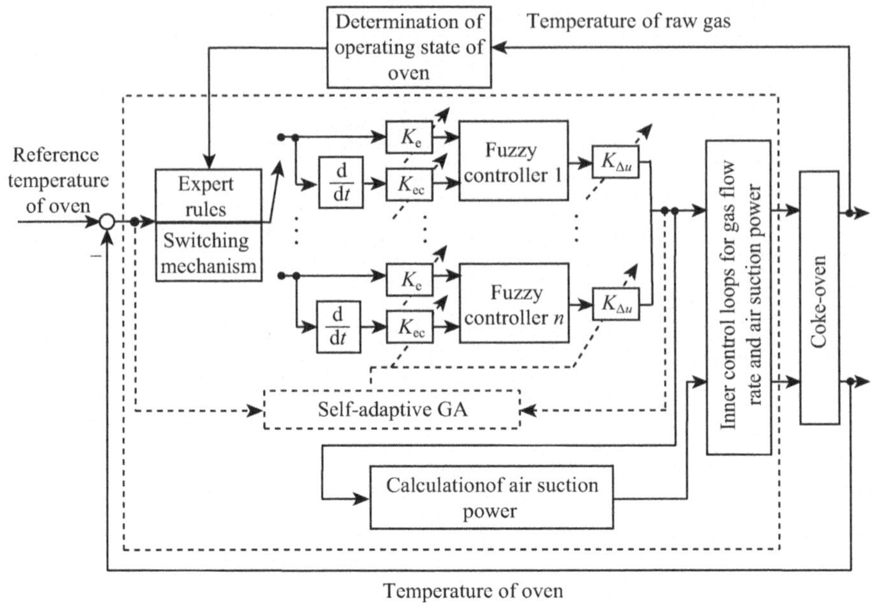

Fig. 3.9 Configuration of optimization and control layer

rational schedule for the combustion process and properly regulating the heat supply. The key points for the optimization and control layer (Fig. 3.9) are as follows:

(1) Optimize the parameters of the controllers.

(2) Choose a controller appropriate to the operating state in a real-time fashion and switch to it, and to use the output of the temperature controller to calculate optimal settings for the gas flow rate and air suction power.

1. Expert Switching

Three types of gas (coke gas, blast furnace gas, a blend of the two) are used in the coking process, and different heating methods are employed for each. There are two heating methods for blended gas: fix the flow rate of coke gas and adjust the flow rate of blast furnace gas, and vice versa. The heating method should be changed when the type of gas is changed; but if the heating method remains unchanged, a change in the type of gas causes the operating state to change.

To obtain good control performance, it is necessary to employ different controllers for various operating states. The following online switching rules were established based on the above classification and determination of the operating state.

R_{ES_i}: **IF** $s = S_i$, **THEN** switch the controller to fuzzy control i.

That is, select fuzzy controller i when the operating state, s, is S_i.

2. Optimization of Parameters of Fuzzy Controller

In fuzzy control, the experience of experts is converted into mathematical models that can be handled by a computer. Since experience is subjective to some extent, there are certain limitations on the parameters of a controller derived directly from such experience. Furthermore, the control system has multiple control objectives (high-quality coke, saving energy, etc.), which requires fine adjustment of the parameters.

As an example, we use the temperature fuzzy controller for blended gas to explain how a fuzzy controller is designed and how its parameters are optimized. Assume that the heating method is to fix the flow rate of coke gas and adjust the flow rate of blast furnace gas. The inputs of the temperature fuzzy controller are the temperature error, e, and its rate of change, ec; and the output is the change in the flow rate of blast furnace gas, Δu. The corresponding fuzzy sets are E, EC, and ΔU; and their linguistic states are { NL (negative large), NM (negative medium), NS (negative small), ZO (zero), PS (positive small), PM (positive medium), PL (positive large) }, { NL, NM, ZO, PS, PM, PL }, and { NL, NM, ZO, PM, PL }, respectively. The ranges of the variables are $[-20, 20]$ for e, $[-15, 15]$ for ec, and $[-200, 200]$ for Δu. The membership functions of E, EC, and U are all chosen to be trapezoidal. The Mamdani method is used for fuzzy inference, and the centroid method is used for defuzzification.

Since the scaling and proportion factors K_e, K_{ec}, and $K_{\Delta u}$ can be used to tune the dynamic characteristics of the closed-loop system, these parameters are determined based on trade-offs among response time, stability, and robustness. However, optimization under these conditions is almost impossible when only the experience of experts is used. Thus, a multiple-objective optimization method was combined with a self-adaptive GA to optimize the parameters of a fuzzy controller.

The optimization targets for the temperature fuzzy controller are to stabilize the oven temperature, to ensure coke quality, to reduce the consumption of gas, and to extend the lifetime of the oven. These considerations yield the following fitness function:

$$F = \frac{1}{w_1 J_1 + w_2 J_2 + w_3 J_3}, \tag{3.20}$$

where w_i $(i = 1, 2, 3)$ is a weight; and J_1, J_2, and J_3 are the performance indices for flue temperature, gas consumption, and response time, respectively. They are given by

$$J_1 = |e|, \tag{3.21}$$

$$J_2 = \sum_{k=0}^{M} u(k), \tag{3.22}$$

$$J_3 = \sum_{k=0}^{N} t_s(k). \tag{3.23}$$

In Eq. (3.22), $u(k)$ is the gas consumption during a sampling period and M is the number of control steps. In Eq. (3.23), $t_s(k)$ is the regulation time and N is the number of steps in the transient response. The constraints are

$$0 \leqslant \quad \Delta u(k) \quad \leqslant \Delta U_{\max}, \tag{3.24}$$

$$0 \leqslant \quad e(k) \quad \leqslant e_{\max}, \tag{3.25}$$

where $\Delta U_{\max}$ is the maximum allowable change in gas flow rate and $e_{\max}$ is the maximum value of e.

A GA is employed to optimize K_e, K_{ec}, and $K_{\Delta u}$ so as to yield the maximum F. K_e, K_{ec}, and $K_{\Delta u}$ are encoded as real numbers; the nth chromosome of the ith generation is $P_n^i = \left[p_{k_e}^n,\ p_{k_{ec}}^n,\ p_{k_{\Delta u}}^n\right]$; and the fuzzy parameters that need to be optimized are $p_{k_e}^n$, $p_{k_{ec}}^n$, and $p_{k_{\Delta u}}^n$.

The elitist strategy [41] and the roulette wheel algorithm [42] are used for selection. When a set of individuals constitutes a larger percentage of the population than a given value, some are selected using the roulette wheel algorithm to maintain the diversity of population; and others are selected using the elitist strategy to preserve the individuals with the best fitness.

The key factors that determine the performance of the GA are the crossover probability, p_c, and the mutation probability, p_m. Since crossover is the primary means of producing new individuals, p_c should be relatively large. However, if it is too large, good patterns will be destroyed; and if it is too small, new individuals will be produced very slowly. p_m has a significant impact on the optimization of the parameters: If it is too large, convergence will be slow; and if it is too small, convergence will be premature. To avoid these problems, this study employed an improved GA [43], in which p_c and p_m are tuned automatically in response to changes in the fitness function as follows:

$$p_c = \begin{cases} \dfrac{k_1(f_{\max} - f_c)}{f_{\max} - f_{\text{avg}}}, & f_c \geqslant f_{\text{avg}}, \\ k_2, & f_c < f_{\text{avg}}, \end{cases} \tag{3.26}$$

$$p_m = \begin{cases} \dfrac{k_3(f_{\max} - f_m)}{f_{\max} - f_{\text{avg}}}, & f_m \geqslant f_{\text{avg}}, \\ k_4, & f_m < f_{\text{avg}}, \end{cases} \tag{3.27}$$

where

$$0 < k_1,\ k_2,\ k_3,\ k_4 < 1,\ \ k_2 > k_1,\ \ k_4 > k_3,$$

$f_{\max}$ is the maximum fitness and f_{avg} is the average, and f_c is the larger of the two individuals selected for crossover and f_m is the fitness of an individual for mutation. Figure 3.10 shows a flowchart of parameter optimization.

3. Calculation of Air Suction Power

In the combustion process, the air suction power determines burning efficiency. It must be changed when the gas flow rate changes to ensure complete combustion. However, since there is no online oxygen sensor, the air–gas ratio cannot be calculated directly. The relationship between air suction power and gas flow is

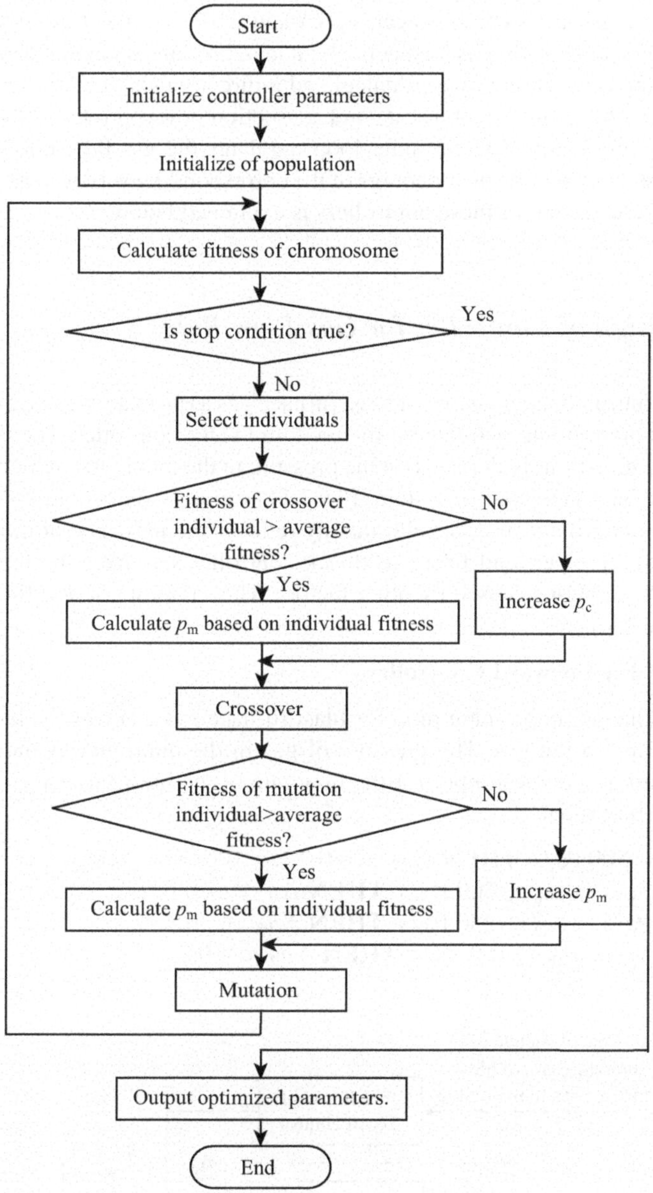

Fig. 3.10 Flowchart of parameter optimization for fuzzy controller

$$u_a(k) = \alpha_0 + \alpha_1 \frac{u_b(k)}{u_b(k-1)} + \alpha_2 \frac{u_c(k)}{u_c(k-1)}, \tag{3.28}$$

where $u_a(k)$ is the air suction power, $u_b(k)$ and $u_c(k)$ are the flow rates of blast furnace gas and coke-oven gas, respectively; and α_0, α_1, and α_2 are constants. In this subsection, the air suction power is calculated indirectly using the above formula.

Using the settings for gas flow rate and air suction power produced by the outer-loop temperature controller, the inner loop regulates the gas flow rate and the air suction power by adjusting the openings of the corresponding valves so as to stabilize the process. The design of these controllers is explained below.

3.3.4 Design of Controller for Gas Flow Rate

Taking the control of the flow rate of blast furnace gas as an example, we explain how to design an appropriate two-degree-of-freedom (TDF) controller. The flow rate of blast furnace gas strongly depends on the pressure in the main pipe, which fluctuates dramatically. The TDF control system (Fig. 3.11) suppresses fluctuations in the flow rate. A feedforward expert controller mainly reduces the influence of the main pipe pressure, and an expert and fuzzy feedback controller stabilizes the flow rate and suppresses fluctuations caused by other factors. The sampling period for the valves was set to be 15 s.

1. Design of Feedforward Controller

A pressure change in the main pipe for blast furnace gas can cause a large change in the flow rate of the gas. The pressure of gas in the main pipe is measured and is fed forward as a control input for the flow rate of the blast furnace gas using the following expert rules:

R_1: **IF** $ec_b > 500$ Pa/s, **THEN** $\Delta u_2 = -0.7$.
R_2: **IF** 250 Pa/s $< ec_b \leqslant 500$ Pa/s, **THEN** $\Delta u_2 = -0.6$.
R_3: **IF** 150 Pa/s $< ec_b \leqslant 250$ Pa/s, **THEN** $\Delta u_2 = -0.4$.
R_4: **IF** 50 Pa/s $< ec_b \leqslant 150$ Pa/s, **THEN** $\Delta u_2 = -0.2$.

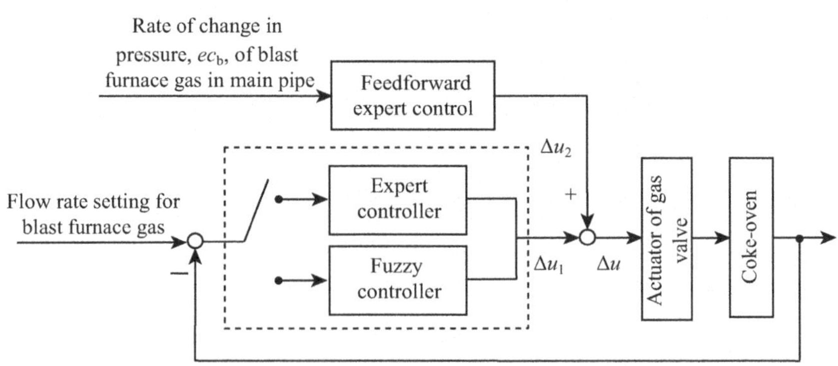

Fig. 3.11 Configuration of TDF control system for flow rate of blast furnace gas

R_5: **IF** -50 Pa/s $\leqslant ec_b < 50$ Pa/s, **THEN** $\Delta u_2 = 0$.
R_6: **IF** -150 Pa/s $\leqslant ec_b < -50$ Pa/s, **THEN** $\Delta u_2 = 0.2$.
R_7: **IF** -250 Pa/s $\leqslant ec_b < -150$ Pa/s, **THEN** $\Delta u_2 = 0.4$.
R_8: **IF** -500 Pa/s $\leqslant ec_b < -250$ Pa/s, **THEN** $\Delta u_2 = 0.6$.
R_9: **IF** $ec_b < -500$ Pa/s, **THEN** $\Delta u_2 = 0.7$.

In the above rules, ec_b is the rate of change in the pressure of the blast furnace gas in the main pipe, and Δu_2 is the change in the valve opening.

2. Design of Feedback Controller

The feedback controller employs both expert and fuzzy control methods. The fuzzy controller precisely adjusts the valve opening when the magnitude of the flow tracking error for the flow rate of blast furnace gas, e_f, is less than a threshold; and the expert controller quickly reduces the tracking error when the magnitude of e_f is greater than the threshold. Since $-3000\ \mathrm{m^3/h} < e_f < 3000\ \mathrm{m^3/h}$ in practice, the threshold was chosen to be 6000 $\mathrm{m^3/h}$. The following expert control rules are used:

R_1: **IF** $e_f > 12000\ \mathrm{m^3/h}$, **THEN** $\Delta u_1 = -2.5$.
R_2: **IF** $6000\ \mathrm{m^3/h} < e_f \leqslant 12000\ \mathrm{m^3/h}$, **THEN** $\Delta u_1 = -2.2$.
R_3: **IF** $-12000\ \mathrm{m^3/h} \leqslant e_f < -6000\ \mathrm{m^3/h}$, **THEN** $\Delta u_1 = 2.2$.
R_4: **IF** $e_f < -12000\ \mathrm{m^3/h}$, **THEN** $\Delta u_1 = 2.5$.

In the above rules, Δu_1 is the change in the valve opening.

The inputs of the fuzzy controller are e_f and ec_f, and the output is the change in the valve opening, Δu_1. The ranges of e_f, ec_f, and Δu_1 are chosen to be $[-6000, 6000]$, $[-2000, 2000]$, and $[-2, 2]$, respectively. Accordingly, the linguistic values of the control error, E_f, its rate of change, EC_f, and the change in the valve opening, ΔU_1, are {NL, NM, NS, ZO, PS, PM, PL}, {NL, NM, ZO, PS, PM, PL}, and {NL, NM, NS, ZO, PS, PM, PL}, respectively. Trapezoidal functions were selected to be the membership functions for the linguistic values of E_f, EC_f, and ΔU_1 because they are insensitive to linguistic values. ΔU_1 is set as shown in Table 3.2. For example, if E is PL and EC is ZO, then the output, ΔU_1, is PL. That means that, when the flow rate of blast furnace gas is much smaller than the reference value and the rate of change in the tracking error is also very small, then the valve should be opened a great deal to quickly increase the flow of gas. The output of the fuzzy controller is obtained by defuzzification. The total change, Δu, in the valve opening is given by

$$\Delta u = \Delta u_1 + \Delta u_2. \tag{3.29}$$

3.3.5 Design of Air Suction Power Controller

The air suction power changes only moderately and is not greatly influenced by any factors. An expert control method is used to control it. Expert rules are derived based on the experience of experts and the analysis of historical data. e_a is the error between

Table 3.2 Fuzzy rules of fuzzy feedback controller in TDF control system for flow rate of blast furnace gas

EC	*E*						
	NL	NM	NS	ZO	PS	PM	PL
NL	NL	NL	NM	NS	NS	ZO	PS
NS	NL	NM	NM	NS	ZO	PS	PM
ZO	NL	NM	NS	ZO	PS	PM	PL
PS	NM	NS	ZO	PS	PM	PM	PL
PL	NS	ZO	PS	PS	PM	PL	PL

the setting and the actual value of the air suction power, and Δu_a is the change in the opening of the air suction valve. Some of the main rules are as follows:

R_1: **IF** $e_\mathrm{a} > 25$ Pa, **THEN** $\Delta u_\mathrm{a} = -2$.
R_2: **IF** 8 Pa $< e_\mathrm{a} \leqslant 25$ Pa, **THEN** $\Delta u_\mathrm{a} = -1.5$.
R_3: **IF** 6 Pa $< e_\mathrm{a} \leqslant 8$ Pa, **THEN** $\Delta u_\mathrm{a} = -1.2$.
R_4: **IF** 4 Pa $< e_\mathrm{a} \leqslant 6$ Pa, **THEN** $\Delta u_\mathrm{a} = -0.9$.
R_5: **IF** 3 Pa $< e_\mathrm{a} \leqslant 4$ Pa, **THEN** $\Delta u_\mathrm{a} = -0.6$.
R_6: **IF** -3 Pa $\leqslant e_\mathrm{a} \leqslant 3$ Pa, **THEN** $\Delta u_\mathrm{a} = 0$.
R_7: **IF** -4 Pa $\leqslant e_\mathrm{a} < -3$ Pa, **THEN** $\Delta u_\mathrm{a} = 0.6$.
R_8: **IF** -6 Pa $\leqslant e_\mathrm{a} < -4$ Pa, **THEN** $\Delta u_\mathrm{a} = 0.9$.
R_9: **IF** -8 Pa $\leqslant e_\mathrm{a} < -6$ Pa, **THEN** $\Delta u_\mathrm{a} = 1.2$.
R_{10}: **IF** -25 Pa $\leqslant e_\mathrm{a} < -8$ Pa, **THEN** $\Delta u_\mathrm{a} = 1.5$.
R_{11}: **IF** $e_\mathrm{a} < -25$ Pa, **THEN** $\Delta u_\mathrm{a} = 2$.

3.4 Operation Planning and Optimal Scheduling of Coking

This section first analyzes the planning and scheduling of coking operations, then describes a configuration of optimal scheduling, finally, presents an optimal scheduling method for different operating states.

3.4.1 Analysis of Operations Planning and Optimal Scheduling of Coking

The coke-oven production process (Fig. 3.12) mainly contains four kinds of car systems: coaling cars, pushers, guide cars, and quenching cars.

The coaling cars deliver weighted raw materials into specified coking chambers. The pushers move to assigned coking chambers and flatten coal in the chambers. Raw materials are carbonized and become coke through the high-temperature coking

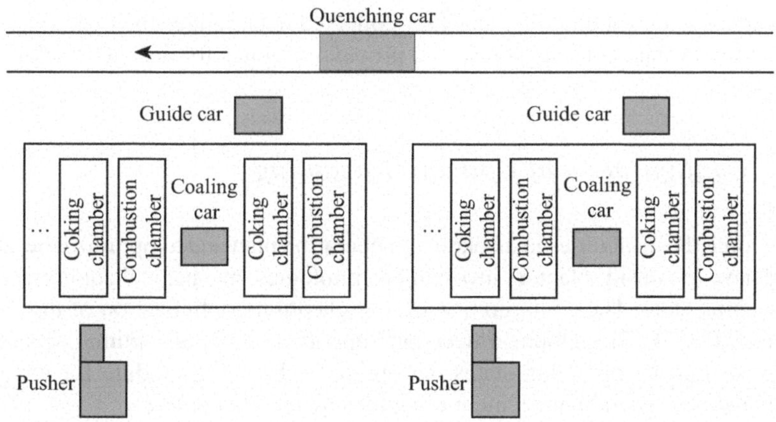

Fig. 3.12 Coke-oven production process

process in the coking chambers. Then, the pushers move precisely to the doors of specified furnaces, and the guide cars move to the doors on the other side of the specified furnaces. The quenching cars move to the front of the corresponding chambers and are ready for the discharge of coke. After that, the pushers and the guide cars open the doors on the two sides, and the pushers push coke to the quenching cars with the help of the guide cars. The quenching cars deliver the coke to a coke quenching tower. Finally, the pushers and the guide cars close the coking doors. These operations repeat during the coking process.

Planning and scheduling of the coking operations is now mainly performed manually to adapt to different operating states. However, this method depends on the experience of workers excessively and is unreliable. So, it is necessary to derive a method of optimizing the scheduling of the coking operations that can adapt to different operating states, and to design a system architecture with good portability and easy implementation of data communication, and automatic backup and recovery of data.

Optimal scheduling of coking operations makes an operating schedule for the next shift based on the maintenance plan, the standard coking time, and the actual pushing time and loading time of each coking chamber in the last turnaround time. Scheduling the coke-pushing operations has to take into account of the turnaround time, the operating time, the maintenance time, and the coking time. The coking time is the time that coal materials stay in a coking chamber for one period. It is the time interval from the beginning of flattening coal (coal loading) to the beginning of pushing coke. The turnaround time refers to the time interval from the beginning of the coke-pushing (or coal loading) to the next beginning of the coke-pushing (or coal loading).

The coke-pushing process has various operating states that give different influence on the scheduling of the coke-oven operations. So, the analysis of the operating states is a fundamental part of the optimal scheduling. The analysis is carried out by a decision analysis module that analyzes collected data, judges the operating state, classifies abnormal operating states, and provides grounds for decision-making.

3.4.2 Configuration of Optimal Scheduling

To overcome the difficulties in a coke-oven operation plan and to optimize the schedule, we have to consider factors involving technologies, mechanical equipment, and coke-pushing state. The configuration of the scheduling optimization of the coking operations (Fig. 3.13) includes a decision support module, an optimal scheduling module for normal operating states, an optimal scheduling module for abnormal operating states, a data management module, and a client module.

For normal operating states, the scheduling of operations is to establish an optimal scheduling model based on the analysis of the coke-pushing process, technological

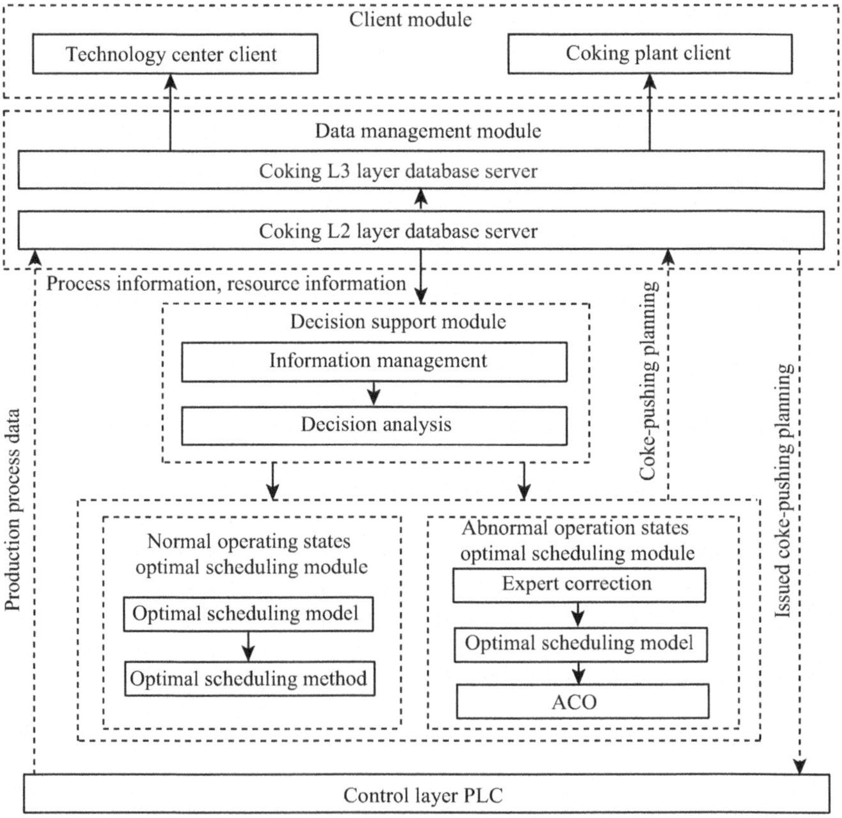

Fig. 3.13 Configuration of optimal scheduling of coking operations

requirements, and constraints. And a reasonable order of coking chambers is arranged in accordance with the order of three shifts for a day. The generated coke-pushing plan includes the number of the furnaces for coke-pushing, the serial number of coking chambers, the corresponding coking time, coaling time, operating time for a single furnace, etc.

For abnormal operating states, the scheduling of operations first needs to classify operating states. An expert correction method is employed when operating states meet requirements of expert correlation rules. Otherwise, an optimal scheduling model for abnormal operating states is established based on coke quality, mechanical wear, operation efficiency, and maintenance time.

The decision support module, composed of information management and decision analysis, stores and deals with the acquired information. It mainly requires three kinds of information: production task, process information, and resource information. The production task contains the indices of coke quality and production. The process information contains the operating time of a single furnace, coking time, coaling time, and coke-pushing time. The resource information contains the states of the coke-oven and four kinds of cars (coaling cars, pushers, guide cars, and quenching cars).

The main function of the data management module is to manage the relevant data of optimizing the schedule of the coking operations. It has the functions of addition, deletion, modification, and printing of the operating plan. It inquires the stored information and the generated schedule of operations according to requirements, and can also manually modify the schedule. The system manages the local information, and also saves information in a data server on the L2 layer of an enterprise through a network.

The client module mainly performs the query and statistics of data, including the operating plan, historical records, the heating system, the gas consumption, the coke-pushing coefficient, etc. The schedule of the coke-pushing for the last three shifts can be viewed from the client module. The historical records can be queried by date or by shift. It is easy to make inquiries about history coke-pushing data including gas flow and pressure of the blast furnace, the mixed gas flow, the gas flow and pressure of the coke-oven, the flue suction and temperature, and the temperature and pressure of gas collectors, as well as the driving coefficient, the coking time, the pushing current, and the coaling coefficient. Many functions are also integrated in the module, for example, the calculation of averages, uniformity coefficients, and stability coefficients.

3.4.3 Optimal Scheduling of Operating States

When the coke-oven is in normal operating states, each coking chamber carries out coaling, pushing, and other operations according to the production schedule. The state of coaling, the status of the coke-pushing equipment, the deformation of the wall of the coking chambers, etc., give big influence on an operating plan, sometimes may even cause that coke-pushing cannot be performed according to a schedule. This leads to various operating states.

In the coking process, the temperature change, mechanical shock, and chemical corrosion caused by repeat operations gradually change the characteristics of the furnace. For example, deformation of the walls of coking chambers in a furnace leads to the increase of pushing resistance. So, we need to extend the coking time of the furnace. Those abnormal furnaces are placed outside a normal sequence when making a schedule of coke-pushing operations. We have to build a circulating table separately for those furnaces according to their operation time. And the coke-pushing operation of the abnormal furnaces should be arranged in the maintenance period as possible to avoid a conflict with normal furnaces.

The standard 5-2 series is used for the schedule of the coke-pushing sequence in normal operating states. And the differences between the scheduled and real coaling time and coke-pushing time are within ±5 min. However, when the state is abnormal, the operations are adjusted under the constraints of resources, energy, the initial conditions, the technological requirements, etc. And the goal is to ensure coke quality and to reduce production cost.

1. Optimal Scheduling for Normal Operating States

Taking the coking process in an iron and steel enterprises as an example, we explain the optimal scheduling of the process for normal operating states. Coke-ovens 1 and 2 use the 5-2 series. The types of Coke-ovens 1 and 2 are a 60-cabinet JN60 and a 55-cabinet JN60, respectively. Each of the coke-oven has two pushers, two coaling cars, a guide car (and a spare), and a coke quenching car.

The implementation of the schedule of the coke-pushing operations should take into account of various factors and meet the following rule for a coking chamber:

$$T_{\mathrm{m}i} + T_{\mathrm{r}i} = T'_{\mathrm{p}i}, \tag{3.30}$$

where $T_{\mathrm{m}i}$ and $T_{\mathrm{r}i}$ are the actual coaling time and coking time of the ith coking chamber in the last turnaround time, respectively; and $T'_{\mathrm{p}i}$ is the planed coke-pushing time of the ith coking chamber in next turnaround time.

The operating time of each operation is maintained at a prescribed value. Coaling starts within eight minutes after the coke-pushing. The guide car, the pushers, and the quenching car start moving simultaneously. They have to meet the following constraints:

$$T'_{\mathrm{m}i} - T'_{\mathrm{p}i} \leqslant 8 \text{ min}, \tag{3.31}$$

$$T'_{\mathrm{p}i} = T'_{\mathrm{s}i}, \tag{3.32}$$

$$T'_{\mathrm{p}i} = T'_{\mathrm{c}i}, \tag{3.33}$$

where $T'_{\mathrm{m}i}$, $T'_{\mathrm{s}i}$, and T'_{ci} are the scheduled coaling time, guiding time, and quenching time of the ith coking chamber in next turnaround time, respectively.

The coking time of Coking chamber i is defined to be

$$T_i = T_{pi} - T_{mi}, \tag{3.34}$$

where T_{mi} is the coaling time of the ith coking chamber, and T_{pi} is the scheduled coke-pushing time of the ith coking chamber.

To ensure coke quality, the coking time needs to be guaranteed for each coking chamber, and the difference between the actual and the standard coking time has to satisfy

$$|T'_{pi} - T_i| \leqslant 5 \text{ min}. \tag{3.35}$$

The difference between the turnaround time, $\tau_{i,\text{cycle}}$, and the actual coking time, T_{ri}, of each coking chamber has to satisfy

$$\tau_{i,\text{cycle}} - T_{ri} \leqslant 15 \text{ min}. \tag{3.36}$$

The maintenance time has to satisfy

$$\tau_{\text{maint}} = \tau_{\text{cycle}} - m \times l, \tag{3.37}$$

$$\tau_{\text{maint}} \geqslant 60 \text{ min}, \tag{3.38}$$

where m is the number of the coking chamber, l is the operation time of a single cabinet, τ_{maint} is the maintenance time, and τ_{cycle} is the turnaround time.

The difference of the coke-pushing time between adjacent two coking chambers has to satisfy

$$|T_{pi} - T_{p(i+1)}| > 120 \text{ min}, \quad |T_{pi} - T_{p(i-1)}| > 120 \text{ min}, \\ i = 2, 3, \ldots, n-1. \tag{3.39}$$

Coke-oven 1 has 60 coking chambers; the operation time of a single cabinet is 10 min. Coke-oven 2 has 55 coking chambers; the operation time of a single cabinet is 9 min. Equation (3.37) gives the turnaround time of 2.75 h. The maintenance time meets the actual production requirement. The coke-pushing plan is strictly carried out according to the prescribed schedule using the 5-2 coke-pushing series. The coking time is approximately equal to the turnaround time, which is generally about 21 h according to the actual situation of the process.

The schedule of a coke-pushing operations for normal operating states is optimized based on expert rules and a model of coke-pushing plan by analyzing the relationship between each time duration and the coke-pushing plan. A circulating table specifies the daily operating time, the baked amount, and the maintenance time.

2. Optimal Scheduling for Abnormal Operating States

When the state of the process is abnormal, the coke-pushing no longer follows the operation plan for normal operating states. We need to optimize the scheduling to gradually restore the schedule to the normal coke-pushing series.

Common abnormal operating states, that is, a disorder of the operating groups, the effects of accidents, and abnormal coking chambers, lead to the disorder of the coke-pushing sequence. Since the impact of these abnormal operating states on the coke-pushing plan can be predicted, the schedule of coke-pushing operations can be optimized for those situations using the expert correction strategy to ensure the quality and production of the coke.

If a delay caused by abnormal operating states is small, then we thin out coking chambers on schedule by one or two for every pushing operation. This gradually restores the normal coke-pushing series. While this method does not lose coking chambers, the adjustment is slow.

If a delay caused by abnormal operating states is short, and the number of affected coking chambers is less than ten, then we employ a coking chamber catch-up method that recovers the production by shortening the operating time and using the maintenance time. It sets the pushing operation for the coking chambers after an accident to a pending state, and performs the pushing operation for the rest ones according to the schedule but tries to shorten the operating time. If the process cannot restore to the normal schedule at the end of the operation, then it uses the maintenance time for pushing operation. If a delay reaches ten coking chambers, then we use a backward approach, that is, skipping the pushing operation of some coking chambers and putting them on a waiting list, to deal with this situation. This method restores the coke-pushing series to the normal one quickly, but it may lose some coking chambers. In addition, to avoid a high-temperature accident, an extension to the coking time should not exceed one-fourth of a specific coking time.

If a delay caused by abnormal operating states is large, and the number of affected coking chambers is larger than ten, then we first use a prescribed normal coke-pushing plan and shorten the operating time as possible. When the saved time is long enough to push a coking chamber, we used the time to push one of the pending coking chambers. And we also use the maintenance time to reduce the number of the pending coking chambers. This shortens the total delay time and makes the coke-pushing plan in the next run can be handled easily. However, if the delay time is too long, we have to discard some of the affected coking chambers.

Some of the expert correlation rules are shown in Table 3.3.

Due to the harsh environment of the process and an influence caused by direct or indirect factors, the abnormal operating states are very complex, and all of them can be corrected by the expert rules. Moreover, a schedule produced by the expert rules may be messy when dealing with several abnormal operating states simultaneously. So, it is necessary to establish a common model to handle the abnormal operating states that the expert correction cannot do.

Optimal scheduling for abnormal operating states not only ensures the quality of coke and reduces the loss, but also meets the constraints of process and mechanical equipment. It is a multi-objective and multi-constrained optimization problem. We used the TSP (see Sect. 1.3.6) as a model for the optimal scheduling, and applied the ACO (see Sect. 1.3.6) to obtain an optimal solution of the problem.

An optimal scheduling problem is described as follows: for given n coking chambers, m maintenance periods; and a coke-pushing penalty between any two of the

Table 3.3 Expert correction rules

Abnormal operating state	Correction rule
Delayed in a group < 10 coking chambers	Skip one coking chamber in the group
Delayed in a group > 10 coking chambers	Extend the coking time of the delayed coking chamber by one fourth of the specified coking time
Number of coking chambers > 10 when accident occurs	Prolong the coking time by adding (accident time)/(number of affected coking chambers), and reduce the operating time of each affected coking chamber by 1 min
Number of coking chambers < 10 when accident occurs	Reduce the operating time of each normal coking chamber by 1 min
Coking chamber is abnormal	Prolong coking time by adding the coke mature time of the chamber
Coke-pushing interval between abnormal and adjacent chambers < 2 h	Set the planned coke-pushing time of the adjacent chamber to 2 h later

chambers, find an optimal coke-pushing sequence that ensures that each coking chamber is pushed once and minimizes the total penalty for all the coking operations of the coking chambers.

While the problem is similar to the TSP, the scheduling is different from the general TSP. For the TSP, a traveler starts from City 1, and must return to the city. A feasible path is a closed loop. However, the scheduling of the coking operations is an open path because each coking chamber can only be pushed once.

To construct a closed loop, we introduce a virtual node that is both starting and ending nodes. A coke-pushing plan starts from and terminates at the virtual node. The penalty between the node and any coking chamber is set to be 0, so as to avoid the influence caused by the introduction of the virtual node. The optimal schedule of the coke-pushing operations is equivalent to an optimal path for the TSP.

Abnormal operating states mainly affect the coking time and ultimately reflect on the change of the coke-pushing sequence. The coke-pushing sequence and coking time seriously affect coke quality and the operating cost. So, the optimal scheduling of the coking operations for abnormal operating states is to find an optimal coke-pushing schedule that performs the production task, ensures the maintenance time, regulates the coking time, guarantees coke quality, and restores the coke operation to normal as soon as possible.

The penalty for two adjacent out-of-order coking chambers is

$$Q_{ij} = Y_{ij} q, \tag{3.40}$$

$$Y_{ij}=\begin{cases}1, & 1<j\leqslant m \text{ \& coking chambers } i \text{ and } j \text{ follow coke-pushing series,}\\ 0, & m<j\leqslant n \text{ \& coking chambers } i \text{ and } j \text{ do not follow coke-pushing series,}\end{cases} \tag{3.41}$$

where $i, j = 0, 1, \ldots, 115,\ i \neq j$, q is the weight of the penalty, m is a constant, and Q_{ij} is the penalty for two adjacent coking chambers that do not follow 5-2 coke-pushing series.

The penalty for the change of the coking time of the ith coking chamber is

$$P_i = e^{t_i} - 1, \tag{3.42}$$

$$t_i = \begin{cases} 0, & T_{\min} - 5 \leqslant T_i \leqslant T_{\max} + 5,\\ \mu\dfrac{T_i - T_{\min}}{5}, & T_i \leqslant T_{\min} - 5,\\ \dfrac{T_{\max} - T_i}{5}, & T_i \geqslant T_{\max} + 5,\end{cases} \tag{3.43}$$

where μ refers to the damage degree of the furnace life and coke quality caused by the change of the coking time; T_i is the scheduled coking time of the ith coking chamber; and $T_{\max}$ and $T_{\min}$ are the maximum and minimum of the standard coking time, respectively.

The total penalty caused by the above two factors for the two adjacent coking chambers is

$$d_{ij} = X_j(Q_{ij} + P_j), \tag{3.44}$$

$$X_j = \begin{cases} 1, & 1 < j \leqslant m \text{ for Coking chamber } j,\\ 0, & m < j \leqslant N \text{ for the } j\text{th maintenance period,}\end{cases} \tag{3.45}$$

$$N = m + m', \tag{3.46}$$

where d_{ij} is the total penalty of two adjacent coking chambers, m is the total number of the coking chambers, and m' is the number of the maintenance periods that is given by

$$m' = \text{Total maintenance time}/120 \text{ min},$$

$$\begin{aligned}\text{Total maintenance time} = {} & \text{Turnaround time} - \text{Operating time of a single cabinet}\\ & \times \text{Number of cabinets in coking chambers.}\end{aligned}$$

To make a schedule of the coke-pushing operations, both the coke-pushing sequence and the insertion position of the maintenance time are essential. Even though the maintenance time does not cause the disorder of the coke sequence, or too long or too short coking time, that is, it does not change the corresponding penalty, its insertion position affects the coking time after the maintenance. So the maintenance period is also considered in the arrangement of the coke sequence. A maintenance period is set to be the same as the coke-pushing time for a coking chamber, but the penalty caused by the maintenance period is set to be zero.

Based on technologies for the scheduling and experience, we divide the constraints into two categories: resource constraints and technological constraints. For some abnormal operating states, the schedule of the coke-pushing operations for state restoration may not satisfy all constraints, but a feasible schedule has to meet the following constraints at least:

(1) The difference of the coke-pushing time between two adjacent coking chambers is not less than 2 h (Eq. (3.39)).

(2) The maintenance time is long enough. A single maintenance time is larger than 1 h (Eq. (3.38)).

The objective function of the optimal scheduling is

$$D = \sum_{k=1}^{N-1} d_{ij}(k). \tag{3.47}$$

Note that City i corresponds to Coking chamber i. Take the penalty of two adjacent coking chambers to be the distance between Cities i and j. The probability that Ant k at Coking chamber i transfers to Coking chamber j is defined to be

$$p_{ij}^{k}(t) = \begin{cases} \dfrac{[\tau_{ij}(t)]^{\alpha}[\eta_{ij}]^{\beta}}{\sum\limits_{j \in allowed_k} [\tau_{ij}(t)]^{\alpha}[\eta_{ij}]^{\beta}}, & j \in allowed_k, \\ 0, & j \notin allowed_k, \end{cases} \tag{3.48}$$

where

$$\eta_{ij} = \frac{1}{d_{ij}}, \tag{3.49}$$

$$\tau_{ij}(t+1) = \rho\tau_{ij}(t) + (1-\rho)\Delta\tau_{ij}^{k}, \tag{3.50}$$

$$\Delta\tau_{ij}^{k} = \frac{Q}{d_{ij}}, \tag{3.51}$$

where $allowed_k$ is the set of the coking chambers that Ant k can go in the next step; α is a control factor for the relative importance of the pheromone concentration; β is a control factor for the relative importance of the heuristic information; η_{ij} is the heuristic concentration for ants from the ith to the jth coking chamber; $\tau_{ij}(t)$ is the pheromone concentration between Ant i and Coking chamber j at time t; $\tau_{ij}(0) =$ Const (Const is the initial pheromone concentration and is equal to a constant); ρ, as a constant ranging from 0 to 1, is the relative importance of the local residual pheromone; $\Delta\tau_{ij}^{k}$ is an increased pheromone concentration on the path of Coking chambers i and j for Ant k at the time between t and $t+1$; and Q is the total amount of pheromone held by each ant.

The following global update rule is used. When all the ants have completed one cycle, the pheromone on all the paths is updated by the ants that produce the global optimal solution:

$$\tau'_{ij}(t+1) = \rho_1 \tau'_{ij}(t) + (1-\rho_1)\Delta\tau'_{ij}, \tag{3.52}$$

$$\Delta\tau'_{ij} = \frac{Q}{d_{Lb_{nc}}}, \tag{3.53}$$

where ρ_1 [$\in (0, 1)$] is the relative importance of the global residual pheromone, and $d_{Lb_{nc}}$ is the sum of the penalty values for the optimal solution in the ncth cycle.

Coke-pushing planning:

Step 1: Build three tables for Ant k: the coking chamber table, G_k; the taboo table, $tuba_k$; and the path table, $allowed_k$. G_k contains N coking chambers. $tuba_k$ stores the visited coking chambers of Ant k. And $allowed_k$ stores the coking chambers that can be visited by Ant k at the next step.

Step 2: Initialize parameters. Set the number of iterations, nc, to be 0; the maximum number of iterations, $nc_{\max}$, to be 500; and the number of ants that complete the task to be 0; $\tau_{ij}(0)$ to be 1; and $\Delta\tau_{ij}$ to be 0. Let the optimal solution of the ncth iteration be Lb_{nc} and the global optimal solution be Lb.

Step 3: Place all ants on the initial point of the algorithm.

Step 4: Empty the set of coking chambers that Ant k has visited, $tuba_k$, and also the set for Ant k's visiting at the next step, $allowed_k$, for all ants.

Step 5: Determine the set that Ant k can visit at the next step.

Step 6: Calculate the transfer probability according to Eq. (3.48), and determine the coking chamber number, s, for the visiting at the next step based on the transfer probability. Add Coking chamber s to $tuba_k$ and delete it from $allowed_k$.

Step 7: Update the pheromone on the path between Coking chambers u and s according to Eq. (3.50).

Step 8: If $tuba_k$ contains all the chambers, it is a coke-pushing plan, Lk. If l is less than the total number of ants, go to Step 5.

Step 9: Find the local optimal solution, Lb_{nc}, and the corresponding optimal ant.

Step 10: Update the pheromone globally on each path passed by the optimal ant according to Eq. (3.52).

Step 11: Finish if $nc = nc_{\max}$ or the algorithm is converged, and output the global optimal solution, Lb, that is, the optimal coke-pushing plan for the next pushing cycle.

3.5 System Implementation and Results of Actual Runs

This section describes the system implementation and results of actual runs of a coking process.

3.5.1 System Implementation

The intelligent integrated optimization and control system was implemented on an industrial control computer. The system consists of the application software, a PLC-based WinCC configuration with an OPC module. WinCC is a communication interface between the application software and the PLC. The application software performs operating-state-based intelligent control with a two-stage decision method to determine the operating state. Once the operating state has been determined, appropriate inner- and outer-loop fuzzy controllers, for which the parameters are optimized offline, are chosen to control the combustion process. Then, the controllers make the gas flow rate and the air suction power track the reference values by regulating the valve openings, thereby ensuring that the oven temperature is stabilized at a given value. The values of the valve openings are sent to the PLC using OPC communication technology to drive the actuators of the valves. The gas flow rate, the air suction power, and the valve openings are measured and sent to the intelligent integrated optimization and control system. Figure 3.14 shows the flow of information and data in the system.

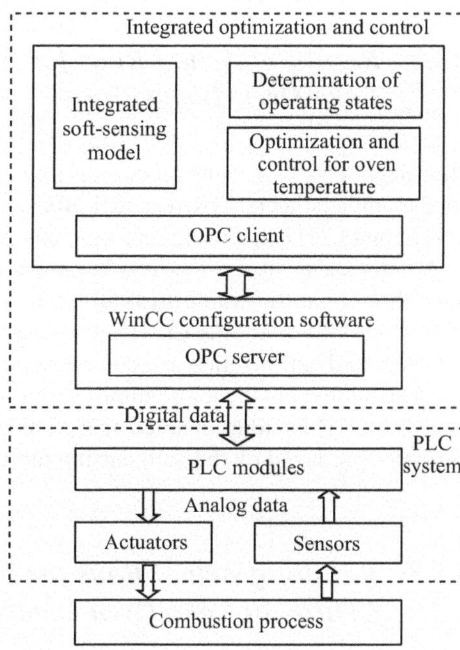

Fig. 3.14 Flow of information and data in integrated optimization and control system for COT

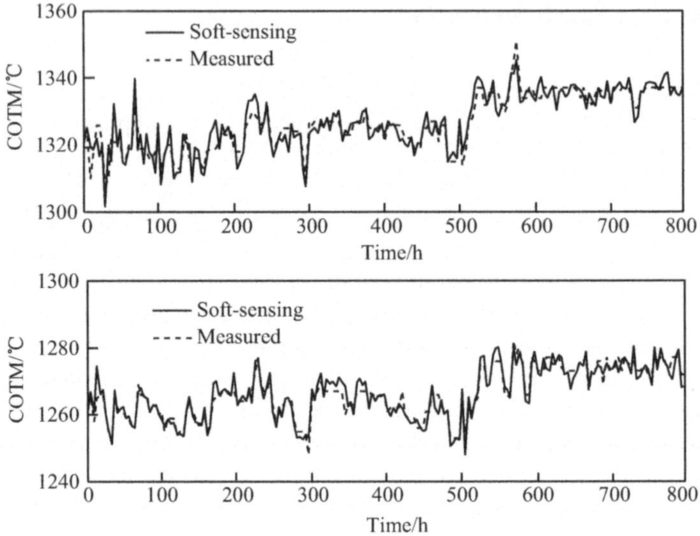

Fig. 3.15 Results of soft-sensing model of COTC and COTM

3.5.2 *Results of Actual Runs for Integrated Soft Sensing of Coke-Oven Temperature*

The integrated soft-sensing model was implemented using Visual C++ language. The OPC module collects TTRMs and TTRCs, and the soft-sensing model estimates the COTM and COTC to enable real-time control of the temperature of the coke-oven.

A statistical analysis of the results obtained with the soft-sensing model (Fig. 3.15) show that, for an allowable error of ±7 °C, the estimation accuracy is 91.2% for the COTM and 88.9% for the COTC. These satisfy the industrial requirements. Before the integrated optimization and control system was implemented, the COT was measured manually; but the long sampling time (4 h) made real-time temperature control impossible. This study enabled real-time control through the use of temperature estimates produced by the soft-sensing model.

3.5.3 *Results of Actual Runs for Intelligent Optimization and Control of Coke-Oven Combustion Process*

The temperature of the coke-oven in the iron and steel company was controlled manually before the intelligent integrated optimization and control system was installed in 2005. Some results of actual runs are shown in Figs. 3.16 and 3.17. It is clear that

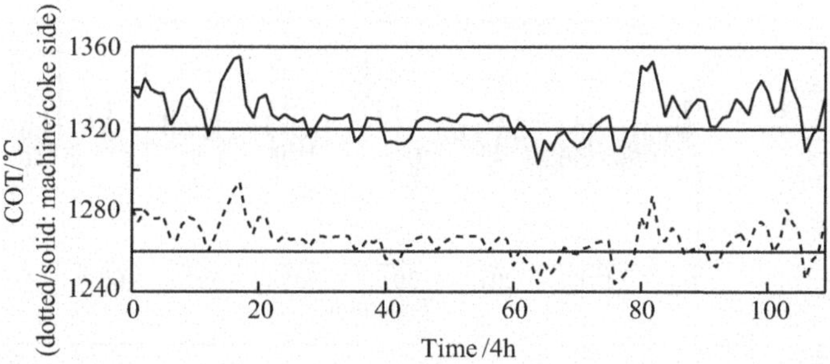

Fig. 3.16 Results of actual runs of COT for manual control

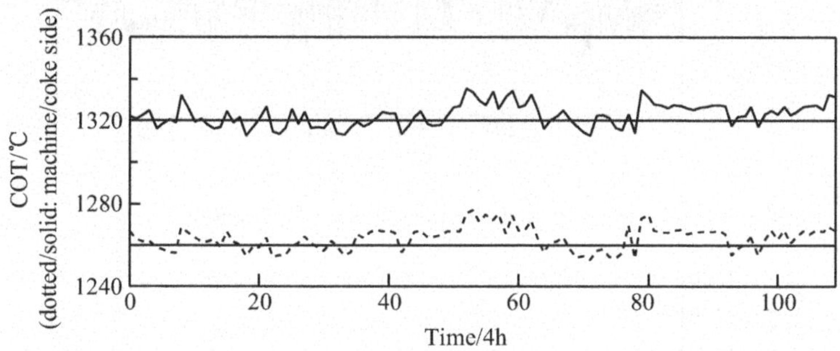

Fig. 3.17 Results of actual runs of COT for integrated optimization and control system

the control system reduced the variation in the oven temperature from ±25 °C to ±10 °C.

Before the system was used, the control of the mixed coal pressures on the machine and coke sides was performed by a simple PID controller. A typical control result is shown in Fig. 3.18. Clearly, the variation in the pressures was larger than 400 Pa. When the mixed coal pressures were influenced by an external disturbance, the controller had to take a long time to adjust them.

The installation of the system greatly improved the control performance of the mixed coal pressures. More than 90% of the deviations of the pressure on the machine side were reduced to a range within ±50 Pa. Strong nonlinear characteristics of the valves make it difficult to adjust them. But the control system ensured that more than 85% of the deviations in the pressures on the machine side were within ±50 Pa (Fig. 3.19).

Before the system was used, the air suction pressures were adjusted manually. It was largely influenced by human factors. Some typical control results are shown in Fig. 3.20. Clearly, it could not meet high accuracy requirements of the process.

After the system was used, the air suction pressures were adjusted by the expert controller. And the control precision was improved greatly. Some control results are shown in Fig. 3.21. It can be seen from the figure that more than 95% of the deviations in the air suction pressures on the coke and machine sides were within ±5 Pa. The

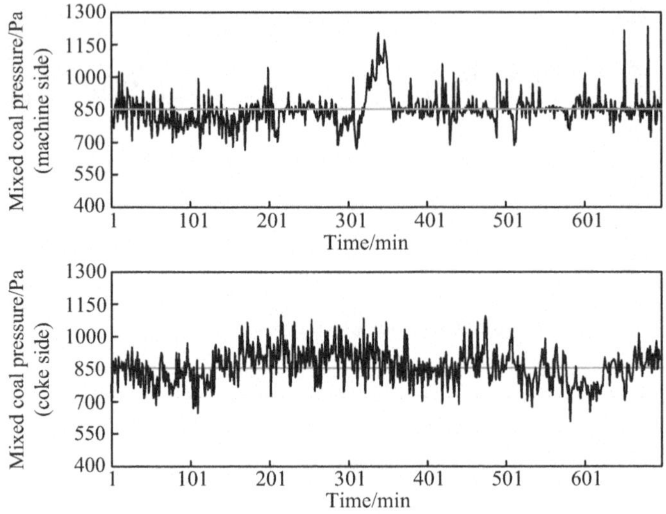

Fig. 3.18 Results of actual runs of mixed coal pressures for PID control

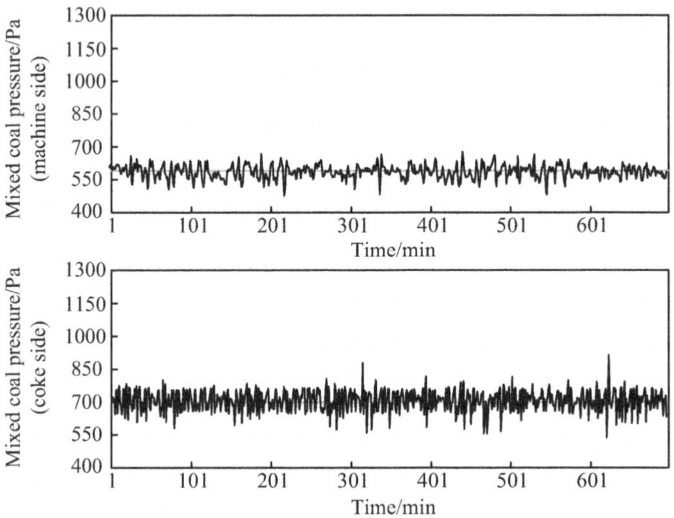

Fig. 3.19 Results of mixed coal pressures for integrated optimization and control system

spikes in the figure show the commutation moments in the coke-oven combustion process.

Coke quality, energy-saving effect, stability coefficient, and uniformity coefficient were chosen as indexes to evaluate the performance.

Two evaluation parameters, the crushing strength (M_{40}) and the abrasion resistance (M_{10}), were employed to measure coke quality. M_{40} is the percentage by weight of coke balls with a diameter greater then 40 mm in 100 kg of coke balls, and M_{10}

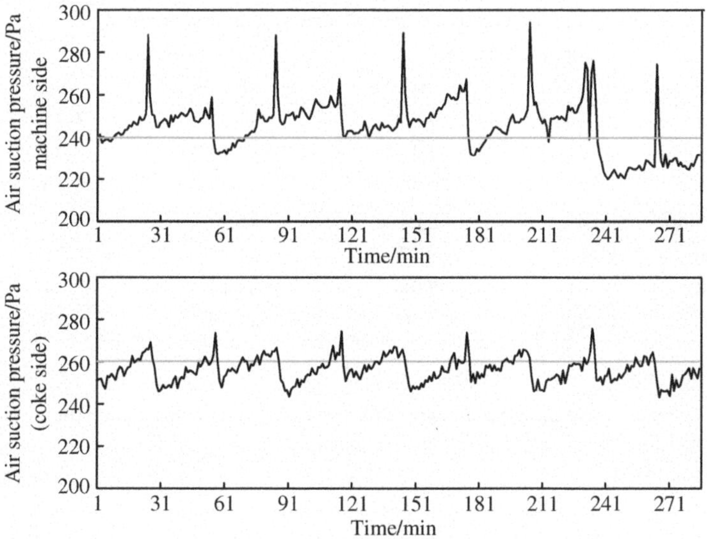

Fig. 3.20 Results of air suction pressures for manual control

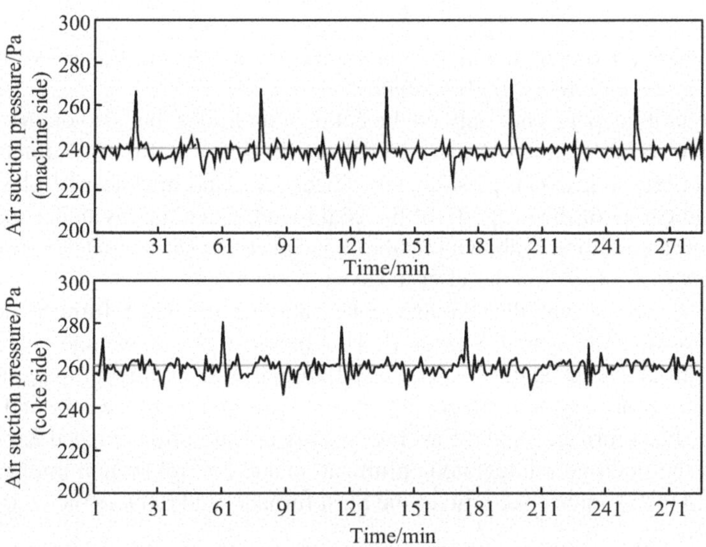

Fig. 3.21 Results of air suction pressures for integrated optimization and control system

is the percentage by weight of coke balls with a diameter less than 10 mm in 100 kg of coke balls. A larger M_{40} and a smaller M_{10} mean better air permeability in iron making, which is desirable. The 6-month values of M_{40} and M_{10} and their average values are shown in Table 3.4. It is clear from Table 3.4 that this system improves

Table 3.4 Comparison of coke quality before and after system was installed

Time	Before installation		After installation	
	$M_{40}/\%$	$M_{10}/\%$	$M_{40}/\%$	$M_{10}/\%$
First month	82.40	7.00	82.65	6.71
Second month	81.77	7.48	82.10	6.74
Third month	81.20	7.70	80.78	7.10
Fourth month	80.80	8.04	81.30	7.06
Fifth month	80.62	8.17	81.70	6.71
Sixth month	80.61	7.98	83.10	6.34
Average	81.20	7.70	82.00	6.90

Table 3.5 Comparison of the coke strength before and after system installation for the similar coal blending condition

Time	Coal blend								Coke strength	
	M_t	A_d	V_{daf}	$S_{t,d}$	D	G	X	Y	M_{40}	M_{10}
Before	11.86	10.27	26.31	0.58	77.68	75.1	30.0	15.4	81.77	7.48
After	11.00	10.37	26.23	0.62	73.85	73.9	29.4	17.5	83.10	6.34

both K_S and K_A. Compared with the parameters for the manual control system, M_{40} is 0.80% larger, and M_{10} is 0.80% smaller.

Coke quality depends not only on the coking conditions, but also on the property of the coal blend. Moisture (M_t), ash (A_d), volatile (V_{daf}), sulfur ($S_{t,d}$), fineness (D), caking property index (G), plastometer indices (X), and thickness of plastic layer (Y) are indicators of the property of the coal blend. Coke quality under the similar coal blending condition is shown in Table 3.5. This table shows that M_{40} was 1.33% larger, M_{10} was 1.14% smaller after the system was used.

Table 3.6 shows that the average coke ratio of the blast furnace decreased 20.950 kg/t after the system was used. This means that the average coke ratio of the blast furnace reduced 4.71%. The installation of this intelligent integrated optimization and control system effectively improved the quality of the coke and the coke ratio of the blast furnace. And the average energy consumption reduced 2.0%. These show that the intelligent integrated optimization and control system improved coke quality and the average coke ratio of the blast furnace, and reduces the consumption of gas.

Oven temperature error is defined to be the difference between the reference value and the average oven temperature. Two variables, the stability coefficient (K_S) and the uniformity coefficient (K_A), are defined to be

$$K_A = \frac{(M - A_{\text{machine}}) + (M - A_{\text{coke}})}{2M}, \tag{3.54}$$

$$K_S = \frac{2N - (A'_{\text{machine}} + A'_{\text{coke}})}{2M}, \tag{3.55}$$

Table 3.6 Comparison of average coke ratio of blast furnace before and after system installation

Time	Before installation/(kg/t)	After installation/(kg/t)
First month	498.003	416.530
Second month	454.280	437.610
Third month	475.210	440.190
Fourth month	454.468	432.040
Fifth month	413.056	428.570
Sixth month	416.069	430.440
Average	451.848	430.898

Table 3.7 Comparison of K_S and K_A before and after system installation

Time	Before installation		After installation	
	K_S	K_A	K_S	K_A
First month	0.48	0.32	0.79	0.80
Second month	0.54	0.25	0.74	0.62
Third month	0.48	0.28	0.75	0.82
Fourth month	0.36	0.36	0.84	0.98
Fifth month	0.45	0.36	0.78	0.88
Sixth month	0.49	0.38	0.68	0.61
Average	0.47	0.33	0.76	0.79

where M is the number of combustion chambers in the coke-oven; A_{machine} is the numbers of combustion chambers on the machine side in which the flue temperatures are outside the error range (±7 °C), and A_{coke}, on the coke side; N is the number of the measurements of the oven temperature; and A'_{machine} is the number of times that the oven temperature error on the machine side is outside the range ±7 °C, and A'_{coke}, on the coke side.

K_S indicates the stability of the oven temperature, and K_A indicates its uniformity. It is clear from Table 3.7 that the intelligent integrated control method improves both K_S and K_A.

3.5.4 Results of Actual Runs for Coke-Oven Operation Planning and Optimal Scheduling

The index of the change in the coking time is defined to be

$$K_{\text{PL}} = \frac{m_1 - a_1}{m_1}, \tag{3.56}$$

Table 3.8 Coefficients before system installation in 1 month

Date	K_{PL}	K_A	K_S
Sept. 1–Sept. 5	0.78	0.66	0.58
Sept. 6–Sept. 10	0.69	0.64	0.59
Sept. 11–Sept. 15	0.82	0.68	0.75
Sept. 16–Sept. 20	0.76	0.72	0.62
Sept. 21–Sept. 25	0.87	0.59	0.65
Sept. 26–Sept. 30	0.79	0.73	0.59

Table 3.9 Coefficients after system installation in 1 month

Date	K_{PL}	K_A	K_S
Nov. 1–Nov. 5	0.94	0.93	0.92
Nov. 6–Nov. 10	0.82	0.89	0.85
Nov. 11–Nov. 15	0.95	0.86	0.9
Nov. 16–Nov. 20	0.98	0.95	0.92
Nov. 21–Nov. 25	0.93	0.88	0.83
Nov. 26–Nov. 30	1	0.94	0.97

where m_1 is the planed number of the furnaces for coke-pushing, a_1 is the number of furnaces for which the coking time is largely different from the standard coking time (the difference is larger than 5 min), and K_{PL} is the coefficient of the planned coke-pushing series. The coke-pushing plan and the situation of the plan execution directly influence the stability of the oven temperature. K_S and K_A were used to measure the performance of actual runs. And the coefficients before and after the installation of the system are shown in Tables 3.8 and 3.9, respectively.

Clearly, the planed coke-pushing, stability, and uniformity coefficients are close to 1 after the system installation. This means that the system satisfied the requirements of actual production.

Actual running results show that the system that optimized comprehensive production goals was stable, and the effect of the system was satisfactory. The multi-objective optimization algorithm may have big effect. The system largely improves the production and quality of coke, and greatly reduces energy consumption.

3.6 Conclusion

A soft-sensing model, which estimates the COT in a real-time fashion, was developed based on the analysis of the characteristics of a coking process. An intelligent integrated hybrid optimization and control system for the temperature of a coke-oven has been developed based on the features of the combustion process. The framework of the control system consists of a decision layer, an optimization and control layer, and a process control layer. The key points of this chapter are as follows:

(1) Based on an analysis of the characteristics of the coking process, two variables, COTM and COTC, were introduced to precisely describe the COT.

(2) An integrated LR model used a linear relationship to describe the dependence of the COTM (COTC) on the TTRMs (TTRCs). An SDNN modeling method with the FCM and the subtractive clustering improved the precision.

(3) An intelligent coordinator balanced the outputs produced by the LR and SDNN models. And a model adaptive unit periodically modified the soft-sensing model to ensure a high estimation precision.

(4) The structure of the control system was based on the analysis of the characteristics of the combustion process, with a special focus on the factors that greatly influenced the operating state and changes in the characteristics of the COT for different operating states.

(5) The temperature control system contained one control loop for temperature and one for gas flow rate and air suction power.

(6) The coke-oven operation planning and optimal scheduling models were established under normal and abnormal operating states.

(7) The optimization and control system stabilized the oven temperature, improved the quality of coke, and reduced energy consumption.

References

1. Smolka J, Slupik L, Fic A, Nowak AJ, Kosyrczyk L (2016) 3-D coupled CFD model of a periodic operation of a heating flue and coke ovens in a coke oven battery. Fuel 165:94–104
2. Wu M, Lei Q, Cao WH, She JH (2011) Integrated soft sensing of coke-oven temperature. Control Eng Pract 19:1116–1125
3. Brosilow CB (1978) Inferential control of process control. AIChE J 24:475–484
4. Ruusunen M, Leivisk K (2004) Fuzzy modeling of carbon dioxide in a burning process. Control Eng Pract 12:607–614
5. Sbarbaro D, Ascencio P, Espinoza P, Felipe M, Guillermo C (2008) Adaptive soft-sensors for on-line particle size estimation in wet grinding circuits. Control Eng Pract 16:171–178
6. Abeykoon C (2014) A novel soft sensor for real-time monitoring of the die melt temperature profile in polymer extrusion. IEEE Trans Ind Electron 61(12):7113–7123
7. Ding Y, Yu JJ, Zhou CH (1994) Quality estimation and supervision for a crude distillation. Pet Process Petrochem 5:23–28 (In Chinese)
8. Bidar B, Sadeghi J, Shahraki F, Khalilipour MM (2017) Data-driven soft sensor approach for online quality prediction using state dependent parameter models. Chemom Intell Lab Syst 162:130–141
9. Wang D, Liu J, Srinvasan R (2009) Data-driven soft sensor approach for quality prediction in a refining process. IEEE Trans Ind Inpormatics 12:113–121
10. James S, Legge R, Budman H (2002) Comparative study of black-box and hybrid estimation methods in fed-batch fermentation. J Process Control 12:113–121
11. Zhu B, Chen ZS, He YL, Yu LA (2017) A novel nonlinear functional expansion based PLS (FEPLS) and its soft sensor application. Chemom Intell Lab Syst 161:108–117
12. Yin S, Zhu XP, Kaynak O (2015) Improved PLS focused on key-performance-indicator-related fault diagnosis. IEEE Trans Ind Electron 62(3):1651–1658

13. Zhao H, Gao S, He Z, Zeng X, Jin W, Li T (2014) Identification of nonlinear dynamic system using a novel recurrent wavelet neural network based on the pipelined architecture. IEEE Trans Ind Electron 61(8):4171–4182
14. Mazinan AH (2013) A new algorithm to AI-based predictive control scheme for a distillation column system. Int J Adv Manuf Technol 66(12):1379–1388
15. Navvab KM, Shahrokh S (2010) A methodology for modeling batch reactors using generalized dynamic neural networks. Chem Eng J 159:195–202
16. Fortuna L, Giannone P, Graziani S, Xibilia MG (2007) Virtual instruments based on stacked neural networks to improve product quality monitoring in a refinery. IEEE Trans Instrum Meas 56:95–101
17. Dufour P, Bhartiya S, Dhurjati PS, Doyle FJ (2005) Neural network-based software sensor: Training set design and application to a continuous pulp digester. Control Eng Pract 13:135–143
18. Jemei S, Hissel D, Péra MC, Kauffmann JM (2008) A new modeling approach of embedded fuel-cell power generators based on artificial neural network. IEEE Trans Ind Electron 55:437–447
19. Khoshgoftaar TM, Hulse JV, Napolitano A (2010) Supervised neural network modeling: an empirical investigation into learning from imbalanced data with labeling errors. IEEE Trans Neural Netw 21:813–830
20. Luo GQ, Wen Z, Chen FH (1998) A flue-oven mathematical model for coke oven. Fuel Chem Process 29:78–82 (In Chinese)
21. Ning FQ, Zhang SF, Yan WF (2004) Generalized predicative control in flue temperature. J UESR China 33:53–55 (In Chinese)
22. Devornique G, Fontchastagner J, Netter D, Takorabet N (2017) Hybrid model: Permeance network and 3-D finite element for modeling claw-pole synchronous machines. IEEE Trans Mag 53(6):7206704-1–7206704-4
23. Yebi A, Ayalew B (2017) Hybrid modeling and robust control for layer-by-layer manufacturing processes. IEEE Trans Control Syst Technol 25(2):550–562
24. Egrioglu E, Aladag CH, Yolcu U, Uslu VR, Basaran MA (2009) A new hybrid approach based on SARIMA and partial high order bivariate fuzzy time series forecasting model. Expert Syst Appl 36:7424–7434
25. Wang X, Chen J, Liu C, Pan F (2010) Hybrid modeling of penicillin fermentation process based on least square support vector machine. Chem Eng Res Des 4:415–420
26. Berkutov NK, Stepanov YV, Popova NK (2007) The relation between coke quality and blast-furnace performance. Steel Transl 37(5):438–441
27. Yurin NI, Morozov OS, Likhacheva OL (2011) Influence of coke quality on blast-furnace performance. Steel Transl 41(11):924–927
28. Uddin MN, Rebeiro RS (2011) Online efficiency optimization of a fuzzy-logic-controller-based IPMSM drive. IEEE Trans Ind Appl 47(2):1043–1050
29. Lei Q, Wu M, She JH (2015) Online optimization of fuzzy controller for coke-oven combustion process based on dynamic just-in-time learning. IEEE Trans Autom Sci Eng 12(4):1535–1540
30. Wang SH, Xu BG, Wang QY, Liu YH (2006) Modified smith predictor and controller for time-delay process with uncertainty. In: Proceedings of the 6th world congress on intelligent control and automation, vol 1, pp 623–627
31. Stratos I, Nikos T, Kimon V (2006) Fuzzy supervisory control of manufacturing systems. IEEE Trans Robot Autom 20(3):379–389
32. Gao XW, Cai XY, Yu XF (2006) Simulation research of genetic neural network base PID control for coke oven heating. In: The sixth world congress on intelligent control and automation, vol 2, pp 21–23
33. Gao XW, Liu H, Zhao YP (2005) Application and research of the fuzzy compound control method in coke oven control system. Control Decis 20(4):434–438 (In Chinese)
34. Romeo LM, Gareta R (2006) Hybrid system for fluling control in biomass boilers. Eng Appl Artif Intell 19(8):919–925
35. Moslehi Z, Taheri M, Mirzaei A, Safayani M (2018) Discriminative fuzzy C-means as a large margin unsupervised metric learning. IEEE Trans Fuzzy Syst 26(6):3534–3544

36. Jiang W, Yang T, Shou Y, Tang Y, Hu W (2018) Improved evidential fuzzy C-means method. J Syst Eng Electron 29(1):187–195
37. Corsini P, Lazzerini B, Marcelloni F (2004) A fuzzy relational clustering algorithm based on a dissimilarity measure extracted from data. IEEE Trans Syst, Man, Cybern Part B: Cybern 34:775–783
38. Huang ZX (1999) Fuzzy k-modes algorithm for clustering categorical data. IEEE Trans Fuzzy Syst 7:446–452
39. Pedrycz W (1998) Conditional fuzzy clustering in the design of radial basis function neural networks. IEEE Trans Neural Netw 9:601–612
40. Chiu SL (1994) Fuzzy model identification based on cluster estimation. J Intell Fuzzy Syst 2:267–278
41. Fujino A, Tobita T, Segawa K, Yoneda K, Togawa A (1997) Elevator group control system with floor-attribute control method and system optimization using genetic algorithms. IEEE Trans Ind Electron 44(4):546–552
42. Mahapatra NK, Bhunia AK, Maiti M (2005) A multiobjective model of wholesaler-retailers' problem VIA genetic algorithm. J Appl Math Comput 19(1–2):397–414
43. Srinival M, Patnaik LM (1994) Adapbive probabilities of crossover and mutation in genetic algorithm. IEEE Trans Syst, Man Cybern 24(4):656–667

Chapter 4
Intelligent Control of Thermal State Parameters in Sintering Process

A sintering process produces sinters as blast furnace materials. The stability of a sintering process and the quality of sinter are closely related to the cost, efficiency, and energy consumption of blast furnace production. And parameters of a thermal state in the process directly affect the stability of the process and the quality of sinter. The stabilizing control of the parameters is the basis for the stabilization of the sintering process.

This chapter explains the characteristics of a sintering process and describes an intelligent control of two thermal state parameters.

4.1 Process Description and Characteristics Analysis

A sintering process is described in this section. The characteristics of the process are presented, and two parameters are introduced to describe the thermal state of the process.

4.1.1 Description of Sintering Process

The sintering process considered in this chapter uses a Dwight-Lloyd sintering machine, which is the type most commonly used in industry.

The sintering of iron ore is an essential part of modern iron making. Sintering prepares ore for a blast furnace by causing the metal in the ore to agglomerate. The characteristics of the agglomerate greatly affect the operating conditions of the blast furnace [1]. In this process, ore is mixed with fine particles of limestone, coke,

M. Wu et al., *Intelligent Optimization and Control of Complex Metallurgical Processes*, Engineering Applications of Computational Methods 3,
https://doi.org/10.1007/978-981-15-1145-5_4

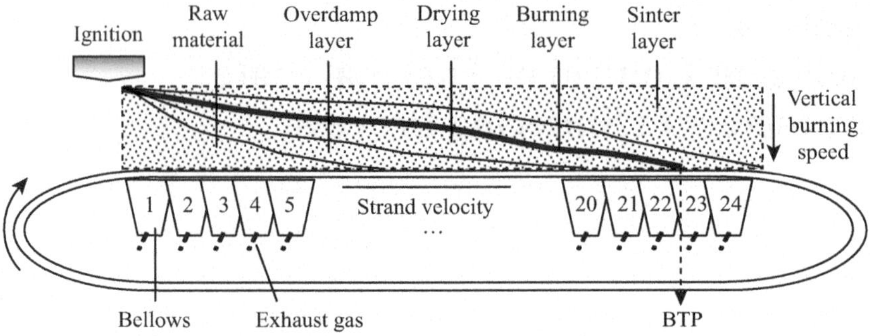

Fig. 4.1 The hierarchy of material layer

and returning sinter; and the heat from the burning coke causes agglomeration. The sintered ore is then broken into small pieces and fed into the blast furnace.

The raw material contains ore, coke, limestone, and returning sinter. After being granulated and moisturized in a mixer, it is fed at a constant rate into a bunker. Then, it passes through a roller feeder and is loaded onto the moving sinter grate to form a bed approximately 700 mm deep. The igniter in the ignition hood ignites the surface of the bed, and the sintering of the ore begins. As the grate moves, bellows beneath the sinter grate supply fresh air to sustain combustion and remove exhaust gas. The temperature of the exhaust gas is measured. Combustion progresses until it reaches the bottom of the bed. The strand velocity has to be set properly so that the raw material is completely sintered before it reaches the end of the strand. The first position on the strand where the material is completely sintered is named as the burn-through point (BTP). It is denoted by L_{BTP} and is specified in terms of a bellows number plus a one-place decimal fraction.

The surface of raw material is ignited under the ignition hood. When the coke in the bed burns, it creates a high-temperature zone. The heat produced induces a series of chemical and physical reactions. As the process continues, various layers are successively formed: a sinter layer, a burning layer, a drying layer, a overdamp layer, and a raw material layer (Fig. 4.1).

The agglomerate is discharged off the edge of the grate, cooled, and crushed. Pieces larger than a certain size are sent to the blast furnace, and the others are sent back to the start of the sintering process as returning sinter.

4.1.2 Characteristic Analysis of Thermal State Parameters in Sintering Process

The sintering process has two important parameters of the thermal state: the BTP (the first position on the sinter strand where the ore is completely sintered) and the ignition temperature. The ignition temperature directly affects the initial molten

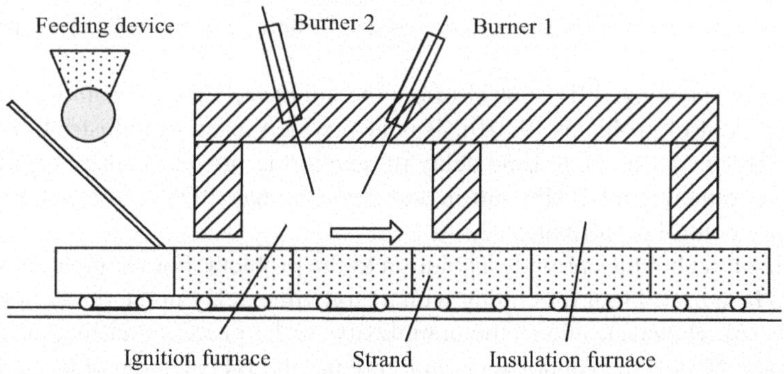

Fig. 4.2 The double-oblique belt-type sintering furnace

state of the mixture, which largely affects the quality of sinters. The stability of the ignition temperature and the BTP directly influence the quality and quantity of sintering agglomerate. If the BTP occurs before an optimal position, the area of the sintering machine is not used effectively. So, the utilization factor is very low. This directly reduces the quantity of agglomerate. On the other hand, if the BTP occurs after the optimal position, some ore remains unsintered. So, the agglomeration rate is very low. This reduces the quantity and degrades the quality of agglomerate, and increases the amount of returning sinter. In addition, frequent changes in the BTP shorten the lifetime of the sintering machine. So, a key problem in the control of a sintering process is to effectively track the BTP to keep it at an optimal value.

The main heat transfer equipment for the sintering ignition process (SIP) is the ignition furnace. A sintering furnace of the double oblique belt type (Fig. 4.2) is explained here as an example. The zone at the place where the flames of Burners 1 and 2 cross has a belt shape and is a high-temperature area on the trolley. The zone should be in the middle of the ignition furnace so as to improve the ignition quality and shorten the ignition time.

Sintering is a complex, nonlinear process with coupled parameters. The BTP cannot be measured online. It is generally estimated by an operator based on observed data, but that value is not very accurate [2]. Many factors affect the BTP, such as the sinter strand velocity, the ignition temperature, the chemical composition of the raw material, and the flow rates of the fresh air blowers. Since most of these factors change frequently and unpredictably, it is difficult to establish a practical, precise mathematical model of the sintering process. The analytical mathematical models [3, 4] that have been devised are very complicated and require hard-to-obtain parameters related to the physical properties of the sinter material. That makes the models unsuitable for process control.

In the last a couple of decades, the research of sintering process for iron ore [5] has made steady progress. Many studies have focused on the prediction and control of the BTP, and a variety of methods, such as PID control [6], predictive control

[7], and neural network (NN)—or fuzzy-based intelligent control [8]—have been devised.

Generally speaking, PID control offers the simplest, effective solution to many real-world control problems [9, 10]. It is also widely used in industrial processes [11, 12]. However, the large time delay of a sintering process makes it difficult to tune PID parameters for BTP control, and conventional PID control may not yield satisfactory control performance.

Intelligent techniques provide a solution to the problem. For example, NNs have proven to be a useful tool for coping with a large time delay in the prediction of the BTP [13–16]. However, due to the complexity of the process and limited process information, NNs alone cannot accurately predict the BTP. On the other hand, predictive control [17, 18] is a promising way to greatly improve control performance, but its use for BTP control is mainly in the simulation stage. Another method is fuzzy control [19]. It involves building a controller based on a fuzzy model of the process, and it employs knowledge representation to transform the experience of experts into a control model that can be understood by a computer [20]. Fuzzy control can handle nonlinearities and uncertainties in a system [21], and has advantages over other control methods [22]. A predictive model of BTP based on fuzzy expert rules is established [23]. The use of fuzzy control for BTP control has been studied [24]. However, most BTP control methods ignore the influence of the level of the charging bunker (hereafter called the bunker level), which is coupled to the BTP via the trolley velocity. So, it is difficult to achieve satisfactory control performance in an actual production process with just one of these methods.

In the ignition temperature control, in order to stabilize the ignition temperature and reduce energy consumption, we need to find an effective identification method to build the mathematical model of SIP. On the basis, an effective advanced control method can be designed to achieve disturbance rejection, and thus to realize the stable control of SIP.

Recently, data-driven modeling [25, 26] has attracted much attention because it makes full use of empirical data. As a typical data-driven modeling method, subspace identification method [27–29] can be used to build a state-space model based on the input and output measurements of a system. Subspace identification method does not need much prior knowledge compared to traditional modeling methods, and the model order can be determined by singular value decomposition (SVD) [30]. Nowadays, the subspace identification method plays an important role in system identification [31] and predictive control [32]. For a SIP, as there exists a time delay that is not considered in traditional subspace modeling methods, the model generated using only traditional subspace identification can hardly match the input–output dynamics accurately. An iterative subspace modeling method is developed to estimate not only the model parameters but also the length of delay time.

At the same time, disturbances in a system not only degrade control performance, but also cause machine malfunctions and other problems. How to reject these disturbances is the key to make the temperature of SIP stable. Some disturbance estimation and compensation methods, such as disturbance observer (DOB) [33–35] and active disturbance rejection control (ADRC) [36–38], have been proposed. One problem with DOB is that it needs to use the inverse dynamics of the plant. ADRC does not require the inverse dynamics of the plant, but a prior information of disturbances is needed. As a typical active rejection method, the equivalent input disturbance (EID) approach was proposed, which, compared to other methods, features that no requirement for a prior information of disturbances [39–41]. It is regarded that an EID exists at the input channel and produces the same impact as the actual disturbances do. It is more reasonable to estimate an equivalent input disturbance than to estimate the disturbance itself because we have to use the control input to improve the disturbance rejection performance. However, all of the above control methods do not consider both the effect of time delay and periodic disturbances. Thus, the performance of periodic disturbance rejection is not as good as expected. Moreover, another key problem, maybe more important, is that the periodic disturbance cannot be completely rejected by conventional EID-based method when the system has an input delay that often degrades the dynamic performance of the control system and sometimes leads the system to be instable. To solve the problem, we modify the configuration of EID method by inserting an appropriate time-delay element into the feedback channel. This improved EID-based method is not only good at compensate unknown periodic disturbance with input time delay, but also easily applied in the practical ignition process of SIP.

Thus, an effective method of controlling the sintering process is urgently needed both to improve the quality and quantity of sintering agglomerate and to promote environmental safety. Recent progress in intelligent integrated modeling [42] and control methods [43], multi-model switching control methods [44, 45], and intelligent coordination strategies [46, 47] provides us with advanced techniques for achieving this objective.

Sintering process is a complex industrial process with nonlinearity, multiple parameters, strong coupling, and multiple working conditions. In order to realize the steady and smooth operation of sintering process, it is necessary to design the intelligent control of sintering thermal state parameters.

4.1.3 Control Requirements

In the ignition temperature control, when the ignition temperature is stable at the target temperature, the quality and quantity of the sinter can be guaranteed. And the control requirement is as follows:

(1) The optimum temperature for the ignition temperature is 1200 °C.

(2) The variation should be at most ± 40 °C.

According to the empirical knowledge of engineers and operators, and statistical data on the process, the optimum position for the BTP is 17.0 in 280 m^2 and 23.0 in 360 m^2. When it is at that position and is relatively stable, the largest quantity of high-quality agglomerate is produced. That is, when the BTP is less than or greater than that, there is less agglomerate and it is of poorer quality.

The total volume of the charging bunker is 100 tons. $h(k)$ should be in the range $[30\,\text{t},\ 60\,\text{t}]$ to ensure continuous operation because there is a danger that the bunker might become empty (if $h(k) < 30$ t) or overflow (if $h(k) > 60$ t).

The following control requirements are demanded by the technology to ensure both safe operation and a large quantity of high-quality agglomerate:

(1) The optimum BTP is 17.0 in $280\,\text{m}^2$ and 23.0 in $360\,\text{m}^2$, and the variation should be at most ± 0.4.

(2) $h(k) \in [30\,\text{t},\ 60\,\text{t}]$.

(3) The strand velocity $(v(k)) \in [1.5\,\text{m/min},\ 3.0\,\text{m/min}]$. This is based on empirical knowledge and data from past runs.

4.2 Intelligent Control of Sintering Ignition Process

The SIP is an important process in sintering. And the control of sintering ignition process is beneficial to reduce the energy consumption of sintering production. This section describes an intelligent control of the SIP.

4.2.1 Control System Architecture

We developed a method of the intelligent control for the SIP based on the optimization of ignition intensity and the characteristics of the SIP. Considering the factors influencing the change of ignition intensity, we choose the quality of sintering ore and the energy consumption in the sintering process to be a comprehensive optimization target. An improved PSO algorithm is used to optimize the target by taking the ignition temperature as a constraint condition. The temperature of the ignition furnace is calculated based on the optimal ignition intensity, and a controller of the ignition temperature is designed to carry out the intelligent self-optimization of the air–fuel ratio. The architecture of the SIP control system is shown in Fig. 4.3, which consists of three parts: a module of the optimization of ignition intensity, an intelligent controller of the ignition temperature, and a self-optimizing controller of the air–fuel ratio.

The setting of the ignition intensity directly determines how much heat is provided to the sintered material, which is the key to determining the quality of the sintering ignition and the normal state of the sintering. The optimal settings of the series of operating parameters including the ignition intensity and the carbon content are used to guide the operation of the SIP and to ensure the achievement of the integrated production goal.

A desired ignition temperature is obtained by the optimization of the ignition intensity. At the same time, the air–fuel ratio self-optimization control method is used to ensure the best combustion efficiency of gas. The intelligent control is also

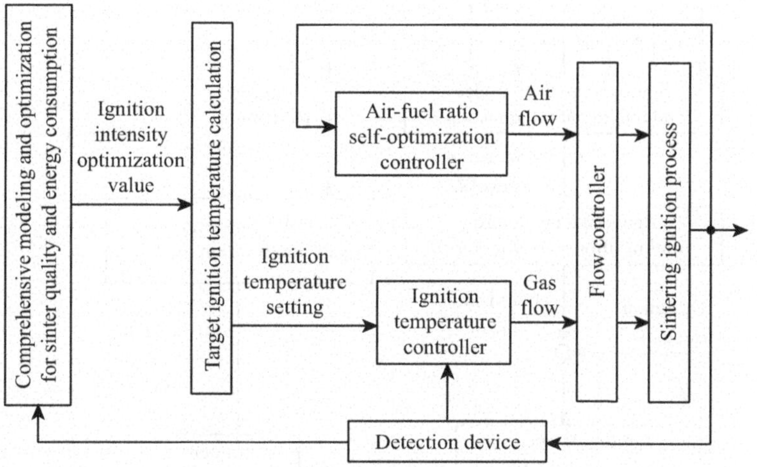

Fig. 4.3 Control system architecture of sintering ignition process

performed to deal with the fluctuations in the gas and the air flow, which are mainly caused by the fluctuations in the blast furnace gas.

4.2.2 *Intelligent Optimization and Control Algorithm*

Based on the analysis of the characteristics of the SIP, we design the structure of the control system and an intelligent control algorithm (Fig. 4.4) for the SIP.

A backpropagation NN (BPNN) model is established using the moisture, the ingredients, the carbon content, the trolley velocity, the ignition intensity, the negative pressure, etc. as the input, and the sinter quality and sintering energy consumption as the output. Actual production data are used to verify the validity of the model.

The ignition temperature of the ignition furnace is calculated based on the optimized ignition intensity, the trolley velocity, the width of the trolley, and other parameters. And a control system is built for the control of ignition temperature. The input and the output of the controller are the tracking error of the temperature and an increasement of the gas flow, respectively.

The air–fuel ratio self-optimization fuzzy controller is designed according to the influence of coal calorific value and pressure fluctuations on the SIP. The optimization of air–fuel ratio under different operating conditions is carried out by the variable step technique, which ensures the quality of combustion and improves the utilization ratio of gas.

The first objective of the intelligent control system for the SIP is to stabilize the ignition temperature. The control system needs to deal with an input time delay. Then, it is desirable to suppresses a periodic disturbance. The design of the control

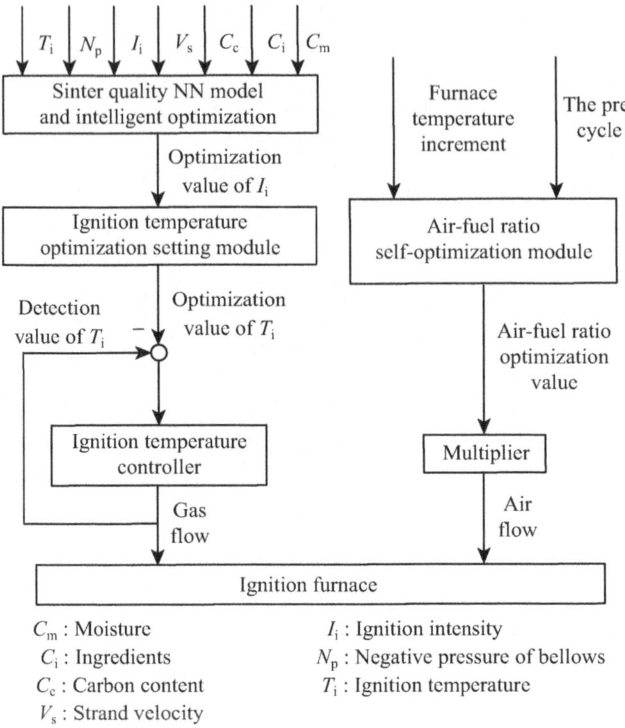

Fig. 4.4 Intelligent control algorithm for sintering ignition process

system for SIP mainly contains subspace modeling and the controller design of the equivalent-input-disturbance (EID)-based periodic disturbance rejection.

4.2.3 Subspace Modeling of Sintering Ignition Process

A stable ignition temperature assures the quality of the sinter. However, the SIP is complex, in which many factors are tightly coupled each other. Moreover, the process contains a large time delay. These make it difficult to obtain a mathematical model. In this subsection, an effective model of sintering ignition is established using a subspace modeling method.

1. Iterative Subspace Modeling

The subspace modeling method builds a state-space model based on the input and output measurements of the system [48]. To apply it to model the SIP with an unknown input time delay, we developed an iteration method.

Considering that the SIP is essentially a combustion process, the SIP can be expressed as a linear space model with a disturbance and an input time delay:

$$\dot{x}(t) = Ax(t) + Bu(t - \tau_p) + Kw(t), \tag{4.1}$$

$$y(t) = Cx(t) + Du(t - \tau_p) + w(t), \tag{4.2}$$

where $u(t - \tau_p) \in \mathbb{R}^m$ is the input with a time delay τ_p; $w(t) \in \mathbb{R}^p$ is a zero mean noise; $x(t) \in \mathbb{R}^n$ is system state; $y(t) \in \mathbb{R}^p$ is the output; A, B, C, and D are the system matrices; and K is the input matrix of the noise. The problem is to find estimates of system matrices A, B, C, and D from finite sequences of the input and output data, $\{u_k\}_{k=0}^N$ and $\{y_k\}_{k=0}^N$.

Assume that the sampling of data is carried out using a sampling period, h, at a time sequence $\{t_k\}_{k=0}^N$, $t_k = t_0 + kh$. The discrete-time input and output data are $\{u_k\}_{k=0}^N$ and $\{y_k\}_{k=0}^N$, respectively, which are sampled from the input and output of the continuous-time dynamic system. In this study, the sampling period is 1 s. The model input is the gas flow, and the model output is the ignition temperature.

Introduction of the complex-variable transform [49]:

$$\lambda = \frac{1}{1 + s\tau}. \tag{4.3}$$

Both forward and inverse multivariable bilinear transforms are special cases of the transform $s = \dfrac{1 - \lambda}{\tau\lambda}$ and can be realized by the state-space Formula:

$$\begin{bmatrix} A_b & B_b \\ C_b & D_b \end{bmatrix} = \begin{bmatrix} (I + \tau A)^{-1} & \tau(I + \tau A)^{-1}B \\ -C(I + \tau A)^{-1} & D - \tau C(I + \tau A)^{-1}B \end{bmatrix}. \tag{4.4}$$

Ignoring the initial conditions and reformulating the linear system equation yield

$$\begin{bmatrix} \xi(t) \\ y(t) \end{bmatrix} = \begin{bmatrix} A_\lambda & B_\lambda \\ C & D \end{bmatrix} \begin{bmatrix} x(t) \\ u(t) \end{bmatrix} + \begin{bmatrix} K_\lambda \\ I \end{bmatrix} w(t), \tag{4.5}$$

where

$$\begin{cases} x(t) = \lambda\xi(t), \\ A_\lambda = I + \tau A, \quad B_\lambda = \tau B, \quad K_\lambda = \tau K. \end{cases} \tag{4.6}$$

Recursion gives

$$\begin{aligned} y(t) &= Cx(t) + Du(t) + w(t) \\ &= CA_\lambda[\lambda x(t)] + CB_\lambda[\lambda u(t)] + Du(t) + CK_\lambda[\lambda w(t)] + w(t) \\ &= \cdots \end{aligned}$$

$$= CA_\lambda^k[\lambda^k x(t)] + \sum_{j=1}^{k} CA_\lambda^{k-j} B_\lambda[\lambda^{k-j+1}u(t)]$$

$$+Du(t) + \sum_{j=1}^{k} CA_\lambda^{k-j} K_\lambda[\lambda^{k-j+1}w(t)] + w(t). \tag{4.7}$$

Then, the extended linear model of the system can be formulated as

$$Y = \Gamma_x X + \Gamma_u U + \Gamma_w W, \tag{4.8}$$

where the input, the output, and the state variable are

$$Y = \begin{bmatrix} \lambda^{i-1}y(t) \\ \lambda^{i-2}y(t) \\ \vdots \\ \lambda y(t) \\ y(t) \end{bmatrix}, \quad U = \begin{bmatrix} \lambda^{i-1}u(t) \\ \lambda^{i-2}u(t) \\ \vdots \\ \lambda u(t) \\ u(t) \end{bmatrix}, \quad X = [\lambda^{i-1}x(t)],$$

and the stochastic disturbance is

$$W = \begin{bmatrix} \lambda^{i-1}w(t) \\ \lambda^{i-2}w(t) \\ \vdots \\ \lambda w(t) \\ w(t) \end{bmatrix}.$$

The parameter matrices of the state variable and the input–output behavior are expressed as

$$\Gamma_x = \begin{bmatrix} C \\ CA_\lambda \\ \vdots \\ CA_\lambda^{i-1} \end{bmatrix} \in \mathbb{R}^{ip\times n}, \tag{4.9}$$

$$\Gamma_u = \begin{bmatrix} D & 0 & \cdots & 0 \\ CB_\lambda & D & \ddots & \vdots \\ \vdots & \vdots & \ddots & 0 \\ CA_\lambda^{i-2}B_\lambda & CA_\lambda^{i-3}B_\lambda & \cdots & D \end{bmatrix} \in \mathbb{R}^{ip\times im}, \tag{4.10}$$

and the matrix of stochastic input–output behavior is

$$\Gamma_w = \begin{bmatrix} I & 0 & \cdots & 0 & 0 \\ CK_\lambda & I & \ddots & \vdots & \vdots \\ \vdots & CK_\lambda & \ddots & 0 & 0 \\ CA_\lambda^{i-3}K_\lambda & \vdots & \ddots & I & 0 \\ CA_\lambda^{i-2}K_\lambda & CA_\lambda^{i-3}K_\lambda & \cdots & CK_\lambda & I \end{bmatrix}. \tag{4.11}$$

It would be the best if we can estimate the time delay and the parameters of system matrices simultaneously. However, this is a difficult task. To make it simple, we estimate the time delay and the parameters separately. Below is a procedure of estimating the state-space parameters of a mathematical model of the SIP.

Step 1: Collect M groups of manufacturing data, which contains the blast furnace gas flow, the coke-oven gas flow, and the ignition temperature of the SIP. N groups of the data are used for modeling. Set the fitting degree F of modeling to be 90% and the input time delay τ_p to be 0.

Step 2: Arrange manufacturing data matrices U_N and Y_N as follows. Note that

$$(\lambda^j u)_k = \lambda^j u(t_k), \quad (\lambda^j y)_k = \lambda^j y(t_k). \tag{4.12}$$

Build

$$Y_N = \begin{bmatrix} (\lambda^{i-1}y)_1 & (\lambda^{i-1}y)_2 & \cdots & (\lambda^{i-1}y)_N \\ (\lambda^{i-2}y)_1 & (\lambda^{i-2}y)_2 & \cdots & (\lambda^{i-2}y)_N \\ \vdots & \vdots & & \vdots \\ (\lambda y)_1 & (\lambda y)_2 & \cdots & (\lambda y)_N \\ y_1 & y_2 & \cdots & y_N \end{bmatrix}, \tag{4.13}$$

$$U_N = \begin{bmatrix} (\lambda^{i-1}u)_1 & (\lambda^{i-1}u)_2 & \cdots & (\lambda^{i-1}u)_N \\ (\lambda^{i-2}u)_1 & (\lambda^{i-2}u)_2 & \cdots & (\lambda^{i-2}u)_N \\ \vdots & \vdots & & \vdots \\ (\lambda u)_1 & (\lambda u)_2 & \cdots & (\lambda u)_N \\ u_1 & u_2 & \cdots & u_N \end{bmatrix}. \tag{4.14}$$

Step 3: Carry out QR factorization of the following data matrices

$$\begin{bmatrix} U_N \\ Y_N \end{bmatrix} = \begin{bmatrix} R_{11} & 0 \\ R_{21} & R_{22} \end{bmatrix} \begin{bmatrix} Q_1 \\ Q_2 \end{bmatrix}. \tag{4.15}$$

Step 4: Carry out the singular value decomposition of the $R_{22} \in \mathbb{R}^{ip \times ip}$

$$R_{22} = \begin{bmatrix} U_n & U_0 \end{bmatrix} \begin{bmatrix} S_n & 0 \\ 0 & S_0 \end{bmatrix} \begin{bmatrix} V_n^{\mathrm{T}} \\ V_0^{\mathrm{T}} \end{bmatrix}. \tag{4.16}$$

and let $\hat{\Gamma}_x = U_n$.

Step 5: Determine the estimates of the system matrices A and C, $\hat{A}$ and $\hat{C}$. $\hat{C}$ can be chosen directly to be the block of the first row in $\hat{\Gamma}_x$. $\hat{A}$ is determined from the shift structure of $\hat{\Gamma}_x$. Denote

$$\underline{\Gamma_x} = \begin{bmatrix} \hat{C} \\ \vdots \\ \hat{C}\hat{A}_\lambda^{i-2} \end{bmatrix} \in \mathbb{R}^{p(i-1)\times n}, \tag{4.17}$$

$$\overline{\Gamma_x} = \begin{bmatrix} \hat{C}\hat{A}_\lambda \\ \vdots \\ \hat{C}\hat{A}_\lambda^{i-1} \end{bmatrix} \in \mathbb{R}^{p(i-1)\times n}, \tag{4.18}$$

where $\underline{\Gamma_x}$ is $\hat{\Gamma}_x$ without the last block in the row and $\overline{\Gamma_x}$ is $\hat{\Gamma}_x$ without the block in the first row; the shift structure implies that

$$\underline{\Gamma_x}\hat{A}_\lambda = \overline{\Gamma_x}. \tag{4.19}$$

Then, we have

$$\hat{A} = \frac{1}{\tau}(\hat{A}_\lambda - I). \tag{4.20}$$

Step 6: Determine estimates $\hat{B}$ and $\hat{D}$ of system matrices B, D from the relationship

$$\hat{\Gamma}_u = R_{21}R_{11}^{-1}. \tag{4.21}$$

Step 7: Compare the model outputs with sample outputs and yield the fitting degree F_i. If $F_i \geqslant F$, stop the iteration. Otherwise, go to Step 8.

Step 8: Keep the output data fixed and the length of the queue to be N. Add the gas flow data of the next sampling time to the tail of the queue and remove the head of the queue. This gives a new input queue. Let $\tau_\mathrm{p} = \tau_\mathrm{p} + 1$, and go back to Step 2.

An optimal fitting model is obtained when the algorithm ends. It serves as a plant model for the design of an EID-based control system.

The above steps yield an extended linear model of the continuous-time system in the state space. Note that the input time delay, τ_p, which is not considered in the model, is estimated below.

2. The Estimated Time-Delay Model

If we perform the above iteration by selecting two inputs and one output, we will obtain a series of models with a different time delay τ_p. In this subsection, the model inputs are the blast furnace gas flow and the coke-oven gas flow, and the model output is the ignition temperature.

Two different sequences of the inputs for one sequence of the output result in two models with two different values of the input time delays, τ_p. A verification of

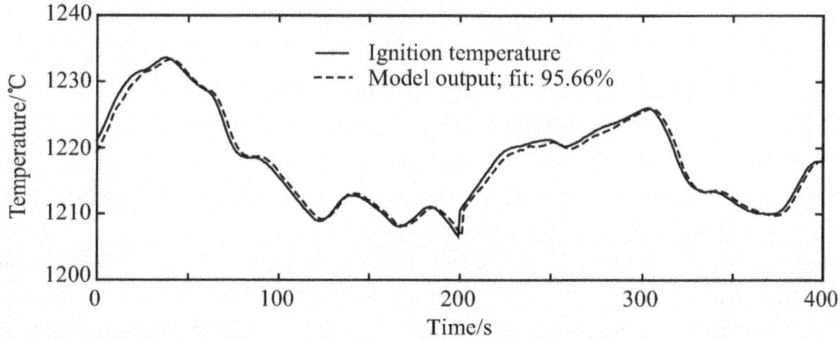

Fig. 4.5 The output of the model with the time delay ($\tau_p = 2$ s)

different values of τ_p shows that $\tau_p = 2$ s had the highest precision, and the prediction error is maintained within ±4 °C (Fig. 4.5). All the parameters of the model are

$$\begin{cases} A = \begin{bmatrix} -41.02 & 0 & -2.2085 \times 10^4 \\ 0 & -10.00 & 4477 \\ 1 & -1 & 0 \end{bmatrix}, \\ B = \begin{bmatrix} 20.3832 & 0 & 0 \\ 0 & 4.2343 & 0 \end{bmatrix}^{\mathrm{T}}, \\ C = \begin{bmatrix} 1 & 0 & 0 \end{bmatrix}, \\ D = 0, \\ \tau_p = 2. \end{cases} \tag{4.22}$$

4.2.4 Periodic Disturbance Rejection Using Equivalent-Input-Disturbance Estimation

Since the fluctuations in the gas pressure in the SIP are periodic, we need to develop a control method that not only ensures stability, but also has the ability to suppress periodic disturbances, even for a plant with an input time delay. At the same time, the control system is also expected to be simple for easy use in practice. This section explains an active disturbance rejection method based on the EID approach.

1. A Modified EID Method Considering Input Time Delay

Since the SIP contains an input time delay, it is difficult to use conventional disturbance rejection methods to handle this kind of plants. Note that, compared to other disturbance rejection methods, the EID-based method has many advantages, for example, it does not need a priori information of disturbances or an inverse model of the plant. To take full advantage of the EID approach, we devised a modified EID (MEID) method for this plant.

A general continuous model with an input disturbance is

$$\begin{cases} \dot{x}(t) = Ax(t) + Bu_{\mathrm{g}}(t - \tau_{\mathrm{p}}) + Bd_{\mathrm{g}}(t - \tau_{\mathrm{p}}), \\ y(t) = Cx(t), \end{cases} \tag{4.23}$$

where $x(t) \in \mathbb{R}^n$ is the state of the plant; $u_{\mathrm{g}}(t - \tau_{\mathrm{p}}) \in \mathbb{R}^m$ is the delayed control input, τ_{p} is a constant describing the time delay; $d_{\mathrm{g}}(t - \tau_{\mathrm{p}}) \in \mathbb{R}^{n_{\mathrm{d}}}$ is a disturbance with the period of T_{d}; $y(t) \in \mathbb{R}^p$ is the output; and A, B, and C are constant matrices of appropriate dimensions. (A, B, C) is assumed to be controllable and observable.

The delayed input disturbance is periodic. It is equivalent to a disturbance acting on the input channel without delay, that is, $d_{\mathrm{e}}(t) = d_{\mathrm{g}}(t - \tau_{\mathrm{p}})$. According to [50], we can produce an estimate of $d_{\mathrm{e}}(t)$, $\hat{d}(t)$, and incorporate it into the system to compensate for the real disturbance. Rewriting the plant Eq. (4.23) gives

$$\begin{cases} \dot{x}(t) = Ax(t) + Bu_{\mathrm{g}}(t - \tau_{\mathrm{p}}) + Bd_{\mathrm{e}}(t), \\ y(t) = Cx(t). \end{cases} \tag{4.24}$$

A full-state observer with an input time delay, τ_{p}, is designed to reproduce the state of the plant:

$$\begin{cases} \dot{\hat{x}}(t) = A\hat{x}(t) + Bu_{\mathrm{f}}(t - \tau_{\mathrm{p}}) + L[y(t) - \hat{y}(t)], \\ \hat{y}(t) = C\hat{x}(t), \end{cases} \tag{4.25}$$

where L is the observer gain; $\hat{x}(t)$ is the estimated state; and $\hat{y}(t)$ is the estimated output.

Assume that

$$\Delta x(t) = x(t) - \hat{x}(t), \tag{4.26}$$

and substituting Eq. (4.26) into Eq. (4.24) yield

$$\dot{\hat{x}}(t) = A\hat{x}(t) + Bu_{\mathrm{g}}(t - \tau_{\mathrm{p}}) + \{Bd_{\mathrm{e}}(t) + [A\Delta x(t) - \Delta\dot{x}(t)]\}. \tag{4.27}$$

Assume that there exists a control input $\Delta d(t)$ that satisfies

$$A\Delta x(t) - \Delta\dot{x}(t) = B\Delta d(t). \tag{4.28}$$

Letting

$$\hat{d}(t) = d_{\mathrm{e}}(t) + \Delta d(t) \tag{4.29}$$

and substituting Eqs. (4.28) and (4.29) into Eq. (4.27) yield

$$\dot{\hat{x}}(t) = A\hat{x}(t) + B[u_{\mathrm{g}}(t - \tau_{\mathrm{p}}) + \hat{d}(t)]. \tag{4.30}$$

Remark 4.1 Equations (4.29) and (4.30) mean that, if we take the state of the plant with an EID to be $\hat{x}(t)$, which is exactly the state of the observer, then the difference between the state of the plant and that of the observer is equivalent to the difference between the exact value and the estimate of the EID. Equation (4.30) plays a key role in the EID estimation.

From Eqs. (4.25) and (4.30), we have

$$\hat{d}(t) = B^{+}LC\Delta x(t) + u_{\mathrm{f}}(t - \tau_{\mathrm{p}}) - u_{\mathrm{g}}(t - \tau_{\mathrm{p}}), \tag{4.31}$$

where

$$B^{+} = (B^{\mathrm{T}}B)^{-1}B^{\mathrm{T}}. \tag{4.32}$$

A low-pass filter, $F(s)$, is used to select the angular frequency band for the disturbance estimate $\hat{d}(t)$. However, since a time delay exists in the control input channel, it causes a phase lag and degrades the disturbance rejection performance. On the other hand, for the periodic disturbance, if we introduce a time delay, τ_{m}, in the feedback channel:

$$\tau_{\mathrm{m}} = nT_{\mathrm{d}} - \tau_{\mathrm{p}}, \tag{4.33}$$

where n is the smallest positive integer that guarantees $\tau_{\mathrm{m}} \geqslant 0$, then the phase difference becomes zero, and the disturbance is completely rejected. From this viewpoint, we define a new EID estimate

$$\tilde{d}(t - \tau_{\mathrm{m}}) = u_{\mathrm{f}}(t) - u_{\mathrm{g}}(t). \tag{4.34}$$

Combining Eqs. (4.31) and (4.34) gives

$$\hat{d}(t) = B^{+}LC\Delta x(t) + \tilde{d}(t - \tau). \tag{4.35}$$

The configuration of the MEID-based control system is shown in Fig. 4.6. It consists of four parts: the plant, an internal model, a state observer, and an MEID estimator.

An internal model is used to guarantee the perfect tracking of a reference input. Its state-space form is

$$\dot{x}_{\mathrm{R}}(t) = A_{\mathrm{R}}x_{\mathrm{R}}(t) + B_{\mathrm{R}}[r(t) - y(t)], \tag{4.36}$$

where $x_{\mathrm{R}}(t)$ is the state; $r(t)$ is the reference input; and A_{R} and B_{R} are the constant matrices, which are given by a known reference input. The state-feedback control law is

$$u_{\mathrm{f}}(t) = K_{\mathrm{R}}x_{\mathrm{R}}(t) + K_{\mathrm{P}}\hat{x}(t), \tag{4.37}$$

where $K = [K_{\mathrm{P}} \ \ K_{\mathrm{R}}]$ represents the state-feedback gains.

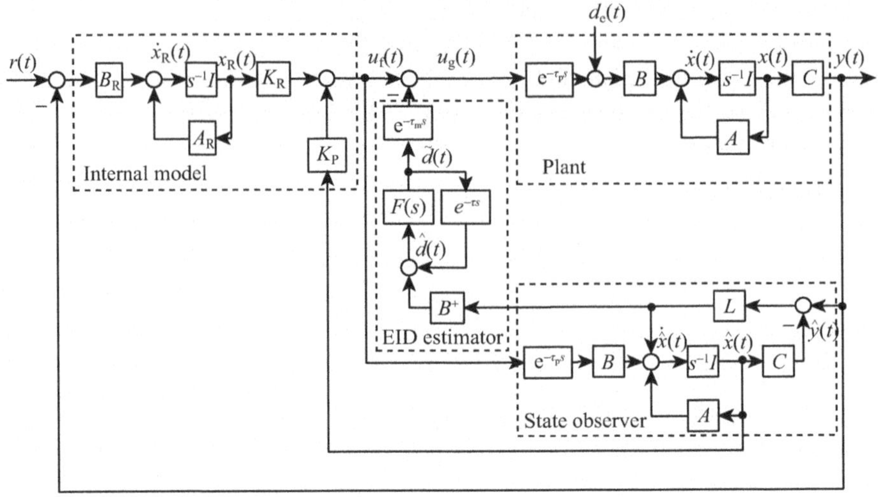

Fig. 4.6 The configuration of the SIP control with EID-based disturbance compensation

2. Stability of the EID-Based Control System

Let $r(t) = 0$, $d_g(t) = 0$. From Eqs. (4.23), (4.25), and (4.34), we have

$$\Delta\dot{x}(t) = (A - LC)\Delta x(t) - B\tilde{d}(t - \tau). \tag{4.38}$$

Based on Eqs. (4.35) and (4.38), we obtain the transfer function from $\tilde{d}(t-\tau)$ to $\hat{d}(t)$.

$$Q(s) = 1 - B^+LC[sI - (A - LC)]^{-1}B. \tag{4.39}$$

Then, redrawing Fig. 4.6 gives Fig. 4.7.

According to the separation theorem, for an appropriately selected K_P and K_R, if the following conditions:

(1) $\|QF\|_\infty < 1$.

(2) $A - LC$ is stable.

hold, then the closed-loop control system is stable.

Based on the above analysis, we obtain the following design steps:

Step 1: Choose the highest angular frequency ω_r for disturbance rejection.

Step 2: Select a first-order low-pass filter and set

$$T \leqslant \frac{1}{5\omega_r}. \tag{4.40}$$

This ensures

$$|F(j\omega)| \approx 1, \quad \omega \in [0, \omega_r]. \tag{4.41}$$

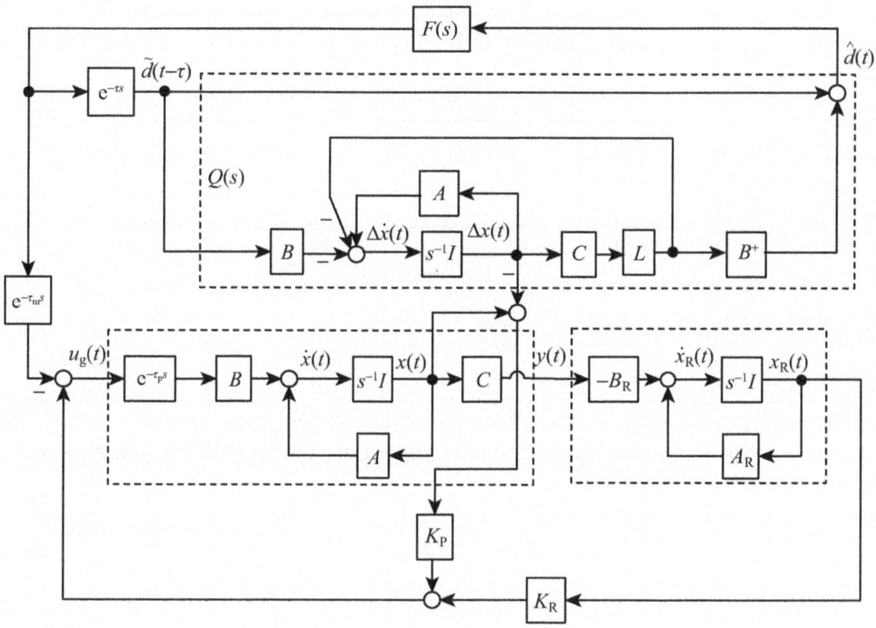

Fig. 4.7 Block diagram of MEID-based control system

Step 3: Determine ρ and calculate the corresponding L_ρ based on the concept of perfect regulation.

Step 4: Plot $|1/Q(j\omega)|$ and $|F(j\omega)|$, and check whether the condition

$$\left|\frac{1}{Q(j\omega)}\right| > |F(j\omega)|, \quad \omega \in [0, \infty) \tag{4.42}$$

is true from the Bode plots. If it is not, change ρ until Eq. (4.42) holds.

Step 5: Design K_P and K_R using a conventional method, for example, LQR.

Remark 4.2 If (A, B, C) is a minimum-phase system, the concept of perfect regulation provides us that a large enough ρ ensures

$$\lim_{\rho\to\infty} [sI - (A - L_\rho C)]^{-1} B = 0. \tag{4.43}$$

Or in other words, it ensures that $Q(j\omega)$ is sufficiently small for all $\omega \in [0, \infty)$.

4.2.5 Experimental Simulation

A numerical example is used to verify the effectiveness of the developed system. A comparison is also carried out between the conventional EID method and this method.

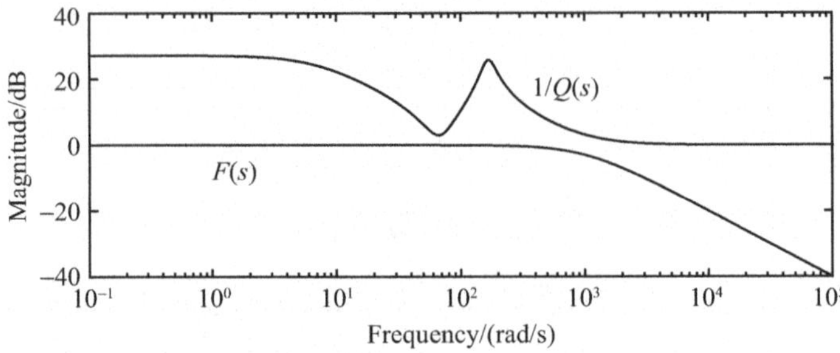

Fig. 4.8 Bode plot of $|1/Q(\mathrm{j}\omega)|$ and $|F(\mathrm{j}\omega)|$

The model in Eq. (4.22) is used for verification. It describes a real process and is a minimum-phase system. The reference temperature is 1200 °C. The parameters of the internal model Eq. (4.36) are

$$A_{\mathrm{R}} = 0, \quad B_{\mathrm{R}} = 1. \tag{4.44}$$

The augmented state-space representation containing the states of the plant and the internal model is

$$\dot{\zeta}(t) = \bar{A}\zeta(t) + \bar{B}u(t - \tau_{\mathrm{p}}), \tag{4.45}$$

where

$$\zeta(t) = \begin{bmatrix} x(t) \\ x_{\mathrm{R}}(t) \end{bmatrix}, \quad \bar{A} = \begin{bmatrix} A & 0 \\ -B_{\mathrm{R}}C & A_{\mathrm{R}} \end{bmatrix}, \quad \bar{B} = \begin{bmatrix} B \\ 0 \end{bmatrix}, \quad \tau_{\mathrm{p}} = 2. \tag{4.46}$$

An optimal observer gain L, which minimizes the performance index, is designed such that

$$\begin{cases} J_L = \displaystyle\int_0^{\infty} \left\{ \rho x_{\mathrm{L}}^{\mathrm{T}}(t) Q_{\mathrm{L}} x_{\mathrm{L}}(t) + R_{\mathrm{L}} u_{\mathrm{L}}^2(t) \right\} \mathrm{d}t, \\ Q_{\mathrm{L}} = \mathrm{diag}\{1,\ 10^{-6},\ 10^{-6}\}, \quad R_{\mathrm{L}} = 1. \end{cases} \tag{4.47}$$

Choosing $\rho = 10^6$ yields

$$L_{\rho} = \begin{bmatrix} 961.0527 & -9.9050 & -0.0559 \end{bmatrix}^{\mathrm{T}}. \tag{4.48}$$

The Bode plot of $|1/Q(\mathrm{j}\omega)|$ and $|F(\mathrm{j}\omega)|$ is depicted in Fig. 4.8. Clearly, the stability conditions are satisfied.

The designed K_{P} and K_{R} are

$$K_{\mathrm{P}} = \begin{bmatrix} -0.2371 & -0.0197 & 0.5906 \\ -0.0041 & -0.2581 & 0.6054 \end{bmatrix}, \quad K_{\mathrm{R}} = \begin{bmatrix} 0.6988 \\ 0.7153 \end{bmatrix}. \tag{4.49}$$

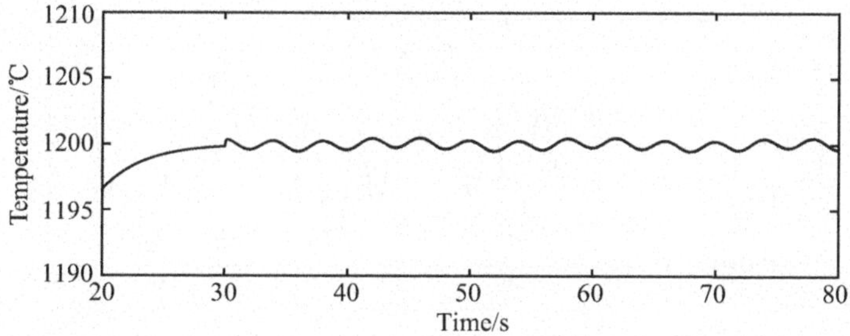

Fig. 4.9 Simulation results of conventional EID-based control system for plant without input time delay for Eq. (4.50)

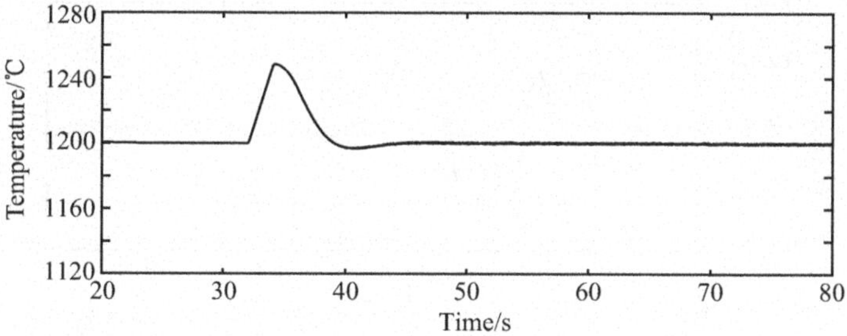

Fig. 4.10 Simulation results of conventional EID-based control system for plant with input time delay for step-type disturbance

Simulation results are shown in Figs. 4.9, 4.10, 4.11, and 4.12. After the system entered the steady state, the following periodic disturbance

$$w(t) = [\,d_{g1}(t)\ d_{g2}(t)\,]^{\mathrm{T}},$$
$$\begin{cases} d_{g1}(t) = 9.95 \times \sin\left(\frac{1}{2}\pi t\right) + 13.83 \times \sin\left(\frac{1}{8}\pi t\right) + e_1(t), \\ d_{g2}(t) = 30.48 \times \sin\left(\frac{1}{2}\pi t\right) + 62.21 \times \sin\left(\frac{1}{8}\pi t\right) + e_2(t), \end{cases} \tag{4.50}$$

where $e_i(t)$ is the zero mean noise, subscripts 1 and 2 denote blast oven gas and coke-oven gas, respectively.

The $w(t)$ was imposed on the system at $t = 30\,\mathrm{s}$. The conventional EID approach [51] had satisfactory disturbance rejection performance for the plant without a input time delay (Fig. 4.9). The conventional EID approach could also reject a step-type disturbance for the plant with a input time delay (Fig. 4.10). However, the conventional EID approach could not reject a periodic disturbance for the plant with a input time delay and had a large steady-state tracking error (peak-to-peak value: 135.42 °C) (Fig. 4.11). On the other hand, the MEID-based control system improved

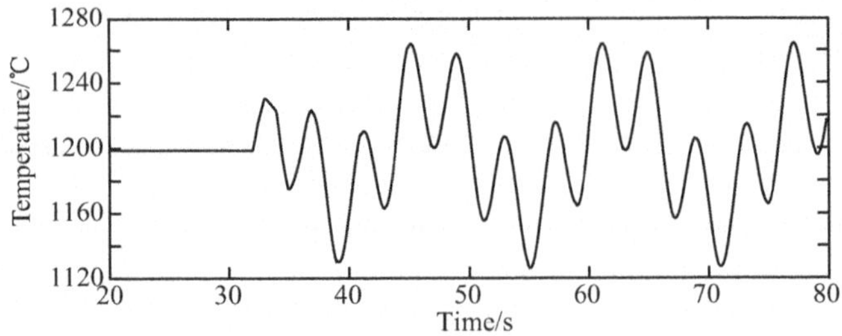

Fig. 4.11 Simulation results of conventional EID-based control system for plant with input time delay for periodic disturbance

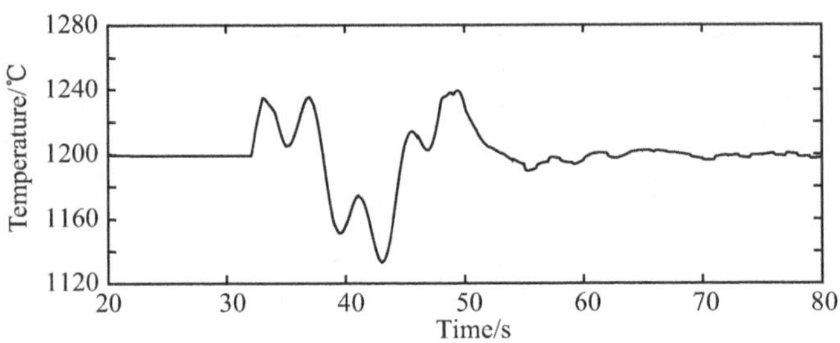

Fig. 4.12 Simulation results of MEID-based control system for plant with input time delay for periodic disturbance

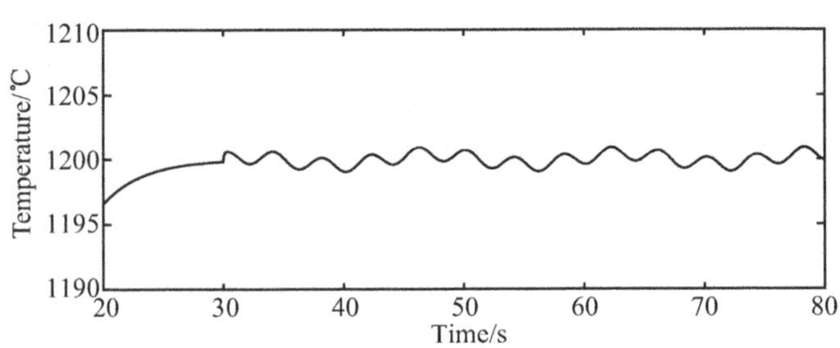

Fig. 4.13 Simulation results of DOB-based control system for plant with input time delay for periodic disturbance

the performance significantly, and the steady-state tracking error dropped to 16.87 °C (peak-to-peak value) (Fig. 4.12).

A disturbance observer (DOB)-based control system [52] was also designed for comparison. K_P, K_R, and L in the system were the same as those given in Eqs. (4.48) and (4.49). The simulation results are shown in Figs. 4.13, 4.14, and 4.15. Clearly, the DOB-based can reject the periodic disturbance for the plant without an input time delay (Fig. 4.13). While it could reject step-type disturbance for the plant with an input time delay [52] (Fig. 4.14), it could not reject the periodic disturbance, and the

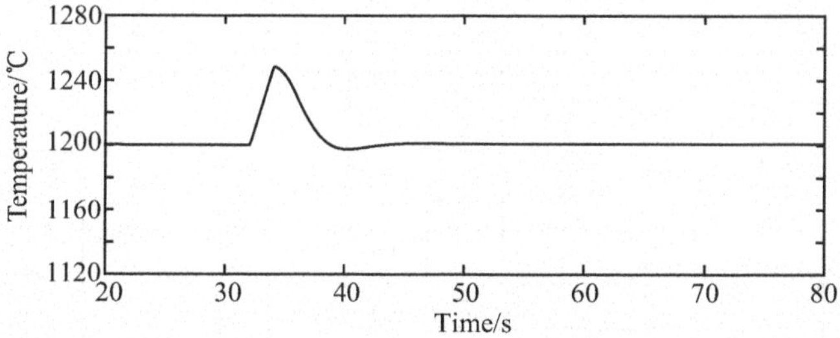

Fig. 4.14 Simulation results of DOB-based control system for plant with input time delay for step-type disturbance

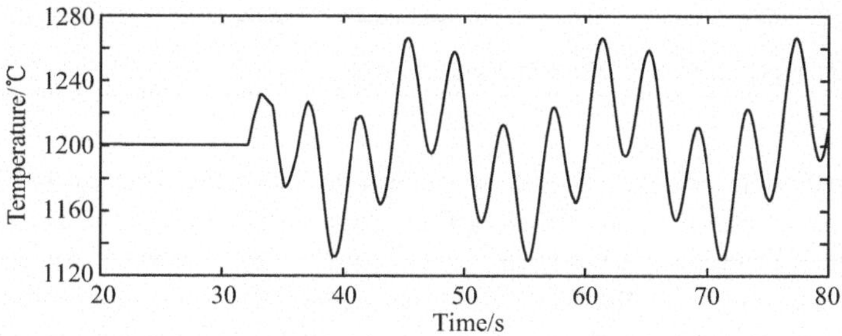

Fig. 4.15 Simulation results of DOB-based control system for plant with input time delay for periodic disturbance

peak-to-peak value of the steady-state error was as large as 138.49 °C (Fig. 4.15). This shows that the DOB-based control system is unable to meet the periodic-disturbance-rejection requirement.

The above simulation results show that, when an additional time delay is added to the feedback channel to compensate for phase difference, the MEID compensator is able to reject a periodic disturbance. So, the MEID provides us a new way to reject a periodic disturbance in the SIP even when the plant has a time delay.

4.3 Intelligent Control System for Burn-Through Point

As a main criterion of determining whether or not the sintering process is normal, the BTP is an important parameter that reflects the sintering state. This section designs an intelligent control of the BTP to achieve the stability of the BTP control.

4.3.1 Control System Architecture

The architecture of the intelligent BTP control system is shown in Fig. 4.16, which consists of seven parts: the sintering process, a BTP soft-sensing model, a BTP

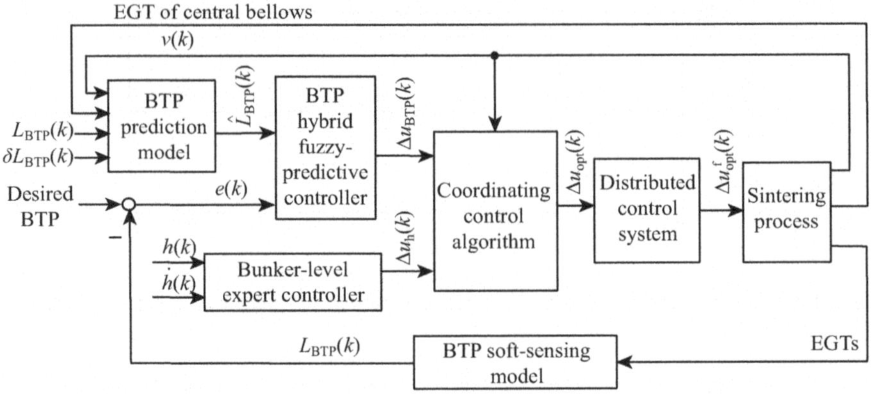

Fig. 4.16 Intelligent control system architecture of BTP

prediction model, a bunker-level expert controller, a BTP hybrid fuzzy-predictive controller, a coordinating control algorithm, and a distributed control system.

First, the BTP prediction model produces $\hat{L}_{BTP}(k+1)$, and the soft-sensing model calculates $L_{BTP}(k)$. Next, the BTP hybrid fuzzy-predictive controller produces a correction to the trolley velocity, $\Delta u_{BTP}(k)$ $(= u_{BTP}(k) - v(k))$, where $u_{BTP}(k)$ is an optimal trolley velocity at the BTP control requirements, and $v(k)$ is a measured trolley velocity. Since the time response of the trolley dynamics is much shorter than that of the sintering process, we do not need to consider the dynamics of the trolley. On the other hand, the bunker-level expert controller also calculates a correction to the trolley velocity, $\Delta u_h(k)$ $(= u_h(k) - v(k))$, where $u_h(k)$ is the optimal trolley velocity determined from the bunker-level control requirements. Finally, based on $\Delta u_{BTP}(k)$, $\Delta u_h(k)$, and $v(k)$, the coordinating control algorithm produces an optimal trolley velocity, $u_{opt}(k)$, that satisfies the synthetic control requirements of the sintering process and sends it to the DCS for tracking control. The DCS generates a corresponding reference input, $u^f_{opt}(k)$, for a converter to carry out the control of the trolley velocity.

The BTP hybrid fuzzy-predictive controller is the core of intelligent control system. It includes a fuzzy-PD controller, a predictive controller, and a soft hand-off controller. The fuzzy-PD controller uses the error between the calculated and the optimal BTP as the input variable. The controller has the advantages of quick response and short setting time. This not only shortens the BTP-adjusting process, but also reduces the fluctuations in the BTP and stabilizes the operating condition. On the other hand, the hysteresis in BTP regulation degrades control performance. In order to solve this problem, we introduce the predictive control in the system to facilitate the adjustment of BTP.

The predictive controller is based on a BTP prediction model. The input of the controller is the error between the predicted and the optimal BTP. The BTP is stabilized in an expected control range by adjusting the trolley speed. However, a longer periodicity of the prediction model results in larger fluctuations in sintering raw material.

The inherent hysteresis and the time-varying of the sintering process lead to frequent changes in the sintering conditions during a sintering cycle, and make it difficult to accurately predict the BTP. Therefore, the hybrid fuzzy-predictive intelligent control strategy, which combines the fuzzy-PD control and the predictive control, is adopted to improve the control precision and enhance the robustness of the system.

The soft hand-off control is the link between the fuzzy-PD control and the predictive control. Meanwhile, the relationship between the central bellows temperature and the BTP is analyzed according to the actual conditions of sintering, and a soft switching model is established. The stability of the BTP is guaranteed by changing the proportional relationship between these two controllers.

4.3.2 Soft Sensing and Prediction of Burn-Through Point

Since the BTP cannot be measured in an online fashion and the system has a large time delay, it is necessary to find ways to determine the BTP and to forecast changes in the BTP. Two models, which are built for these purposes, are the foundation of the BTP control.

1. Soft Sensing of BTP

Soft sensing solves the problem of how to measure the BTP. Since the BTP occurs where the exhaust gas temperature (EGT) is a maximum, the EGT is measured at the leading edge of each bellows by a thermocouple. Figure 4.17 shows the typical relationship between the EGT and the bellows number.

Statistical data and simulation results revealed that the relationship between the EGT and the bellows number could be approximated by the piecewise quadratic functions [53]:

$$T_i(k) = A_i(k)i^2 + B_i(k)i + C_i(k), \tag{4.51}$$

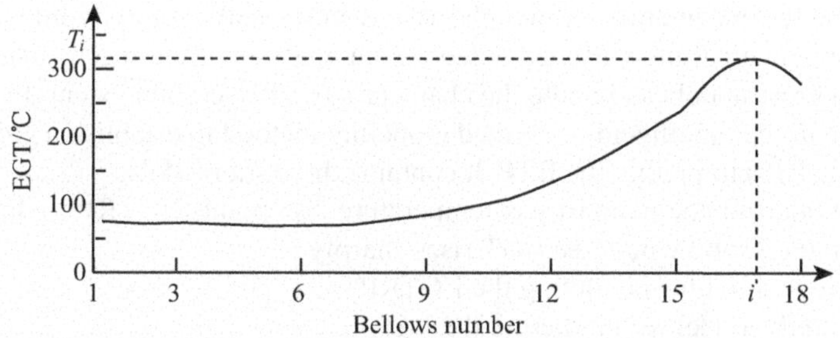

Fig. 4.17 Relationship between EGT and bellows number

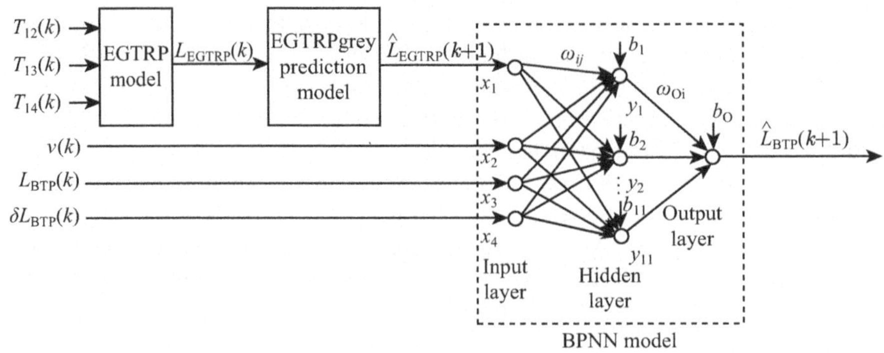

Fig. 4.18 BTP prediction model

where i is the bellows number; k is the sampling time; $T_i(k)$ is the EGT measured at the ith bellows; and $A_i(k)$, $B_i(k)$, and $C_i(k)$ are the binomial coefficients of the quadratic polynomial around the ith bellows.

For the EGTs $(T_{i-1}(k), T_i(k), T_{i+1}(k))$ of three adjacent bellows $(i-1, i, i+1)$, the parameters of the quadratic polynomial in Eq. (4.51) are

$$\begin{cases} A_i(k) = \dfrac{T_{i+1}(k) - 2T_i(k) + T_{i-1}(k)}{2}, \\ B_i(k) = T_i(k) - T_{i-1}(k) - (2i-1)A_i(k), \\ C_i(k) = T_{i-1}(k) - (i-1)^2 A_i(k) - (i-1)B_i(k). \end{cases} \tag{4.52}$$

The soft sensing of the BTP is used to analyze and preprocess the detection of parameters. It corrects detected abnormal temperature and yields accurate BTP.

2. Prediction of BTP

The sintering process has the features of nonlinearity, a large time delay, and incomplete parameter information. Although these features make it impossible to build a precise mathematical model based solely on an analysis of the mechanism, an NN model or a grey model can handle the characteristics to a certain extent. In this subsection, we use an intelligent integrated modeling method to establish a grey BPNN model (Fig. 4.18) to predict the BTP. It contains three submodels:

(1) A model for the exhaust gas temperature rise point (EGTRP), which is the position on the strand where the EGT rises sharply.

(2) A grey model for predicting the EGTRP.

(3) A BPNN model for predicting the BTP.

Note that the EGTRP is specified by a bellows number plus a decimal fraction, just as the BTP is.

The complexity of the sintering process and the lack of complete information on it make it necessary to use all available parameter information directly reflecting the state and trend of the BTP to obtain a satisfactory prediction accuracy.

In the actual process, if the trolley slows down a great deal, the EGTs of the bellows in the middle rise rapidly, thereby causing the BTP to occur too early. In contrast, if the trolley speeds up a great deal, the EGTs of the bellows in the middle rise slowly, thereby causing the BTP to occur too late. So, we focus on the EGT of Bellows 13. A change in it portends a change in BTP and is an important variable for predicting the BTP.

The relationship between the EGT and Bellows i near the EGTRP can also be described by a quadratic function. Since the EGTRP is usually 13.0, we choose the EGTs $(T_{12}(k), T_{13}(k), T_{14}(k))$ of Bellows 12, 13, and 14 at the present sampling time to fit the quadratic polynomial in Eq. (4.51) as follows:

$$\begin{cases} A_{\mathrm{EGTRP}}(k) = \dfrac{T_{14}(k) - 2T_{13}(k) + T_{12}(k)}{2}, \\ B_{\mathrm{EGTRP}}(k) = T_{13}(k) - T_{12}(k) - 25A_{\mathrm{EGTRP}}(k), \\ C_{\mathrm{EGTRP}}(k) = T_{12}(k) - 144A_{\mathrm{EGTRP}}(k) - 12B_{\mathrm{EGTRP}}(k). \end{cases} \tag{4.53}$$

The experience of experts tells us that the optimal EGT for the EGTRP is $T_{\mathrm{EGTRP}} = 160\,^{\circ}\mathrm{C}$. Substituting it into

$$A_{\mathrm{EGTRP}}(k)L^2_{\mathrm{EGTRP}}(k) + B_{\mathrm{EGTRP}}(k)L_{\mathrm{EGTRP}}(k) + C_{\mathrm{EGTRP}}(k) = 160, \tag{4.54}$$

and solving this equation yield the position of the EGTRP, $L_{\mathrm{EGTRP}}(k)$. This is the EGTRP calculation model.

To adjust the BTP in a timely manner, we built a grey GM(1,1) model that predicts the EGTRP at the next sampling time. Assume that the time sequence for the EGTRP calculated from Eq. (4.54) is

$$\begin{aligned} g^{(0)} &= \left\{g^{(0)}(1), \ldots, g^{(0)}(n-1), g^{(0)}(n)\right\} \\ &= \left\{L_{\mathrm{EGTRP}}(k-n+1), \ldots, L_{\mathrm{EGTRP}}(k-1), L_{\mathrm{EGTRP}}(k)\right\}, \end{aligned} \tag{4.55}$$

and let

$$g^{(1)}(j) = \sum_{i=1}^{j} g^{(0)}(i), \quad j = 1, 2, \ldots, n, \tag{4.56}$$

where n is the dimension of the sequence, which is generally in the range [4, 6]. We chose $n = 5$ in this study.

The movement of $g^{(1)}(j)$ $(j = 1, 2, \ldots, n)$ is described by the first-order differential equation

$$\frac{\mathrm{d}g^{(1)}(t)}{\mathrm{d}t} + ag^{(1)}(t) = r, \tag{4.57}$$

where a is a development coefficient, and r is a grey action quantity. They are calculated using the method of a least-squares estimation [54] and are

$$\hat{a} = [a,\ r]^{\mathrm{T}} = \left(X^{\mathrm{T}}X\right)^{-1} X^{\mathrm{T}}Y, \tag{4.58}$$

where

$$X = \begin{bmatrix} -0.5\left(g^{(1)}(1) + g^{(1)}(2)\right) & 1 \\ -0.5\left(g^{(1)}(2) + g^{(1)}(3)\right) & 1 \\ -0.5\left(g^{(1)}(3) + g^{(1)}(4)\right) & 1 \\ -0.5\left(g^{(1)}(4) + g^{(1)}(5)\right) & 1 \end{bmatrix},$$
$$Y = \left[g^{(0)}(2),\ g^{(0)}(3),\ g^{(0)}(4),\ g^{(0)}(5)\right]^{\mathrm{T}}.$$

The time response of Eq. (4.57) in discrete form is

$$\hat{g}^{(1)}(j+1) = \left[g^{(0)}(1) - \frac{r}{a}\right]\mathrm{e}^{-aj} + \frac{r}{a},\ j = 1, 2, \ldots, n. \tag{4.59}$$

The restored values of $g^{(0)}(j+1)$ $(j = 1, 2, \ldots, n)$ are

$$\begin{cases} \hat{g}^{(0)}(1) = g^{(0)}(1), \\ \hat{g}^{(0)}(j+1) = \hat{g}^{(1)}(j+1) - \hat{g}^{(1)}(j), \quad j = 1, 2, \ldots, n. \end{cases} \tag{4.60}$$

So, a grey GM(1,1) model is

$$\begin{aligned} \hat{L}_{\mathrm{EGTRP}}(k+1) &= \hat{g}^{(0)}(n+1) \\ &= \hat{g}^{(1)}(n+1) - \hat{g}^{(1)}(n) \\ &= \left[g^{(0)}(1) - \frac{r}{a}\right]\left[\mathrm{e}^{-an} - \mathrm{e}^{-a(n-1)}\right] \\ &= \left[g^{(0)}(1) - \frac{r}{a}\right]\left[\mathrm{e}^{-5a} - \mathrm{e}^{-4a}\right]. \end{aligned} \tag{4.61}$$

It predicts the EGTRP at the next sampling time.

Based on the predicted EGTRP and other information, we build a three-layer BPNN model to predict the BTP. The numbers of input and output neurons are set to be 4 and 1, respectively. The trial and error determined the number of hidden neurons to be 11. Defining the inputs to be x_j $(j = 1, 2, 3, 4)$ yields the BPNN model

$$\hat{L}_{\mathrm{BTP}}(k+1) = \sum_{i=1}^{11} \omega_{\mathrm{O}i}\,\mathrm{tansig}\left(\sum_{j=1}^{4} \omega_{ij}x_j + b_i\right) + b_{\mathrm{O}}, \tag{4.62}$$

where, for $i = 1, 2, \ldots, 11$ and $j = 1, 2, 3, 4$, ω_{ij} is the weight of the signal from the jth input neuron to the ith hidden neuron, b_i is the bias of the ith hidden neuron, $\omega_{\mathrm{O}i}$ is the weight of the signal from the ith hidden neuron to the output neuron, and b_{O} is the bias of the output neuron. These weights are determined by training the BPNN.

Summarizing the above results gives us the steps for BTP prediction:

Step 1: Calculate $A_{\mathrm{EGTRP}}(k)$, $B_{\mathrm{EGTRP}}(k)$, and $C_{\mathrm{EGTRP}}(k)$ using Eq. (4.53) and $T_{12}(k)$, $T_{13}(k)$, and $T_{14}(k)$, which are the EGTs of Bellows 12, 13, and 14 at the present sampling time. Then, solve Eq. (4.54) for $L_{\mathrm{EGTRP}}(k)$.

Step 2: Estimate the EGTRP for the next sampling time, $\hat{L}_{\mathrm{EGTRP}}(k+1)$, using the grey GM(1,1) model Eq. (4.61).

Step 3: Put $\hat{L}_{\mathrm{EGTRP}}(k+1)$, $v(k)$, $L_{\mathrm{BTP}}(k)$, and the change in BTP, $\delta L_{\mathrm{BTP}}(k) = L_{\mathrm{BTP}}(k) - L_{\mathrm{BTP}}(k-1)$, into Eq. (4.62) and predict the BTP for the next sampling time, $\hat{L}_{\mathrm{BTP}}(k+1)$.

4.3.3 Hybrid Fuzzy-Predictive Controller

Sintering is a continuous industrial production process. The operating parameters and raw material parameters affect the control of the state parameters with a large time delay. And the coupling between the factors is strongly nonlinear. Conventional control methods based on a linear mathematical model are difficult to meet actual process requirements.

A hybrid fuzzy-predictive controller for the BTP is designed based on fuzzy control and prediction control to produce a correction to the desired-BTP input, Δu_{BTP}. The controller (Fig. 4.19) contains a fuzzy-PD controller, a predictive controller, and a soft hand-off controller. The model switches smoothly between the fuzzy-PD and the predictive controllers. Regarding the control strategy, when the BTP is relatively stable, the fuzzy-PD controller is mainly used to quickly track the BTP and suppress fluctuations arising from prediction error. On the other hand, when the variation in the BTP is relatively large, the predictive controller is mainly used to compensate for the influence of the large time delay and to adjust the input in advance.

1. Fuzzy-PD Controller

The input variables for the controller are the error between the calculated BTP and the optimal BTP. The optimal BTP is 17.0 for a sintering machine 280 m^2 and 23.0 for one 360 m^2. Take a sintering machine 280 m^2 as an example, $e(k) = L_{\mathrm{BTP}}(k) - 17.0$. Let the rate of change in the error be $\dot{e}(k)$, and the output is a correction to the trolley velocity be $\Delta u_{\mathrm{BTP}}^{\mathrm{F}}(k)$.

When the error, $e(k)$, is relatively large and is outside the range $[-1.0, 1.0]$, which means that the BTP is either less than 16.0 or greater than 18.0, expert rules based on the experience of veterans are employed to coarsely tune the trolley velocity so as to quickly force the error inside the range $[-1.0, 1.0]$. The rules are

R_{e1}: **IF** $e(k) > 1.0$, **THEN** $\Delta u_{\mathrm{BTP}}^{\mathrm{F}}(k) = 0.2\,\mathrm{m/min}$.

R_{e2}: **IF** $e(k) < -1.0$, **THEN** $\Delta u_{\mathrm{BTP}}^{\mathrm{F}}(k) = -0.2\,\mathrm{m/min}$.

R_{e1} means that, if the BTP is before the optimal position and is less than 16.0, then the trolley velocity is increased by $0.2\,\mathrm{m/min}$. R_{e2} says that, if the BTP is after the optimal position and is greater than 18.0, then the trolley velocity is reduced by $0.2\,\mathrm{m/min}$.

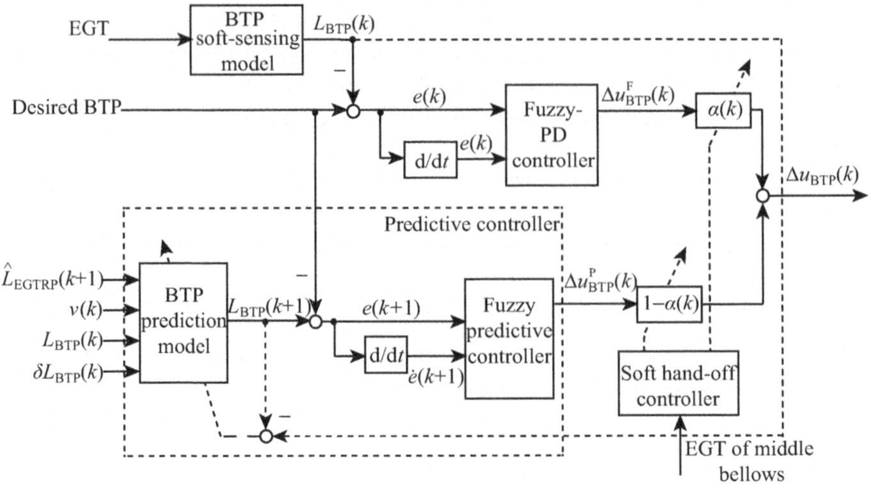

Fig. 4.19 Structure of BTP hybrid fuzzy-predictive controller

Table 4.1 Rule base of fuzzy-PD controller (inputs: E^F and $\dot{E}^F$, output: U^F_{BTP})

$\dot{E}^F$	E^F						
	NB	NM	NS	ZO	PS	PM	PB
NB	PB	PB	PB	PM	PM	ZO	ZO
NM	PB	PB	PM	PM	PM	ZO	ZO
NS	PM	PM	PM	PS	ZO	NS	NS
ZO	PM	PS	PS	ZO	NS	NM	NM
PS	PS	PS	ZO	NM	NM	NM	NM
PM	ZO	ZO	NM	NB	NB	NB	NB
PB	ZO	ZO	NM	NB	NB	NB	NB

When the error is in the range $[-1.0, 1.0]$, the fuzzy-PD controller carries out fine control of the BTP. This controller has seven linguistic labels containing values from negative big to positive big for $e(k)$, $\dot{e}(k)$, and $\Delta u^F_{BTP}(k)$. Their universes of discourse are defined to be $e(k) \in [-1.0, 1.0]$, $\dot{e}(k) \in [-0.2, 0.2]$, and $\Delta u^F_{BTP}(k) \in [-0.15, 0.15]$. And the corresponding fuzzy domains, E^F, $\dot{E}^F$, and U^F_{BTP}, are $\{-6, -5, -4, -3, -2, -1, 0, 1, 2, 3, 4, 5, 6\}$, $\{-6, -5, -4, -3, -2, -1, 0, 1, 2, 3, 4, 5, 6\}$, and $\{-7, -6, -5, -4, -3, -2, -1, 0, 1, 2, 3, 4, 5, 6, 7\}$, respectively. The membership function for U^F_{BTP} was chosen to be a triangular function, and the membership functions for E^F and $\dot{E}^F$ were chosen to be bell-shaped functions so as to improve the adaptive capacity of the controller. The rule base (Table 4.1) is based on the experience of experts. The center of gravity method is used to defuzzify the inferred output fuzzy subset.

Table 4.2 Rule base of fuzzy-predictive controller (inputs: E^{P} and $\dot{E}^{\mathrm{P}}$, output: $U_{\mathrm{BTP}}^{\mathrm{P}}$)

$\dot{E}^{\mathrm{P}}$	E^{P}						
	NB	NM	NS	ZO	PS	PM	PB
NB	PB	PB	PB	PB	PM	ZO	ZO
NM	PB	PB	PB	PB	PM	ZO	ZO
NS	PM	PM	PM	PM	ZO	NS	NS
ZO	PM	PM	PS	ZO	NS	NM	NM
PS	PS	PS	ZO	NS	NM	NM	NM
PM	ZO	ZO	NM	NM	NM	NB	NB
PB	ZO	ZO	NM	NB	NB	NB	NB

2. Predictive Controller

The predictive controller consists of the BTP prediction model described in Sect. 4.3.2 and a fuzzy-predictive controller. The steps in predictive control for the well-trained BPNN are as follows:

Step 1: Use the BPNN to predict the BTP for the next step, $\hat{L}_{\mathrm{BTP}}(k+1)$.

Step 2: Calculate the difference $e(k+1) = \hat{L}_{\mathrm{BTP}}(k+1) - 17.0$ and its rate of change $\dot{e}(k+1)$. Input them to the fuzzy-predictive controller and use fuzzy inference to determine the change in trolley velocity, $\Delta u_{\mathrm{BTP}}^{\mathrm{P}}(k)$.

Step 3: Return to Step 1 until the sintering process is finished.

The structure of the fuzzy-predictive controller is the same as that of the fuzzy-PD controller, but the universes of discourse are defined to be $e(k+1) \in [-1.0, 1.0]$, $\dot{e}(k+1) \in [-0.3, 0.3]$, and $\Delta u_{\mathrm{BTP}}^{\mathrm{P}}(k) \in [-0.15, 0.15]$. The corresponding fuzzy domains, E^{P}, $\dot{E}^{\mathrm{P}}$, and $U_{\mathrm{BTP}}^{\mathrm{P}}$, are $\{-6, -5, -4, -3, -2, -1, 0, 1, 2, 3, 4, 5, 6\}$, $\{-6, -5, -4, -3, -2, -1, 0, 1, 2, 3, 4, 5, 6\}$, and $\{-5, -4, -3, -2, -1, 0, 1, 2, 3, 4, 5\}$, respectively. In addition, the membership functions for the fuzzy domains E^{P}, $\dot{E}^{\mathrm{P}}$, and $U_{\mathrm{BTP}}^{\mathrm{P}}$ were chose to be triangular functions. The rule base (Table 4.2) is based on the experience of experts.

3. Soft Hand-off Control

The key point for switching between the fuzzy-PD and predictive controllers is determining whether or not the BTP is relatively stable. Since the variation in the EGTs of the middle bellows directly reflects changes in the process state and the BTP, they can be used as the basis for a criterion for a soft hand-off between the two controllers. Two variables are used for the criterion:

$$\begin{cases} \overline{T}(k) = \dfrac{1}{3} \displaystyle\sum_{i=11,13,14} |T_i(k) - T_{\mathrm{d}i}|\,, \\ \delta\overline{T}(k) = \dfrac{1}{3} \displaystyle\sum_{i=11,13,14} |T_i(k) - T_i(k-1)|\,, \end{cases} \tag{4.63}$$

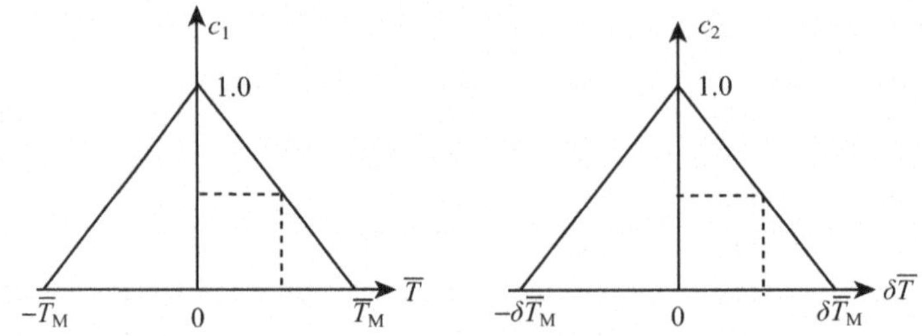

Fig. 4.20 Determination of coefficients for soft hand-off

where the subscript i is bellows number; $\overline{T}(k)$ is the average absolute value of the error between the measured EGTs, $T_i(k)$, and the optimum EGTs, T_{di}, in the middle of the strand; and $\delta\overline{T}(k)$ is the average of the absolute values of the changes in EGT in the middle of the strand between the present and previous steps. Based on empirical data on the process, we chose

$$T_{d11} = 100\,°\text{C}, \quad T_{d13} = 150\,°\text{C}, \quad T_{d14} = 200\,°\text{C}.$$

Let $\overline{T}_M$ and $\delta\overline{T}_M$ be the maxima of $\overline{T}(k)$ and $\delta\overline{T}(k)$, respectively. Based on an analysis of the data from the actual runs, we set

$$\overline{T}_M = 50\,°\text{C}, \quad \delta\overline{T}_M = 10\,°\text{C}.$$

The coefficients $c_1(k)$ and $c_2(k)$ for the soft hand-off are determined by the fuzzy mathematics (Fig. 4.20):

$$c_1(k) = 1 - \frac{\overline{T}(k)}{\overline{T}_M}, \quad c_2(k) = 1 - \frac{\delta\overline{T}(k)}{\delta\overline{T}_M}. \tag{4.64}$$

Let

$$\alpha(k) = \max\{c_1(k), c_2(k)\}. \tag{4.65}$$

The output of the BTP hybrid fuzzy-predictive controller is given by

$$\Delta u_{BTP}(k) = \alpha(k)\Delta u^F_{BTP}(k) + [1 - \alpha(k)]\Delta u^P_{BTP}(k). \tag{4.66}$$

When the EGTs in the middle of the strand are near the optimum values and the variations are small, both $c_1(k)$ and $c_2(k)$ are close to 1. This means that the BTP is relatively stable. In this case, the output of the fuzzy-PD controller, $\Delta u^F_{BTP}(k)$, plays the main role in BTP control; otherwise, the output of the fuzzy-predictive controller, $\Delta u^P_{BTP}(k)$, plays the main role.

While the hybrid fuzzy-predictive controller inherits the advantages of fast prediction, it does not have a steady-state tracking error to achieve a smooth transition from one control mode to another. When a large disturbance affects the process in steady state, the fuzzy-PD controller suppresses it and enhances the robustness of the control system.

4.3.4 Bunker-Level Expert Controller

Since the bunker continuously feeds raw material to the trolley, safe, uninterrupted operation of the process requires that the bunker does not become empty or overflow. An expert controller is designed to control the bunker level by adjusting the trolley velocity. The input variables are the current bunker level, $h(k)$, and its rate of change, $\dot{h}(k)$. The output is a correction to the trolley velocity, $\Delta u_{\mathrm{h}}(k)$. The objectives of bunker-level control are to

(1) ensure that the amount of material in the bunker is in the safe range [30 t, 60 t];

(2) keep the amount relatively stable.

The basic idea behind the expert rules is now explained. When the bunker level is either very high or is increasing at a high rate, the strand should be sped up to increase the flow of material out of the bunker while keeping the depth of the bed constant. On the other hand, when the bunker level is either very low or is decreasing at a very high rate, the strand should be slowed down to reduce the flow of material out of the bunker while again keeping the depth of the bed constant. The expert rules have the form of IF-THEN statements, with the ith rule being indexed as $R_{\mathrm{h}i}$. For example,

$R_{\mathrm{h}1}$: **IF** $(0\,\mathrm{t} \leqslant h(k) \leqslant 30\,\mathrm{t})$ **OR** $(\dot{h}(k) < 0)$, **THEN** $-0.3\,\mathrm{m/min} \leqslant \Delta u_{\mathrm{h}}(k) \leqslant -0.15$ $\mathrm{m/min}$.

4.3.5 Coordinating Control Algorithm

Note that, since the BTP and the bunker level are both controlled by means of the trolley velocity, they are coupled. An satisfactory solution principle (SSP)-based coordinating control algorithm is employed to handle this situation and to produce a satisfactory reference input for the trolley velocity. SSP attempts to optimize multiple performance indices rather than just one. Since solutions to real-world control problems usually necessitate the satisfaction of multiple requirements at the same time, this concept is very practical [55].

To balance $u_{\mathrm{BTP}}(k)$ and $u_{\mathrm{h}}(k)$, we first need to determine satisfactory degree functions (SDFs) for the BTP and bunker level. We employ a triangular function for the SDF of the BTP and a trapezoidal function for the SDF of the bunker level (Fig. 4.21). In the figure, u_{BTP} is the optimal trolley velocity output by the BTP controller, $u_{\mathrm{BTP}}(k) = \Delta u_{\mathrm{BTP}}(k) + v(k)$. u_{hm} and u_{hM} are the minimal and maximal trolley velocities, respectively, determined by the bunker-level expert controller, $u_{\mathrm{h}}(k) = \Delta u_{\mathrm{h}}(k) + v(k)$. And u_{m} and u_{M} are the minimal and maximal trolley velocities determined by the technology requirements.

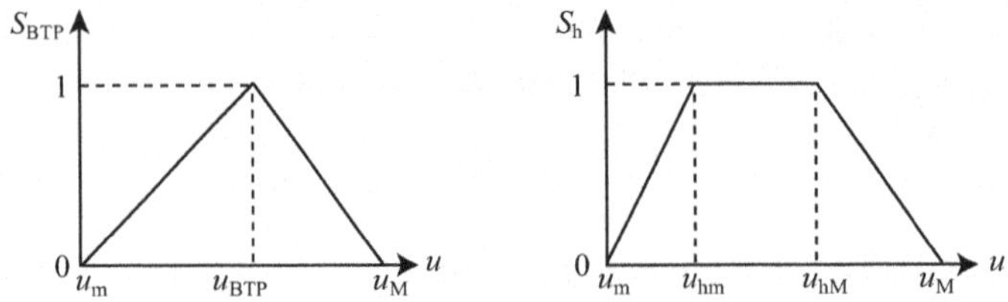

Fig. 4.21 Satisfactory degree functions of BTP and bunker level

Let $S_{BTP}(u)$ be the satisfactory degree of the BTP, and $u(k)$ be the strand velocity reference input to be determined.

$S_{BTP}(u)$ is defined to be

$$S_{BTP}(u) = \begin{cases} 0, & u(k) \leqslant u_m, \\ S_{BTP}^{(1)}(u), & u_m < u(k) < u_{BTP}(k), \\ 1, & u(k) = u_{BTP}(k), \\ S_{BTP}^{(2)}(u), & u_{BTP}(k) < u(k) < u_M, \\ 0, & u(k) \geqslant u_M, \end{cases} \tag{4.67}$$

where

$$S_{BTP}^{(1)}(u) = \frac{u(k) - u_m}{u_{BTP}(k) - u_m}, \quad S_{BTP}^{(2)}(u) = \frac{u_M - u(k)}{u_M - u_{BTP}(k)}.$$

And the satisfactory degree of the bunker level, $S_h(u)$, is

$$S_h(u) = \begin{cases} 0, & u(k) \leqslant u_m, \\ S_h^{(1)}(u), & u_m < u(k) < u_{hm}, \\ 1, & u_{hm} \leqslant u(k) \leqslant u_{hM}, \\ S_h^{(2)}(u), & u_{hM} < u(k) < u_M, \\ 0, & u(k) \geqslant u_M, \end{cases} \tag{4.68}$$

where

$$S_h^{(1)}(u) = \frac{u(k) - u_m}{u_{hm} - u_m}, \quad S_h^{(2)}(u) = \frac{u(k) - u_{hM}}{u_M - u_{hM}}.$$

We use the linear-weighted-sum method to determine the synthetic satisfactory degrees of the BTP and the bunker level:

$$S(u) = \gamma(k) S_{BTP}(u) + [1 - \gamma(k)] S_h(u), \tag{4.69}$$

$$\gamma(k) = 1 - \frac{|h(k) - h_{mid}|}{h_{mid}}, \tag{4.70}$$

where h_{mid} is the middle level of the bunker, for example, $h_{\text{mid}} = 45$ t in this subsection. In Eq. (4.69), $\gamma(k)$ and $1 - \gamma(k)$ are weight coefficients describing the importance of the BTP and bunker level. When the bunker is either nearly empty or about to overflow, the output of the bunker-level controller is given the greatest importance, and $\gamma(k)$ is set to a relatively small value. So, $\gamma(k)$ is defined to be Eq. (4.70).

According to the technical requirements, the problem of coordinating the control of the BTP and the bunker level is defined to be

$$\begin{aligned} &\max S(u) = \max\left\{\gamma(k)S_{\text{BTP}}(u) + [1-\gamma(k)]S_{\text{h}}(u)\right\}, \\ &\text{subject to } u_{\text{m}} \leqslant u(k) \leqslant u_{\text{M}}. \end{aligned} \tag{4.71}$$

We calculate a strand velocity, $u_{\text{opt}}(k)$, that maximizes $S(u)$ under the constraint in Eq. (4.71). The optimization problem is solved for three situations:

(1) **IF** $u_{\text{hm}} \leqslant u_{\text{BTP}}(k) \leqslant u_{\text{hM}}$, **THEN** $u_{\text{opt}}(k) = u_{\text{BTP}}(k)$.

(2) When $u_{\text{BTP}}(k) < u_{\text{hm}}$, from Eqs. (4.67), (4.68), and (4.69), we obtain

$$S(u) = \begin{cases} 0, & u(k) \leqslant u_{\text{m}}, \\ S_1(u), & u_{\text{m}} < u(k) \leqslant u_{\text{BTP}}(k), \\ S_2(u), & u_{\text{BTP}}(k) < u(k) \leqslant u_{\text{hm}}, \\ S_3(u), & u_{\text{hm}} < u(k) \leqslant u_{\text{hM}}, \\ S_4(u), & u_{\text{hM}} < u(k) \leqslant u_{\text{M}}, \\ 0, & u(k) > u_{\text{M}}, \end{cases} \tag{4.72}$$

where

$$\begin{aligned} S_1(u) &= \gamma(k)S_{\text{BTP}}^{(1)}(u) + [1-\gamma(k)]S_{\text{h}}^{(1)}(u), \\ S_2(u) &= \gamma(k)S_{\text{BTP}}^{(2)}(u) + [1-\gamma(k)]S_{\text{h}}^{(1)}(u), \\ S_3(u) &= \gamma(k)S_{\text{BTP}}^{(2)}(u) + [1-\gamma(k)] \cdot 1, \\ S_4(u) &= \gamma(k)S_{\text{BTP}}^{(2)}(u) + [1-\gamma(k)]S_{\text{h}}^{(2)}(u). \end{aligned}$$

Optimize the above $S(u)$ for $u(k)$ in different ranges. Choose the largest $S(u)$ from among the solutions, and set the corresponding $u(k)$ to be the solution $u_{\text{opt}}(k)$.

(3) When $u_{\text{BTP}}(k) > u_{\text{hM}}$, the solution method is logically identical to that of the second case. The only difference is that $S(u)$ changes to

$$S(u) = \begin{cases} 0, & u(k) \leqslant u_{\text{m}}, \\ S_1(u), & u_{\text{m}} < u(k) \leqslant u_{\text{hm}}, \\ S_5(u), & u_{\text{hm}} < u(k) \leqslant u_{\text{hM}}, \\ S_6(u), & u_{\text{hM}} < u(k) \leqslant u_{\text{BTP}}(k), \\ S_4(u), & u_{\text{BTP}}(k) < u(k) \leqslant u_{\text{M}}, \\ 0, & u(k) > u_{\text{M}}, \end{cases} \tag{4.73}$$

where

$$S_5(u) = \gamma(k)S^{(1)}_{\mathrm{BTP}}(u) + [1-\gamma(k)]\cdot 1,$$
$$S_6(u) = \gamma(k)S^{(1)}_{\mathrm{BTP}}(u) + [1-\gamma(k)]S^{(2)}_{\mathrm{h}}(u).$$

We use an example to illustrate the solution method. Consider the case $u_{\mathrm{BTP}}(k) = 2.85\,\mathrm{m/min}$, $u_{\mathrm{hm}} = 2.6\,\mathrm{m/min}$, $u_{\mathrm{hM}} = 2.75\,\mathrm{m/min}$, and $h(k) = 25\,\mathrm{t}$. Since $u_{\mathrm{BTP}}(k) > u_{\mathrm{hM}}$, we obtain $S(u)$ from Eqs. (4.67)–(4.73):

$$S(u) = \begin{cases} 0, & u(k) \leqslant 1.5, \\ 0.81548u(k) - 1.22322, & 1.5 < u(k) \leqslant 2.6, \\ 0.41185u(k) - 0.17375, & 2.6 < u(k) \leqslant 2.75, \\ 5.59823 - 1.80815u(k), & 2.75 < u(k) \leqslant 2.85, \\ 16.448 - 5.4827u(k), & 2.85 < u(k) \leqslant 3.0, \\ 0, & u(k) > 3.0. \end{cases} \tag{4.74}$$

Calculating the maximum $S(u)$ for different intervals yields $S_{\max}(1.5) = 0$, $S_{\max}(2.6) = 0.8970$, $S_{\max}(2.75) = 0.9588$, $S_{\max}(2.75 < u(k) \leqslant 2.85) = 0.6258$, $S_{\max}(2.85 < u(k) \leqslant 3.0) = 0.8223$, and $S_{\max}(u(k) > 3.0) = 0$. So, the global maximum is $S_{\max} = 0.9588$ and the corresponding strand velocity is $u_{\mathrm{opt}} = 2.75\,\mathrm{m/min}$.

The intelligent control system is being used for the real-time control of the BTP of the sintering process of an iron and steel plant. The results of actual runs demonstrate its validity.

4.4 Industrial Implementation and Results of Actual Runs

This section describes the industrial implementation of the intelligent control method. It has been applied to the actual sintering process of sintering machines 280 m^2 and 360 m^2.

4.4.1 Industrial Implementation

The intelligent control system consists of an industrial computer, an operator station & DDE server, an engineering station, and the PLCs (Fig. 4.22).

The intelligent controller at the intelligent control level of the intelligent control system was implemented on an industrial computer, and the application software was written in the Visual C++ language. The software package contains an HMI module, a control algorithm module, and a DDE client. The HMI module displays process data and information on an industrial computer, analyzes real-time data and the process states, sends the reference input $u_{\mathrm{opt}}(k)$ to the DCS, manages the security of the

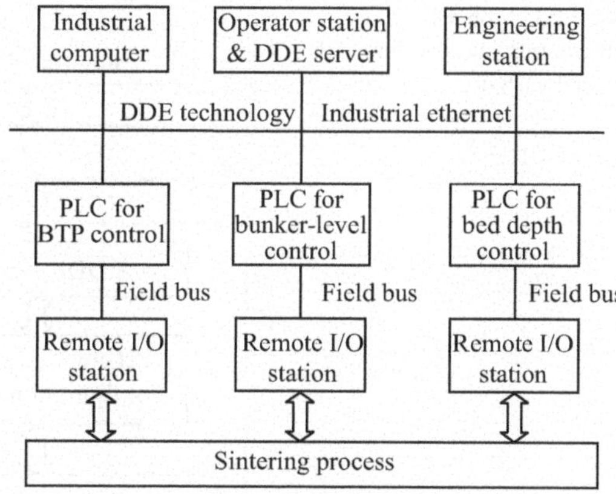

Fig. 4.22 Hardware architecture of intelligent control system

system, and issues warning alarms. The control methods were implemented in the algorithm module. The DDE client handles communications between the industrial computer and the DCS.

The basic automation level of the intelligent control system was implemented in the DCS. The DCS contains, a DDE server, cimplicity HMI configuration software, and PLC modules (Modicon TSX Quantum PLC of Schneider Electric Ltd.) which carry out process control. The intelligent controller generates a strand velocity reference input, which is sent to the PLC module through DDE data communication technology. More specifically, the DDE server transfers data between the DDE client and the cimplicity HMI configuration software. The PLC modules send those data to the frequency converter (Schneider Electric ATV-58F), which controls the motor drive of the strand. At the same time, it sends the actual strand velocity back to the intelligent controller. The cimplicity HMI configuration software controls and monitors the sintering process, collects data in a real-time fashion, analyzes historical curves, and produces reports and tables. Figure 4.23 shows the flow of information and data in the control system.

Based on the reference input produced by the intelligent controller, $u_{\mathrm{opt}}(k)$, the PLC control module produces a corresponding $u^{\mathrm{f}}_{\mathrm{opt}}(k)$ and uses it to regulate the frequency of the converter of the motor drive to adjust the strand velocity.

4.4.2 Results of Actual Runs

The intelligent control system has been in operation since May 2006. The process runs 24 h a day. It is divided into three 8-h shifts. Figure 4.24 shows BTP control results for 15 days (360 h). In the sintering machine 280 m^2, the optimal BTP was set to 17.0. Figure 4.24a shows the results for manual control, which was the method used before the intelligent control system was implemented, and Fig. 4.24b shows the results for the intelligent control system. Clearly, the intelligent control system

Fig. 4.23 Flow of information and data in BTP control system

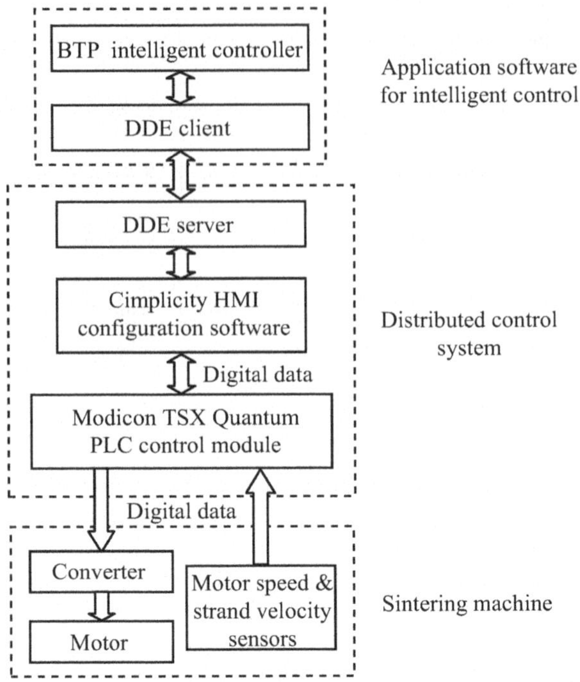

suppressed the variation in the BTP and kept it under ±0.4 at all times. This satisfies the technological requirement.

Table 4.3 compares the control results for manual control and intelligent control system in the sintering machine 280 m^2. We compared statistical data on the variation in BTP, the amount of agglomerate produced, and the tumbler index for one year. The BTP variation, ϵ_{BTP}, was calculated using peak-to-peak values during a given period

$$\epsilon_{\text{BTP}} = \frac{L_{\text{BTP}}^{\max} - L_{\text{BTP}}^{\min}}{17.0} \times 100\%. \tag{4.75}$$

The tumbler index is a measure of sinter quality. To determine it, a sample of agglomerate is put in a special tumbler, and the tumbler is rotated for 8 min. Then, the tumbler index is calculated by dividing the weight of the pieces of agglomerate with a diameter greater than 6.3 mm by the total weight of the sample. The larger the tumbler index is, the higher the mechanical strength of the agglomerate is. Clearly, the intelligent control system suppressed the variation in the BTP much better than manual control and kept it in a small range, increased the amount of sinter produced, and improved the quality of the sinter. More specifically, it reduced the variation in the BTP by 6.41%, increased the yearly production of agglomerate by 1.1%, and improved the tumbler index by 3.2%. The resulting economic benefits were significant.

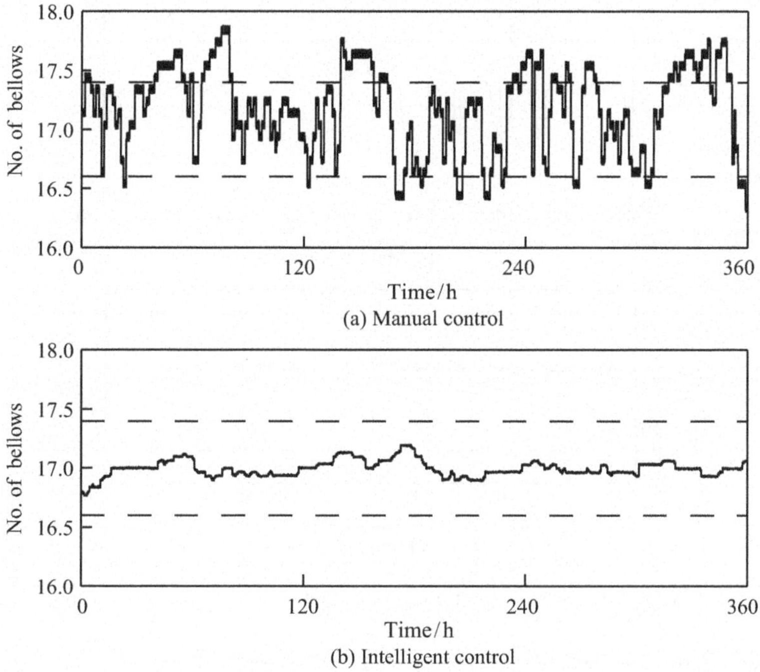

Fig. 4.24 BTP control results for manual control and intelligent control system in 280 m^2 sintering machine

Table 4.3 Comparison of control results between manual control and intelligent control system in 280 m^2 sintering machine

Control method	ϵ_{BTP} (%)	Production (t/h)	Tumbler index (%)
Manual control	9.29	464.7	78.1
Intelligent control system	2.88	469.6	81.3

The intelligent control system is also able to be used in sintering machine 360 m^2, and the optimal BTP was 23.0. Figure 4.25 shows the following:

(1) The results from manual control, which was the method used before the intelligent control system was implemented.

(2) The results from the intelligent control system.

Clearly, the intelligent control system suppressed the variation in the BTP. To analyze the effect of the intelligent control system for BTP control, the utilization factor of sinter machine, BTP fluctuation, and the screening index of agglomerate are chosen to be the three key indicators. Results' comparison of the indexes is shown in Table 4.4.

The intelligent control system greatly improved the quantity and quality of sinters. Compared to manual control, the utilization factor of sinter machine is increased by

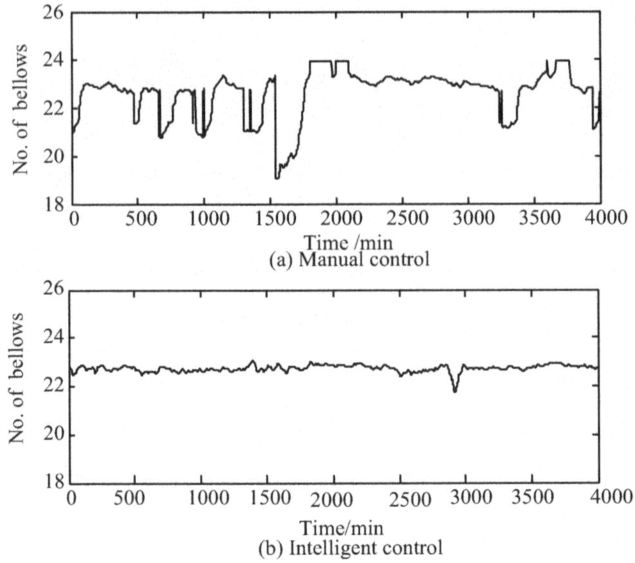

Fig. 4.25 BTP control results for manual control and intelligent control system in 360 m^2 sintering

Table 4.4 Statistic analysis of control results in 360 m^2 Sintering machine

Control method	Utilization factor (t/(h ·m^2))	BTP fluctuation (%)	Screening index (%)
Manual control	1.31	22.16	5.03
Intelligent control system	1.46	10.66	4.84

0.15 $t/(h \cdot m^2)$; the BTP variation is decreased by 11.5%; and the screening index is reduced by 0.19%.

4.5 Conclusion

The sintering ignition process is an important part of steel sintering production. The sintering ignition process directly affects the BTP and ultimately affects the sinter quality and sintering energy consumption. We propose a sintering ignition intelligent control method based on the ignition intensity optimization setting. The main contributions include the following aspects:

(1) A comprehensive modeling and intelligent optimization method for sinter quality and sintering energy consumption is proposed, which realizes the comprehensive optimization of sinter quality and sintering energy consumption.

(2) An improved sintering ignition temperature intelligent control method is proposed to realize the intelligent control of sintering ignition temperature.

We introduce an intelligent control system to control the BTP of the sintering process of an iron and steel plant. The characteristics of the intelligent control are as follows:

(1) A control scheme based on a hierarchical configuration was developed that provides a practical solution to the problem of controlling the BTP of a sintering process.

(2) An intelligent control strategy employing the prediction of the BTP was devised based on the characteristics of the sintering process. The system contains a BTP soft-sensing model, a BTP prediction model, a BTP hybrid fuzzy-predictive controller, a bunker-level expert controller, and a coordinating control algorithm.

(3) The BTP soft-sensing model produces a value for the BTP in a timely manner. The BTP prediction model uses an intelligent integrated modeling method based on grey theory and BPNNs. It also uses the EGTRP. These two models are the foundation of BTP control.

(4) A BTP hybrid fuzzy-predictive controller was established. It consists of a controller, a predictive controller, and a soft hand-off controller. It combines the advantages of fuzzy control and predictive control to find a suitable strand velocity control input.

(5) An intelligent coordinating control algorithm was devised that employs SSP to handle the coupling between BTP control and bunker-level control.

A real-world implementation of the system showed that it sufficiently suppressed the variation in BTP and improved the quantity and quality of sintering agglomerate. This system not only provides effective control but also ensures continuous, safe operation of the sintering process. This method provides a practical way of controlling the sintering process.

In conclusion, the intelligent control of sintered thermal state parameters can improve the sinter production and reduce the production cost, which is of great significance to realize energy saving and emission reduction.

References

1. Kwon WH, Kim YH, Lee SJ, Paek KN (1999) Event-based modeling and control for the Burnthrough Point in sintering processes. IEEE Trans Control Syst Technol 7(1):31–41
2. Terpark J, Dorcak L, Kostial I, Pivka L (2005) Control of burn-through point for agglomeration belt. Metalurgija 44(4):281–284
3. Giri BK, Roy GG (2012) Mathematical modelling of iron ore sintering process using genetic algorithm. Ironmak Steelmak 39(1):59–66
4. Yang W, Choi S, Choi ES, Ri DW, Kim S (2006) Combustion characteristics in an iron ore sintering bed-evaluation of fuel substitution. Combust Flame 145(3):447–463
5. Chen X, Hu J, Wu M, Cao WH (2017) T-S fuzzy logic based modeling and robust control for Burning-Through Point in sintering process. IEEE Trans Ind Electron 64(12):9378–9388
6. Gao XQ, Shang C, Huang DX, Yang F (2017) A novel approach to monitoring and maintenance of industrial PID controllers. Control Eng Pract 64:111–126
7. Zhang F, Wu X, Shen J (2017) Extended state observer based fuzzy model predictive control for ultra-supercritical boiler-turbine unit. Appl Therm Eng 118:90–100

8. Mai A, Commuri S (2016) Intelligent control of a prosthetic ankle joint using gait recognition. Control Eng Pract 49:1–13
9. Li C, Zhang N, Lai X, Zhou J, Xu Y (2017) Design of a fractional-order PID controller for a pumped storage unit using a gravitational search algorithm based on the Cauchy and Gaussian mutation. Inf Sci 396:162–181
10. Kumar A, Kumar V (2017) Evolving an interval type-2 fuzzy PID controller for the redundant robotic manipulator. Expert Syst Appl 73:161–177
11. Yu DL, Chang TK, Yu DW (2005) Fault-tolerant control of multivariable processes using auto-tuning PID controller. IEEE Trans Syst Man Cybern 35(1):32–43
12. Pereira RDO, Veronesi M, Visioli A, Normey-Rico JE, Torrico BC (2017) Implementation and test of a new autotuning method for PID controllers of TITO processes. Control Eng Pract 58:171–185
13. Wang CS, Wu M (2011) Closed-loop identification method of sintering process with part of the inputs controlled. Procedia Eng 15:1910–1915
14. Wu M, Duan P, Cao WH, She JH, Xiang J (2012) An intelligent control system based on prediction of the burn-through point for the sintering process of an iron and steel plant. Expert Syst Appl 39(5):5971–5981
15. Wu M, Xu CH, She JH, Cao WH (2012) Neural-network-based integrated model for predicting burn-through point in lead Czinc sintering process. J Process Control 22(5):925–934
16. Wang CS, Wu M (2013) Hierarchical intelligent control system and its application to the sintering process. IEEE Trans Ind Inform 9(1):190–197
17. Wu M, Wang CS, Cao WH, Lai XZ, Chen X (2012) Design and application of generalized predictive control strategy with closed-loop identification for burn-through point in sintering process. Control Eng Pract 20(10):1065–1074
18. Rodrigues EMG, Godina R, Pouresmaeil E, Ferreira JR, Catalão JPS (2017) Domestic appliances energy optimization with model predictive control. Energy Convers Manag 42:402–413
19. Bouhoune K, Yazid K, Boucherit MS, Chériti A (2017) Hybrid control of the three phase induction machine using artificial neural networks and fuzzy logic. Appl Soft Comput 55:289–301
20. Precup RE, Preitl S, Tar JK, Tomescu ML, Takacs M, Korondi P, Baranyi P (2008) Fuzzy control system performance enhancement by iterative learning control. IEEE Trans Ind Electron 55(9):3461–3475
21. Yang DD, Cai KY (2010) Finite-time quantized guaranteed cost fuzzy control for continuous-time nonlinear systems. Expert Syst Appl 37(10):6963–6967
22. Soyguder S, Alli H (2010) Fuzzy adaptive control for the actuators position control and modeling of an expert system. Expert Syst Appl 37(3):2072–2080
23. Wu M, Xu CH, Du YX (2006) Intelligent optimal control for lead-zinc sintering process state. Trans Nonferrous Metals Soc China 16(4):975–981
24. Chen XL, Fan XH, Wang Y, Long HM, Jiang T, Shi J, Song QY, Yang XD (2009) Control guidance system for sintering burn through point. Ironmak Steelmak 36(3):209–211
25. Hu J, Wu M, Chen X, She JH, Cao WH, Chen LF, Ding HF (2017) Hybrid prediction model of carbon efficiency for sintering process. IFAC-PapersOnLine 50(1):10238–10243
26. Hill DJ, Minsker BS (2010) Anomaly detection in streaming environmental sensor data: a data-driven modeling approach. Environ Model Softw 25(9):1014–1022
27. Chen ZX, Fang HJ, Chang Y (2016) Weighted data-driven fault detection and isolation: a subspace-based approach and algorithms. IEEE Trans Ind Electron 63(5):3290–3298
28. Verhaegen M, Hansson A (2016) N2sid: Nuclear norm subspace identification of innovation models. Automatica 72:57–63
29. Hu YS, Dai Y, Liang Q, Wang Y (2015) Subspace-based continuous-time model identification using nuclear norm minimization. IFAC PapersOnline 48(28):338–343
30. Golub GH, Reinsch C (1970) Singular value decomposition and least squares solution. Numer Math 14:403–420
31. Kim D, Cai J, Braun JE, Ariyura KB (2018) System identification for building thermal systems under the presence of unmeasured disturbances in closed loop operation: theoretical analysis and application. Energy Build 167:359–369

32. Liu X, Cui J (2018) Economic model predictive control of boiler-turbine system. J Process Control 66:59–67
33. Ding ZT (2015) Consensus disturbance rejection with disturbance observers. IEEE Trans Ind Electron 62(9):5829–5837
34. Sariyildiz E, Ohnishi K (2014) A guide to design disturbance observer. J Dyn Syst Meas Control 136(2):729–736
35. Wang L, Su JB (2015) Robust optimal disturbance observer design for the non-minimum phase system. Asian J Control 17(4):1453–1457
36. Chu Z, Sun Y, Wu C, Sepehri N (2018) Active disturbance rejection control applied to automated steering for lane keeping in autonomous vehicles. Control Eng Pract 74:13–21
37. Huang Y, Xue WC (2014) Active disturbance rejection control: methodology and theoretical analysis. ISA Trans 53(4):963–976
38. Sun L, Li DH, Hu KT, Lee KY, Pan FP (2016) On tuning and practical implementation of active disturbance rejection controller: a case study from A regenerative heater in a 1000 MW power plant. Ind Eng Chem Res 55(23):6686–6695
39. Chen X, Jiao W, Wu M, Cao WH (2018) EID-estimation-based periodic disturbance rejection for sintering ignition process with input time delay. Asian J Control 20(5):1–14
40. She JH, Zhang AC, Lai XZ, Wu M (2012) Global stabilization of 2-DOF underactuated mechanical systems C an equivalent-input-disturbance napproach. Nonlinear Dyn 69(1–2):495–509
41. Yu P, Wu M, She JH, Lei Q (2016) Robust repetitive control and disturbance rejection based on two-dimensional model and equivalent-input-disturbance approach. Asian J Control 18(6):2325–2335
42. Wu M, Xu CH (2007) An intelligent integrated predictive method based on gas temperature profile for burn-through point. Acta Autom Sin 33(12):1313–1320 (In Chinese)
43. Wu M, Cao WH, He CY, She JH (2009) Integrated intelligent control of gas mixing-and-pressurization process. IEEE Trans Control Syst Technol 17(1):68–77
44. Wang YJ, Zhang CB, Chen ZH (2015) A method for state-of-charge estimation of Li-ion batteries based on multi-model switching strategy. Energy Procedia 137:427–434
45. Battistelli G, Mosca E, Tesi P (2014) Adaptive memory in multi-model switching control of uncertain plants. Automatica 50(3):874–882
46. Li SY, Hu CF (2007) Two-step interactive satisfactory method for fuzzy multiple objective optimization with preemptive priorities. IEEE Trans Fuzzy Syst 15(3):417–425
47. Khan R, Khan SU (2017) Energy saving through intelligent coordination among daily used fixed and mobile devices. Sustain Comput Inform Syst 14:43–57
48. Bergamascoa M, Lovera M (2011) Continuous-time predictor-based subspace identification using laguerre filters. IET Control Theory Appl 5(7):856–867
49. Johansson R (2010) Continuous-time model identification and state estimation using non-uniformly sampled data. In: Proceedings of the 19th international symposium on mathematical theory of networks and systems, 2010, pp 347–354
50. She J, Xin X, Yamaura T (2006) Analysis and design of control system with equivalent-input-disturbance estimation. In: Proceedings of IEEE international conference on control applications, pp 1463–1469
51. Wu M, Xu BG, Cao WH (2014) Aperiodic disturbance rejection in repetitive-control systems. IEEE Trans Control Syst Technol 22(3):1044–1051
52. Sun L, Li DH, Lee KY (2015) Enhanced decentralized PI control for fluidized bed combustor via advanced disturbance observer. Control Eng Pract 42:128–139
53. Li MH, Wang J (2006) The research for soft measuring technique of sintering burning through point. In: IEEE conference on industrial electronics and applications, pp 1–4
54. Yao TX, Liu SF, Xie NM (2009) On the properties of small sample of GM(1,1) model. Appl Math Model 33(4):1894–1903
55. Zhang G, Wang ZQ, Han XL, Hu SS. Research on satisfactory control theory and its application in fault-tolerant technology. In: The 5th world congress on intelligent control and automation, 2004, pp 1521–1524

Chapter 5
Intelligent Decoupling Control of Gas Collection and Mixing-and-Pressurization Processes

Coke is an important raw material in the metallurgy industry [1]. In the coking process, large amount of by-product gas will be generated from coke-ovens [2], and the process of recycling by-product gas is called gas collecting process. Gas collection involves using gas collectors to collect the gas produced by coke-ovens and sending it where it will be used. Generally, several coke-ovens are in operation at the same time. After the purification of the coal gas, the gas mixing-and-pressurization process of clean coal gas together with clean coal gas is a very important step in the production of steel and nonferrous metals.

In the gas collecting process, the pressures of the gas collectors are important technological parameters of the thermodynamic system of coking and gas collection [3, 4]. The stability of the pressures directly influences the gas quality, the lifetime of the ovens, and the operating conditions [5, 6]. However, many factors like disturbances and couplings may deteriorate the pressure [7].

If the pressure of a gas collector is too low, air is sucked into the coke-oven, causing the coke to burn. This reduces the quality of the coke and shortens the lifetime of the oven. It also raises the temperature of the gas, which imposes a large load on the cooling system, leading to excessive energy consumption. On the other hand, if the pressure is too high, the coke may ignite and a gas leak will occur. This produces environmental pollution and wastes energy. Furthermore, ignition heats the wall of the oven, thereby reducing its stiffness, which shortens the lifetime of the oven [8]. So, it is very important to keep the pressure of a gas collector within the proper range.

Coal gas is an important industrial raw material in the production of steel and nonferrous metals, and gas mixing and pressurization is a key step in the production process [9, 10]. In this step, two kinds of gas delivered separately from a coke-oven and a blast furnace through valved pipelines are first mixed and then pressurized at a pressurization station before use. To meet production requirements, the calorific value and pressure of the mixed gas must each be above a certain value. These parameters are important in determining the cost-effectiveness of the process and the quality and quantity of the product. More specifically, frequent fluctuations in the pressure shorten the lifetime of a reheating furnace, and a low calorific value

M. Wu et al., *Intelligent Optimization and Control of Complex Metallurgical Processes*, Engineering Applications of Computational Methods 3,
https://doi.org/10.1007/978-981-15-1145-5_5

reduces the heating capability and may even lead to the extinguishing of the fire in a reheating furnace, which could cause a serious accident [11]. So, stabilizing the calorific value and pressure of the mixed gas at specified values is extremely important in a production process.

Gas collection and mixing-and-pressurization processes are two important links in the iron and steel production, which not only save energy but also reduce environmental pollution. They directly affect the quality and quantity of the iron and steel. Therefore, it is a vital significance to study on the control of gas collecting and mixing-and-pressurization processes.

The gas collection and mixing-and-pressurization processes are highly complex industrial processes. Their characteristics of multivariable, strong coupling, nonlinearity, time-varying, and hard in modeling are the major difficulties in the control of gas collection and mixing-and-pressurization processes. Many control schemes have been developed for the control of gas collection and mixing-and-pressurization processes.

In the past, most pressure control systems regulated the pressure of just a single coke-oven, but the strong coupling between ovens led to unsatisfactory results. In the last decades, a variety of methods (PID control, feedforward control, fuzzy control, expert control, fuzzy decoupling control) have been used to control the pressures in the gas collection process for multiple coke-ovens.

Expert control is a common, practical approach in control engineering [12, 13], and is used to control many processes [14, 15]. It is able to manage different kinds of information and employ both open- and closed-loop control strategies to handle a time-varying and/or uncertain plant, as well as random disturbances [16]. Since expert control is flexible with regard to control method and knowledge expression, it is a powerful tool for the control of complex processes [17–19]. A knowledge-based expert pressure control strategy for gas collectors was developed in [20] with a focus on the characteristics of expert control. However, several problems with the system were left unsolved: the pressure variation was relatively large, the reliability was not high enough, and disturbances significantly degraded performance.

Fuzzy control employs knowledge representation to transform the experience of experts into a control model that can be understood by a computer, thus yielding effective control [21–23]. Since it is simple, and since the design and implementation are easy to carry out with commercially available programming tools, it has become an important tool for nonlinear control [24]. The control of knowledge-based systems using linguistic variables, which do not have precise values, is very practical because it allows us to use previously developed heuristic procedures and experience in designing control systems [25]. While some experience can be translated into a fuzzy characterization involving conventional numerical probabilities [26], much of it cannot be expressed numerically. So, fuzzy control relies more on information from experts and less on the mathematical model of a physical system than is possible with conventional control approaches. Many researchers have investigated the advantages of fuzzy control over other control methods, e.g., [27–30]. Specifically, [31] and [32] studied the fuzzy control of gas collection in coke-ovens. Even though those systems regulate pressures quickly enough, the disturbance rejection performance is not adequate. References [21] and [33] devised a hybrid intelligent, coordinated strategy for pressure control that integrates PID control and feedforward control with expert control. However, these methods are only suitable for two symmetrically arranged coke-ovens of the same capacity.

Decoupling control is based on modern control theory [34]. Over the past few decades, a considerable number of studies have been devoted to adaptive decoupling control [35, 36], fuzzy-compensated decoupling control [37], neural network decoupling control [38, 39], and their application. Reference [23] devised a decoupling control algorithm based on a correlativity analysis and combined it with a gain-scheduling PI controller. Reference [40] designed an intelligent controller by combining expert rules with fuzzy control for the pressure system of the gas collectors of multiple coke-ovens to achieve parallel decoupling among ovens. Reference [41] described a multivariable decoupling control algorithm based on a PID neural network and a radial-basis-function fuzzy neural network. The hierarchical intelligent coordinated control method in [42] uses a hierarchical structure to handle the complexity of the process and divides the pressure control system for gas collectors into three levels: basic control, decoupling, and coordination. These control methods employ useful ideas and are rather effective in solving the decoupling problem for multiple coke-ovens. However, they are all aimed at symmetric ovens and cannot handle ovens with different capacities and/or pipelines with different layouts.

Nowadays, most pressurization stations are manually controlled. Manual control means not only that changes are made slowly and that large fluctuations in calorific value and pressure occur, but also that workers have to be rather strong physically. To improve working conditions and the quality of products, attempts to automatically control pressurization stations have been made [43–46], mainly by three control methods: a pressure-and-calorific-value double-control-loop regulation method, a feedforward decoupling control method, and an intelligent decoupling control method. Double-control-loop regulation [44] uses a variable-matching strategy, which carries out decoupling control by creating a suitable match between control inputs and outputs. This method provides neither high control precision nor satisfactory disturbance rejection performance. In feedforward decoupling control [45], the pressures of the pipelines from the blast furnace and coke-oven, and the pressure differences between the valves in those pipelines are measured and fed forward to the system as disturbances to perform decoupling control. The drawback of this method is that its effectiveness deteriorates when the fluctuations become relatively large. As for intelligent decoupling control, even though the control strategies that have been reported over the last decade are promising ways of greatly improving control performance, this method is mainly at the stage of simulations [43, 46].

This chapter presents a hierarchical intelligent decoupling control method for the gas collection process of multiple asymmetric coke-ovens of an iron and steel company and an integrated intelligent decoupling control method for the gas mixing-and-pressurization process of a steel mill. First, the characteristics of the two processes are analyzed. Then the control system configurations of the two processes are, respectively, devised. A hierarchical decoupling method is presented that weakens the effects of asymmetry on system interactions and progressively decouples the ovens for the gas collection process. A control method intelligently carries out decoupling control between the calorific value and pressure of the mixed gas for the gas mixing-and-pressurization process. The results of real-world applications of the methods validate the effectiveness of the developed methods. Finally, some concluding remarks are made.

5.1 Process Description and Characteristic Analysis

The configuration of the gas collection and mixing-and-pressurization processes is shown in Fig. 5.1. It mainly includes two parts. One is the gas collection process and the other is the gas mixing-and-pressurization process.

In gas collection process, we concern three coke-ovens in the coking plant of an iron and steel company: two large ones (Coke-oven 1 has 55 apertures, and Coke-oven 2 has 60) and a small one (Coke-oven 3 has 42 apertures). Coke-ovens 1 and 2 are about 60 m apart, but they are about 600 m from Coke-oven 3. Each gas collector has two pressure measurement points and a butterfly valve. The pressure of a gas collector is defined to be the average of the pressures measured at the two measurement points. The gas produced in a coke-oven flows through a gas collector into the main pipe, where it mixes with the gas from the other collectors. The gas is cooled by three primary coolers and sent through a blast blower to the purification process including the desulfurization, the ammonia washing, the final colding, and the benzene washing processes.

After purifying the gas, the gas is sent to the users. A portion of the clean coal gas is sent back to the cove ovens for coking. When the remaining clean coal gas is too much for the gas mixing-and-pressurization station, a part of it is stored in gasometers, just as illustrated in Fig. 5.1. But when the remaining clean coal gas is insufficient for the gas mixing-and-pressurization station, the clean coal gas stored in gasometers could be used to the gas mixing-and-pressurization process. Here, we use the first situation to explain the gas collection and mixing-and-pressurization processes. As shown in Fig. 5.1, the blast furnace gas from blast furnace together with the clean coke-oven gas is used in the gas mixing-and-pressurization process.

In the gas mixing-and-pressurization process (Fig. 5.1), gas from a blast furnace with a low calorific value (typically $3150-4180\ \mathrm{kJ/m^3}$) and gas from a coke-oven with a high calorific value (typically $5900-18300\ \mathrm{kJ/m^3}$) are first mixed and then pressurized in a pressurization station.

5.1.1 Description and Analysis of Gas Collection Process

This section describes the gas collection process and analyzes the factors that affect the pressures of the gas collectors.

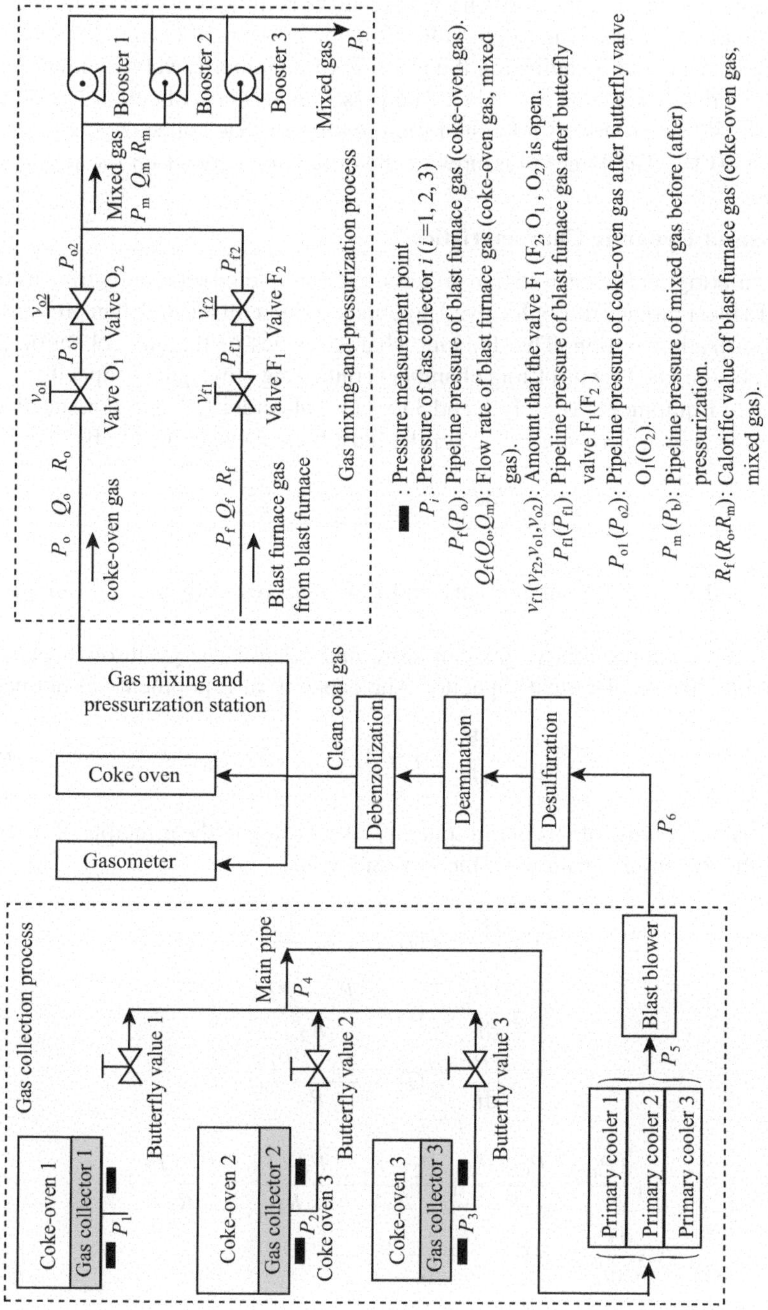

Fig. 5.1 Gas collection and gas mixing-and-pressurization processes

1. Process Description

The pressures of the gas collectors vary with changes in the process variables and also with the demand for gas. During gas collection, the pressure of each of the three gas collectors is stabilized by adjusting the position of its butterfly valve, which means the angle of the disk within the valve. The pressures of Gas collectors 1 and 2, P_1 and P_2, should be around 100 Pa, and the pressure of Gas collector 3, P_3, should be around 30 Pa. The target pressures of the collectors depend on the production schedule.

2. Analysis of Dynamic Characteristics

In the gas mixing-and-pressurization process, gases delivered separately from a coke-oven and a blast furnace through valved pipelines are first mixed and then pressurized at a pressurization station. The dynamic characteristics of the gas collection process can be simulated by throttling chambers containing atmosphere capacitors. The resistance of Equipment i ($i = 1, 2$, and 3 for Gas collectors 1, 2, and 3, respectively; $i = 4$ for the primary coolers; $i = 5$ for the blast blower) is defined to be

$$R_i = \frac{\mathrm{d}P_i}{\mathrm{d}G_i}, \quad i = 1, 2, \ldots, 5, \tag{5.1}$$

where P_i and G_i are the gas pressure and flow rate, respectively, of Equipment i (Fig. 5.1).

Due to its compressibility, gas can store and release energy through pressure changes, just like an electrical capacitor. The capacity of Equipment i is defined to be

$$C_i = \frac{\mathrm{d}V_i}{\mathrm{d}P_i}, \quad i = 1, 2, \ldots, 5, \tag{5.2}$$

where V_i is the volume of gas in Equipment i. According to the principle of material balance, the dynamic equations of the system are

$$C_1 \frac{\mathrm{d}P_1}{\mathrm{d}t} = G_1 - \frac{P_1 - P_4}{R_1}, \tag{5.3}$$

$$C_2 \frac{\mathrm{d}P_2}{\mathrm{d}t} = G_2 - \frac{P_2 - P_4}{R_2}, \tag{5.4}$$

$$C_3 \frac{\mathrm{d}P_3}{\mathrm{d}t} = G_3 - \frac{P_3 - P_4}{R_3}, \tag{5.5}$$

$$C_4 \frac{\mathrm{d}P_4}{\mathrm{d}t} = \frac{P_1 - P_4}{R_1} + \frac{P_2 - P_4}{R_2} + \frac{P_3 - P_4}{R_3} - \frac{P_4 - P_5}{R_4}, \tag{5.6}$$

$$C_5 \frac{\mathrm{d}P_5}{\mathrm{d}t} = -G - \frac{P_5 - P_6}{R_5}, \tag{5.7}$$

where G_i is the volume of gas in Coke-oven i ($i = 1, 2, 3$); G is the volume of gas to be used; and P_6 is the pressure of the outlet gas.

Clearly, this is an MIMO system, and the pressures of the gas collectors, P_i $(i = 1, 2, 3)$ and the suction power before the primary coolers, P_4, are coupled to each other. The parameters R_i and C_i $(i = 1, 2, \ldots, 5)$ are closely related to the positions of the butterfly valves, the lengths and diameters of the pipelines, the rotational speed of the blast blower, the flow rate of the circulating ammonia water, etc. These parameters are time-varying and uncertain. So, it is difficult to construct a precise mathematical model of the process. And a main source of nonlinearity is the characteristics of the butterfly valves. The following qualitative results were obtained from the analysis, equations, and production technology explained above.

(1) Influence of coking operations on pressures of the three gas collectors: Coaling suddenly increases the gas pressure of a coke-oven, and a sudden change in gas flow rate directly affects the pressure of a gas collector.

(2) Influence of suction power before the primary coolers, P_4, on the pressures of the gas collectors: Eqs. (5.6) and (5.7) show that changes in the volume of gas to be used, G, the outlet pressure, P_6, and/or the rotational speed of the blast blower change the suction power before the blast blower, P_5, and that before the primary coolers, P_4. A change in P_4 directly influences P_1, P_2, and P_3; since the pipeline from Gas collector 3 to the primary coolers is much longer than those from Gas collectors 1 and 2 to the coolers, the influence is weaker on P_3 than on P_1 or P_2.

(3) Coupling among the three coke-ovens: From Eqs. (5.3)–(5.5), it is clear that, when G_1 increases, Butterfly valve 1 should open more to stabilize P_1. However, this causes an increase in P_4, and thus an increase in P_2 and P_3. So, the operation of Butterfly valve 1 changes not only the pressure of Gas collector 1, but also the pressures of Gas collectors 2 and 3. Since Gas collectors 1 and 2 are very close to each other and the pipelines from them to the primary coolers are very short, the coupling between Gas collectors 1 and 2 is strong, and the couplings between Gas collector 3 and Gas collectors 1 and 2 are weak.

(4) Nonlinearity of actuators: The characteristics of the butterfly valves are highly nonlinear. They have a dead zone and exhibit output saturation. As a result, their characteristics can only be described by a piecewise linear function in the operating range.

5.1.2 Description and Analysis of Gas Mixing-and-Pressurization Process

The control objective for the gas mixing and pressurization process is to stabilize the calorific value and pressure of the mixed gas after pressurization at the values 12000 $\mathrm{kJ/m^3}$ and 13 kPa, respectively. The pipeline from the blast furnace and that from the coke-oven each contain two butterfly valves. Adjusting these four valves regulates the flow rate ratio of the gases, thus enabling tracking control of the calorific value of the mixed gas before pressurization. Then, the mixed gas is pressurized by three boosters, and the pressure is stabilized at the specified value by adjusting the converters of the boosters.

The law of conservation of energy gives us the following equation regarding calorific value:

$$R_f Q_f + R_o Q_o = R_m Q_m. \tag{5.8}$$

where Q_m is determined by the demands of the process. When one of the input flow rates (Q_f or Q_o) changes, the other (Q_o or Q_f) has to be adjusted to ensure that R_m has a suitable value.

The relationship between the flow rate of the gas passing through a butterfly valve, Q (m^3/s), and the pressure difference across the valve, ΔP (Pa), is

$$Q = \frac{aA}{\sqrt{\xi}}\sqrt{\frac{\Delta P}{\rho}}, \tag{5.9}$$

where a is a constant, A (m^2) is the effective cross-sectional area of the valve, ξ is the coefficient of friction, and ρ (kg/m^3) is the density of the gas. From the above equation it follows that, when $\Delta P/\rho$ is constant, a decrease (or increase) in the coefficient of friction causes an increase (or decrease) in the flow rate of the gas. This allows us to regulate the flow rate of a gas by changing the coefficient of friction, which involves adjusting the core traveling schedule of a butterfly valve [47].

5.2 Intelligent Decoupling Control of Gas Collection Process

The gas collection process is highly nonlinear and time-varying, and the pressures of the gas collectors are strongly coupled to each other. Many factors affect the pressures: coaling, the discharge of coke, variations in coking time, changes in heating methods, the suction power before the blast blower, etc. So, it is almost impossible to establish a precise mathematical model of the process, and it is difficult to control the process by conventional methods.

This section presents a hierarchical intelligent decoupling control method for the gas collection process of multiple asymmetric coke-ovens of an iron and steel company. Based on an analysis of the couplings between the ovens, a hierarchical decoupling method is presented that weakens the effects of asymmetry on system interactions and progressively decouples the ovens. We first present the structure of an intelligent grouping–decoupling control system and then discuss decoupling control rules for the different interactions between and within groups. Furthermore, to suppress pressure fluctuations and to eliminate the influence of intra- and inter-group coupling, expert control, fuzzy control, and feedforward control are combined with decoupling control in the design of multilayer decoupling controllers. Finally, expert valve control and multirate sampling strategies are described that suit the characteristics of the actuators so as to guarantee the control performance.

Throughout this section, for a variable x, $\delta x(k)$ is the change in x between two sampling points, i.e., $\delta x(k) = x(k) - x(k-1)$, and we omit k and simply use the notation δx when no confusion may arise. X means a fuzzy linguistic variable of x, and Δu is a correction to the control input u produced by a controller.

5.2.1 *Intelligent Decoupling Control Based on Coupling Degree Analysis*

The analytical results in the previous section show that the gas collection process is a strongly intercoupled, time-varying, highly nonlinear, and dramatically changing MIMO system. There are also many disturbances in the system. Decoupling control is an effective way to achieve the desired control performance. However, conventional decoupling control methods require a precise mathematical model of the plant. This makes it difficult to use them directly for this process. On the other hand, human operators have accumulated a wealth of experience that can be exploited to achieve effective decoupling control. With this in mind, we designed an intelligent decoupling control system for the gas collection process of multiple asymmetric coke-ovens. Our method combines advanced control techniques, such as fuzzy control, expert control, fuzzy decoupling, feedforward control, and multirate sampling.

There is a serious coupling relationship among the three coke-ovens. To eliminate the coupling effects between the three coke-ovens, and also to simplify the control algorithm to facilitate the system realization, we first analyze the coupling degree among the three coke-ovens. Then we apply a layered structure in Fig. 5.2 to control the gas collection process of the asymmetric coke-ovens. The control structure is divided into three layers: the basic layer, decoupling layer, and coordination layer. The basic layer is for a single coke, where a controller is designed for a coke. The decoupling layer is the core of the control system. The module for analysis of the coupling degree is introduced, and an analysis for a coupling degree of the position of butterfly valves and the pressures of gas collectors is conducted. Then a grouped decision is made based on the coupling degree. Some advanced control techniques, like fuzzy control, expert control, and fuzzy decoupling, are combined to design intragroup fuzzy decoupling controller and intergroup fuzzy decoupling controller. The coordination layer is responsible for the coordination between the inputs and outputs of the system and to deal with the external disturbance acting on the system and to optimize the set point of the pressure. Then the hierarchical decoupling control of the gas collection process is implemented. The layers are almost independent on each other. This helps to implement the modularization of the algorithm, which is easy to combine with the hard and software platform and easy to maintain, use, and upgrade, and helps to improve the reliability of the whole physical system.

In order to realize the decoupling control of the gas collection process, it is necessary to analyze the coupling relationship among the variables in the process of collecting the gas, so as to find the degree of coupling of each variable in the gas

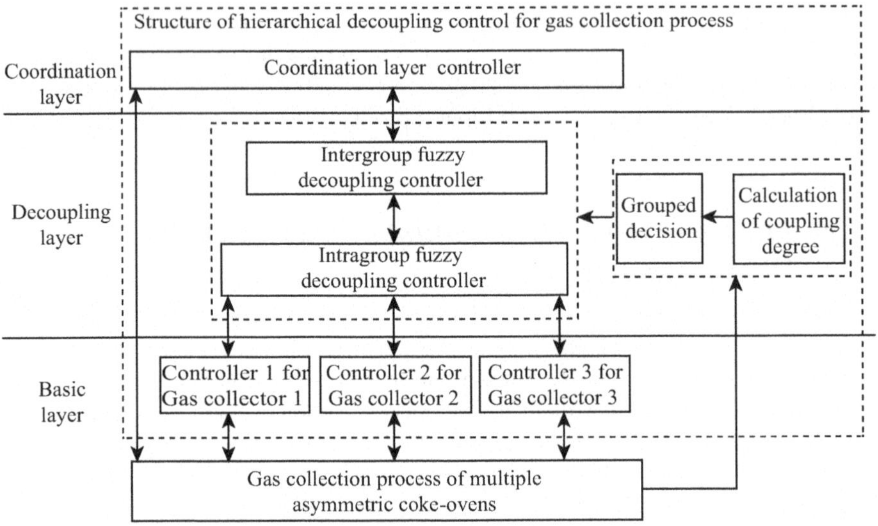

Fig. 5.2 Hierarchical decoupling control structure for gas collection process

collection process, and then put forward the corresponding decoupling scheme to achieve good decoupling control. There are a variety of methods for evaluating the degree of coupling of multivariable systems. The most widely used is the static and dynamic relative gain matrix theory and the block relative gain matrix theory.

Bristol [48] defined a measure-called relative gain to describe the interaction between the inputs and outputs of an MIMO system. For a system with N inputs and N outputs, the relative gain matrix, RGM, is given by the following formula:

$$RGM = \begin{bmatrix} \lambda_{11} & \lambda_{12} & \dots & \lambda_{1N} \\ \lambda_{21} & \lambda_{22} & \dots & \lambda_{2N} \\ \vdots & \vdots & & \vdots \\ \lambda_{N1} & \lambda_{N2} & \dots & \lambda_{NN} \end{bmatrix}, \tag{5.10}$$

$$\lambda_{ij} = \frac{\Phi_{ij}}{P_{ij}}, \tag{5.11}$$

$$\Phi_{ij} = \left. \frac{\partial c_i}{\partial m_j} \right|_{m_r=\text{const}(r \neq j)}, \tag{5.12}$$

$$P_{ij} = \left. \frac{\partial c_i}{\partial m_j} \right|_{c_r=\text{const}(r \neq j)}, \tag{5.13}$$

where $i = 1, 2, \dots, N$; Φ_{ij} is defined as the first magnification factor, denoting the gain of the channel from m_j to c_i when $m_\text{r} = [m_1 \ \cdots \ m_{j-1} \ m_{j+1} \ \cdots \ m_N]^\text{T}$ is invariable, that is to say, the other loops all being open, and P_{ij} is defined as the second magnification factor, denoting the gain of the channel from m_j to c_i when $c_r = [c_1 \cdots c_{j-1}\ c_{j+1} \cdots c_N]^T$ is invariable, namely, the other loops all being open.

Relative gain λ_{ij} reflects the coupling relationship:

(1) When λ_{ij} is close to 1, for example, $0.9 < \lambda_{ij} < 1.1$, which indicates that the association of other channels to the channel is small, it is not necessary to take a decoupling strategy; when $\lambda_{ij} = 1$, it indicates that the control loop from m_j to c_i has nothing to do with the other channels, so the control loop is not coupled, because the open-loop gain of the channel from m_j to c_i is not affected by other loops no matter they are closed or open loops.

(2) When $\lambda_{ij} < 0$ or is close to 0, which indicates that the regulator of this channel cannot obtain a good control effect, that is to say, the regulator should be reselected; when $\lambda_{ij} = 0$, which indicates that c_i is not affected by m_j, namely, the loop is super-coupled. So we cannot use m_j to control c_i.

(3) When λ_{ij} is between 0.3 and 0.7 or greater than 1.5, which indicates that there is a serious coupling, it is necessary to decouple the system; when $\lambda_{ij} \gg 1$, which is due to that the denominator of λ_{ij} tends to zero, indicating that the presence of other loops makes c_i free from m_j, so we cannot use m_j to control c_i.

The above calculation of the coupling degree that is based on the definition of probability statistics needs not to know the transfer function of the system, which just needs to know the input and output of the system and does not care about the internal structure of the system. The definition of coupling degree is only a basic problem of multivariable decoupling control. The key to the establishment of multivariable decoupling control system based on the appropriate coupling degree analysis is variable pairing and hierarchical decoupling.

Considering the multivariable and strong coupling characteristics between the pressure of the gas collectors, the coupling degree analysis module is designed based on the butterfly valve positions, and the historical data of the pressure, and probabilistic statistical methods. Then, the hierarchical decision-making system for the coke-oven gas collectors is built. Eventually, layer by layer, the grouped decoupling is achieved.

Since the iron and steel enterprises have three coke-ovens, input variables of the gas collection system are the set points of positions of the butterfly valves and output variables y_1, y_2, y_3 are detected pressure values corresponding to each gas collector.

Due to the serious coupling relation in the gas collection system, the change of the position of the butterfly valve of a coke-oven will affect pressures of all the other gas collectors, and the change of the pressure of a gas collector is a comprehensive result by all the other butterfly valves of the gas collectors. When the positions, x_1, x_2, x_3, act on the pressure of a gas collector y_i, the impacts on the pressures of the gas collectors y_i $(i = 1, 2, 3)$ are

$$\begin{aligned} p(y_1) &= a_{11}p(x_1) + a_{12}p(x_2) + a_{13}p(x_3) + c, \\ p(y_2) &= a_{21}p(x_1) + a_{22}p(x_2) + a_{23}p(x_3) + c, \\ p(y_3) &= a_{31}p(x_1) + a_{32}p(x_2) + a_{33}p(x_3) + c. \end{aligned} \tag{5.14}$$

The equation is based on the probability of statistical algorithm. a_{ij} is the impact factor of the position of the j ($j = 1, 2, 3$) butterfly valve to the pressure of i ($i = 1, 2, 3$) gas collector; c stands for other effect elements (such as the offset of the positions of each butterfly valve, etc.). Similarly, when the position x_j of a butterfly valve acts on all the pressures of the gas collectors, y_1, y_2, and y_3, the impact of the position of a butterfly valve x_j on all the pressures of gas collectors, y_1, y_2, and y_3, is expressed as

$$\begin{aligned} p(x_1) &= b_{11}p(y_1) + b_{21}p(y_2) + b_{31}p(y_3) + d, \\ p(x_2) &= b_{12}p(y_1) + b_{22}p(y_2) + b_{32}p(y_3) + d, \\ p(x_3) &= b_{13}p(y_1) + b_{23}p(y_2) + b_{33}p(y_3) + d, \end{aligned} \tag{5.15}$$

where b_{ij} is the contribution of the position of the j butterfly valve to the pressure of the i gas collector; d stands for other impact elements.

Define the coupling degree between the position of the i butterfly valve and the pressure of the j gas collector to be $\lambda_{ij} = a_{ij}b_{ij}$. Take $n = 4$ groups of the positions of the butterfly valves at any period of time as the inputs, and record the corresponding four groups of pressure values of the gas collectors. Using the technique that is used to solve the ternary linear regression problem in probability method, set the four sample observations to be $(x_{k1}, x_{k2}, x_{k3}, y_{ki})$, where $k = 1, 2, 3, 4$; x_{kj} represents the kth input value of x_j; y_{ki} represents the kth output value of y_i $(i, j = 1, 2, 3)$. Introduce the following notation:

$$l_{ij} = \sum_{k=1}^{4} (x_{ki} - \bar{x}_i)(x_{kj} - \bar{x}_j), \tag{5.16}$$

$$f_{ij} = \sum_{k=1}^{4} (x_{ki} - \bar{x}_i)(y_{kj} - \bar{y}_j), \tag{5.17}$$

where $\bar{x}_i$ is the mean of four input values for the i input variable; $\bar{x}_j$ is the mean of four input values for the j input variable; and $\bar{y}_j$ is the mean of four output values for the j output variable. Calculating $\bar{x}_1$, $\bar{x}_2$, $\bar{x}_3$ and $\bar{y}_1$, $\bar{y}_2$, $\bar{y}_3$ gets the following equations:

$$\begin{cases} l_{11}a_{11} + l_{12}a_{12} + l_{13}a_{13} = f_{11}, \\ l_{21}a_{21} + l_{22}a_{22} + l_{23}a_{23} = f_{22}, \\ l_{31}a_{31} + l_{32}a_{32} + l_{33}a_{33} = f_{33}. \end{cases} \tag{5.18}$$

Denote the coefficient matrix as $L = (l_{ij})$, $F = (f_{ij})$, and $A^{\mathrm{T}} = L^{-1}F$, and obtain

$$A = \begin{bmatrix} a_{11} & a_{12} & a_{13} \\ a_{21} & a_{22} & a_{23} \\ a_{31} & a_{32} & a_{33} \end{bmatrix}.$$

In the same way, introduce the following notation:

$$h_{ij} = \sum_{k=1}^{4} (y_{ki} - \bar{y}_i)(y_{kj} - \bar{y}_j), \tag{5.19}$$

$$g_{ij} = \sum_{k=1}^{4} (x_{ki} - \bar{x}_j)(y_{kj} - \bar{x}_j), \tag{5.20}$$

where $\bar{y}_i$ is the mean of four output values for the i output variable. And we obtain $B^{\mathrm{T}} = H^{-1}G$, where

$$B = \begin{bmatrix} b_{11} \ b_{12} \ b_{13} \\ b_{21} \ b_{22} \ b_{23} \\ b_{31} \ b_{32} \ b_{33} \end{bmatrix}.$$

As a result, we obtained the coupling degrees λ_{ij}. The coupling degree between the positions of Butterfly valves 1 and 2 and the pressures of their own gas collector was about 0.6. The coupling degree between the position of Butterfly valve 3 and the pressure of Gas collector 3 is greater than 0.9. The coupling degree between the position of Butterfly valve 1 and the pressure of Gas collector 2 was about 0.4, and the coupling degree between the position of Butterfly valve 2 and the pressure of Gas collector 1 was also about 0.4. The coupling degrees between the position of Butterfly valve 3 and the pressures of Gas collections 1 and 2 both are less than 0.1, and the coupling degrees between the position of Butterfly valves 1 and 2 and the pressures of Gas collection 3 were less than 0.1.

As we can see above, the coupling degrees between the positions of Butterfly valves 1 and 2 and the pressures of Gas collections 1 and 2 are strong, while the coupling degree between the position and the pressure of Butterfly valve 3 and Coke-ovens 1 and 2 are weak. Moreover, according to the analysis of the distribution and process characteristics of the coke-ovens in the steel enterprise, we find that Coke-ovens 1 and 2 layout symmetrically and the lengths and sizes of the pipelines of the two coke-ovens are also roughly the same. While the Coke-oven 3 is relatively far away from Coke-ovens 1 and 2, the lengths and sizes of their pipelines are far away the same. So, first divide Gas collections 1 and 2 into a group, and conduct an intragroup decoupling control. Then together with the Gas collection 3, carry out intergroup decoupling control. By now, the parallel decoupling control has been achieved.

5.2.2 *Configuration of Intelligent Decoupling Control System*

Based on the locations of and interrelationships among the coke-ovens, the collectors are divided into two groups: Gas collectors 1 and 2 constitute Group 1, and Gas collector 3 by itself constitutes Group 2. An intragroup decoupling controller is designed to rapidly decouple Gas collectors 1 and 2. Then, an intergroup decoupling controller is designed to decouple the two groups.

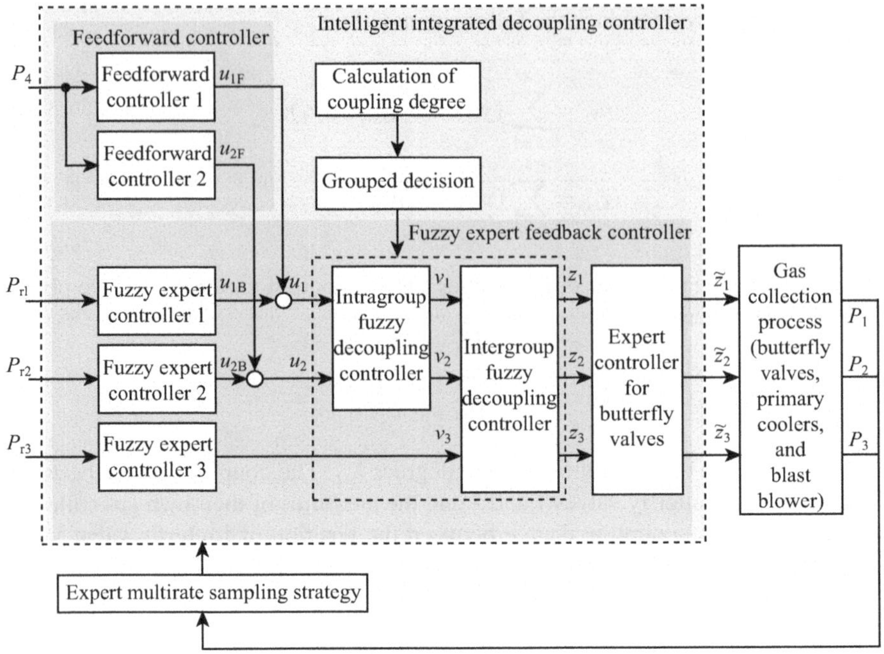

P_{ri}: referfence pressure of Gas collector i ($i = 1, 2, 3$)

Fig. 5.3 Intelligent decoupling control system for gas collection process

The configuration of the control system is shown in Fig. 5.3. Before decoupling, Fuzzy expert controllers 1–3 (one for each gas collector) stabilize the pressure of the collectors and cooperate with Feedforward controllers 1 and 2 to suppress the influence of the suction power before the primary coolers on Gas collectors 1 and 2. Decoupling is carried out by intragroup and intergroup fuzzy decoupling controllers. The fuzzy expert controllers monitor the pressures of the gas collectors and select suitable control methods. Pressure changes are suppressed by adjusting the valve positions by the amounts $\Delta\tilde{z}_1$, $\Delta\tilde{z}_2$, and $\Delta\tilde{z}_3$. The control inputs of the intragroup fuzzy decoupling controller, intergroup decoupling controller, and butterfly expert controller are calculated one after another. By gradually adjusting the positions of the butterfly valves, the pressure interactions among the gas collectors are decoupled. The tracking error caused by the nonlinearities of the actuators is also taken into account. An expert multirate sampling strategy is employed to quickly suppress pressure fluctuations in the gas collectors.

5.2.3 Decoupling Control Strategies

The intelligent integrated decoupling controller has two degrees of freedom and consists of five main parts:

(1) To suppress pressure fluctuations, three fuzzy expert controllers, one for each gas collector, were designed based on the pressure characteristics of the collectors.

(2) Feedforward controllers were also designed for Gas collectors 1 and 2 to compensate for the large influence of the suction power before the primary coolers.

(3) Intra- and intergroup decoupling controllers were designed to hierarchically decouple the pressures of the gas collectors.

(4) An expert valve controller was designed to handle the nonlinear characteristics of the butterfly valves so as to provide good control performance.

(5) An expert multirate sampling strategy was employed: A long sampling period is used for pressure stabilization, and a short sampling period is used for an expert high-pass filter that monitors pressure so that large, rapid pressure fluctuations can be captured and suppressed.

5.2.4 Design of Intelligent Decoupling Control System

To suppress pressure fluctuations and to eliminate the influence of intra- and intergroup coupling, expert control, fuzzy control, and feedforward control are combined with decoupling control in the design of multilayer decoupling controllers. Furthermore, expert valve control and multirate sampling strategies are described that suit the characteristics of the actuators so as to guarantee the control performance.

1. Fuzzy Expert Controllers for Gas Collectors

The pressure of a gas collector is affected by the gas flow rate, the suction power before the blast blower, the resistances of the primary coolers and the blast blower, the pressure of the outlet gas, etc. Since these factors are dynamic and uncertain, it is difficult to establish a precise mathematical model to describe them. For this reason, conventional PID control does not yield satisfactory results. So, this study employed a fuzzy expert control method, and feedforward control is also used for Gas collectors 1 and 2. This subsection explains fuzzy and expert control, taking Gas collector 1 as an example.

(1) Expert Controller

Fuzzy expert controller 1 for Gas collector 1 contains three fuzzy controllers. The following expert rules are used to choose a suitable controller for Gas collector 1, based on the pressure characteristics of the collector:

R_{e1} : **IF** $|e_1| < 20$ Pa,
THEN employ Fuzzy controller 1.

R_{e2} : **IF** 20 Pa $\leq |e_1| \leq$ 40 Pa,
THEN employ Fuzzy controller 2.

R_{e3} : **IF** $|e_1| > 40$ Pa,
THEN employ Fuzzy controller 3.

In the above rules, e_1 is the tracking error between the reference pressure, P_{r1}, and the actual pressure, P_1, of Gas collector 1. The design of Fuzzy controller i ($i = 1, 2, 3$) is explained below.

(1) Since small fluctuations in P_1 (magnitude < 10 Pa) are allowable and since a butterfly valve has a dead zone, the change in the control input produced by Fuzzy controller 1 is set to zero when the magnitude of the pressure tracking error is less than 20 Pa.

(2) Since the resistances of the gas collectors, R_i ($i = 1, 2, \ldots, 5$), are relatively large, the influence of other pressures on P_1 appears relatively slowly. For this reason, conventional control methods, which employ time-invariant linear controllers, do not produce good control results. So, Fuzzy controller 2 utilizes a fine-adjustment fuzzy control method.

(3) Coaling and other operations that produce sudden changes in the suction power before the primary coolers and/or in the outlet pressure cause a large variation in P_1. Fuzzy controller 3 employs a fast-adjustment fuzzy control method to quickly suppress this kind of variation.

(2) Fuzzy Controller

The inputs of Fuzzy controller i ($i = 1, 2, 3$) in Fuzzy expert controller 1 are e_1 and its rate of change, and the output is the change in the position of Butterfly valve 1. As an example, in Fuzzy controller 2, the numbers of linguistic values that E_1, its rate of change, $\dot{E}_1$, and the output, ΔU_{B1}, can take on were set to 7, 5, and 7, respectively. Note that the number of linguistic values is strongly related to the technological requirements. Generally speaking, strict technological requirements require sophisticated adjustment and thus a large number of linguistic values. The commonly used triangular function was selected as the membership function of $\dot{E}_1$ and of ΔU_{B1}, and a trapezoidal function was selected as the membership function for the linguistic values of E_1 because it is insensitive to linguistic values.

When the tracking error, e_1, of the pressure, P_1, is small (20 Pa $\leq |e_1| \leq$ 40 Pa), fine-adjustment fuzzy control rules based on the experience of experts are employed to adjust the tracking error (Table 5.1). For example, if E_1 is PB and $\dot{E}_1$ is ZO, then the output, ΔU_{B1}, is NM. This means that, when the pressure is much less than the reference and is almost unchanging, then the butterfly valves should be set to the middle position to reduce the gas flow and increase the pressure. On the other hand, when e_1 is large ($|e_1| > 40$ Pa), the rules for fast-adjustment fuzzy control (Table 5.2) are employed. These rules quickly shift the pressure of Gas collector 1 into the range $|e_1| \leq 40$ Pa.

2. Feedforward Controllers

Since a change in the suction power before the primary coolers, P_4, causes large variations in P_1 and P_2, information on P_4 is fed forward to suppress this influence. Feedforward controller 1 employs the following control law:

Table 5.1 Fuzzy rules of fine-adjustment fuzzy control method used in Fuzzy controller 2 of Fuzzy expert controller 1 (inputs: E_1 and $\dot{E}_1$, output: ΔU_{B1})

E_1	$\dot{E}_1$				
	NB	NM	ZO	PM	PB
NB	PB	PB	PM	ZO	ZO
NM	PB	PM	PS	ZO	ZO
NS	PM	PS	ZO	ZO	ZO
ZO	PS	ZO	ZO	ZO	NS
PS	ZO	ZO	ZO	NS	NM
PM	ZO	ZO	NS	NM	NB
PB	ZO	ZO	NM	NB	NB

Table 5.2 Fuzzy rules for fast-adjustment fuzzy control method used in Fuzzy control 3 of Fuzzy expert controller 1 (inputs: E_1 and $\dot{E}_1$, output: ΔU_{B1})

E_1	$\dot{E}_1$				
	NB	NM	ZO	PM	PB
NB	PB	PM	PM	PS	ZO
NM	PB	PM	PS	PS	ZO
ZO	PM	PS	ZO	NS	NM
PM	ZO	NS	NM	NM	NB
PB	ZO	NS	NS	NM	NB

$$\Delta u_{1\mathrm{F}} = K_{\mathrm{F}}(k)\dot{P}_4(k-1), \tag{5.21}$$

$$K_{\mathrm{F}}(k) = \begin{cases} 0, & |\dot{P}_4(k-1)| < \dot{P}_{4\mathrm{m}}, \\ K, & |\dot{P}_4(k-1)| \geqslant \dot{P}_{4\mathrm{m}}, \end{cases} \tag{5.22}$$

where $\dot{P}_4(k-1)$ is the rate of change in P_4 at the sampling point $k-1$, and $\dot{P}_{4\mathrm{m}}$ is the threshold for the rate of change in P_4 (Feedforward controller 1 is employed when P_4 is above the threshold).

Thus, the integrated output of Fuzzy expert controller 1 and Feedforward controller 1 is given by

$$\Delta u_1(k) = \Delta u_{1\mathrm{B}}(k) + \Delta u_{1\mathrm{F}}(k), \tag{5.23}$$

where $\Delta u_{1\mathrm{B}}(k)$ and $\Delta u_{1\mathrm{F}}(k)$ are the corrections to the control input u_1 produced by Fuzzy expert controller 1 and Feedforward controller 1, respectively.

3. Intelligent Decoupling Controllers

Decoupling is carried out based on the control inputs, u_1, u_2, and u_3. As shown in the structure of the control system (Fig. 5.3), the three gas collectors are divided into two groups, based on the analysis of the relationships among them: Gas collectors 1 and 2 constitute Group 1, and Gas collector 3 constitutes Group 2. Gas collectors 1 and 2 are strongly coupled to each other, and this coupling is the key to decoupling control of the system. To exploit these relationships, intelligent decoupling is carried out in two steps: First, an intragroup decoupling controller removes the interaction between Gas collectors 1 and 2. Then, an intergroup decoupling controller removes the interaction between the two groups. This is explained below.

(1) Intragroup Fuzzy Decoupling Controller

Due to the proximity of Gas collectors 1 and 2, the coupling between them is the strongest in the system. A change in the pressure of one of them is immediately reflected in the pressure of the other. The first step in decoupling control is to design an intragroup fuzzy decoupling controller to handle the interaction between these two collectors.

The inputs of the controller are $u_1(k)$ and $u_2(k)$, and the outputs are the corrections $\Delta v_{1d}(k)$ and $\Delta v_{2d}(k)$. Based on an analysis of a large number of data from actual runs and a consideration of control precision, the numbers of linguistic values of U_1, U_2, ΔV_{1d}, and ΔV_{2d} were all set to 5 ({NB, NS, ZO, PS, PB}).

Fuzzy decoupling has been shown to be very effective in MIMO systems. The main reason is that the design of fuzzy decoupling rules is very simple and does not require difficult decoupling theory. In this study, the decoupling rules were established based on an analysis of the coupling relationships and the experience of experts. For example, if Gas collector 1 has a higher pressure than Gas collector 2 does, then gas flows from 1 to 2, thereby reducing the pressure of 1 and increasing the pressure of 2. This situation should be dealt with by subtracting a suitable positive amount from $u_1(k)$ and adding it to $u_2(k)$. The following fuzzy decoupling rule describes this situation in the IF-THEN form:

IF $(U_1 = \text{PB})$ AND $(U_2 = \text{NB})$,
THEN $(\Delta V_{1d} = \text{NB})$ AND $(\Delta V_{2d} = \text{PB})$.

Table 5.3 lists all the rules for fuzzy decoupling control. Note that the integrated outputs of the feedforward controllers and the fuzzy expert controllers are the changes in the butterfly valve positions, $\Delta u_1(k)$ and $\Delta u_2(k)$. In addition, the outputs of the intragroup fuzzy decoupling controller, $\Delta v_{1d}(k)$ and $\Delta v_{2d}(k)$, are the first corrections to the butterfly valve positions. The total changes to those positions are

$$\Delta v_1(k) = \Delta u_1(k) + \Delta v_{1d}(k), \tag{5.24}$$

$$\Delta v_2(k) = \Delta u_2(k) + \Delta v_{2d}(k). \tag{5.25}$$

(2) Intergroup Fuzzy Decoupling Controller

Since the pipeline between the two groups is very long, the coupling between them is much weaker than that within Group 1. This makes it possible to handle the intergroup decoupling independently on the basis of the results of the intragroup decoupling. Intergroup decoupling involves calculating a second set of corrections to the butterfly valve positions.

If we let the inputs for intergroup decoupling control be

$$\Delta v_{12}(k) = \frac{\Delta v_{1d}(k) + \Delta v_{2d}(k)}{2}, \tag{5.26}$$

$$\Delta v_3(k) = \Delta u_3(k), \tag{5.27}$$

Table 5.3 Rules for intragroup fuzzy decoupling control (inputs: U_1 and U_2; outputs: ΔV_{1d} and ΔV_{2d})

U_1	U_2				
	NB	NS	ZO	PS	PB
NB	$\Delta V_{1d} = $ ZO	$\Delta V_{1d} = $ PS	$\Delta V_{1d} = $ PS	$\Delta V_{1d} = $ PS	$\Delta V_{1d} = $ PB
	$\Delta V_{2d} = $ ZO	$\Delta V_{2d} = $ NS	$\Delta V_{2d} = $ ZO	$\Delta V_{2d} = $ NS	$\Delta V_{2d} = $ NB
NS	$\Delta V_{1d} = $ NS	$\Delta V_{1d} = $ ZO	$\Delta V_{1d} = $ ZO	$\Delta V_{1d} = $ PS	$\Delta V_{1d} = $ PS
	$\Delta V_{2d} = $ PS	$\Delta V_{2d} = $ ZO	$\Delta V_{2d} = $ ZO	$\Delta V_{2d} = $ NS	$\Delta V_{2d} = $ NS
ZO	$\Delta V_{1d} = $ ZO	$\Delta V_{1d} = $ ZO	$\Delta V_{1d} = $ ZO	$\Delta V_{1d} = $ ZO	$\Delta V_{1d} = $ ZO
	$\Delta V_{2d} = $ PS	$\Delta V_{2d} = $ ZO	$\Delta V_{2d} = $ ZO	$\Delta V_{2d} = $ ZO	$\Delta V_{2d} = $ NS
PS	$\Delta V_{1d} = $ NS	$\Delta V_{1d} = $ NS	$\Delta V_{1d} = $ ZO	$\Delta V_{1d} = $ ZO	$\Delta V_{1d} = $ PS
	$\Delta V_{2d} = $ PS	$\Delta V_{2d} = $ PS	$\Delta V_{2d} = $ ZO	$\Delta V_{2d} = $ ZO	$\Delta V_{2d} = $ NS
PB	$\Delta V_{1d} = $ NB	$\Delta V_{1d} = $ NS	$\Delta V_{1d} = $ NS	$\Delta V_{1d} = $ NS	$\Delta V_{1d} = $ ZO
	$\Delta V_{2d} = $ PB	$\Delta V_{2d} = $ PS	$\Delta V_{2d} = $ ZO	$\Delta V_{2d} = $ PS	$\Delta V_{2d} = $ ZO

and set the outputs to be the changes in the control inputs, $\Delta z_{11}(k)$ and $\Delta z_{22}(k)$, then the rules for intergroup fuzzy decoupling are easy to build and are very similar to those in Table 5.2. The total changes to the butterfly valve positions are

$$\begin{aligned}\Delta z_1(k) &= \Delta v_1(k) + \Delta z_{11}(k) \\ &= \Delta u_1(k) + \Delta v_{1d}(k) + \Delta z_{11}(k), \end{aligned} \tag{5.28}$$

$$\begin{aligned}\Delta z_2(k) &= \Delta v_2(k) + \Delta z_{11}(k) \\ &= \Delta u_2(k) + \Delta v_{2d}(k) + \Delta z_{11}(k), \end{aligned} \tag{5.29}$$

$$\begin{aligned}\Delta z_3(k) &= \Delta v_3(k) + \Delta z_{22}(k) \\ &= \Delta u_3(k) + \Delta z_{22}(k). \end{aligned} \tag{5.30}$$

4. Expert Controller for Butterfly Valves

The relationship between butterfly valve position and flow rate is highly nonlinear and can be approximately described by a piecewise linear function. In other words, a given change in valve position produces different results depending on the initial position:

(1) Due to the thickness of the plate of a butterfly valve, the plate is still in contact with the sealing ring when the position is in the range 0–5% open. So, there is no flow.

(2) Flow occurs when the valve position is greater than 5% open. The flow rate changes linearly with position when the position is in the range 5–30% open.

(3) The flow rate changes linearly with position when the position is in the range 30–85% open, but the gradient is larger than that for the range 5–30% open.

(4) The valve does not exert a regulatory function when the position is in the range 85–100% open because the flow rate is at a maximum.

The following expert controller was designed to deal with this strong nonlinearity. For $i = 1, 2, 3$,

R_{b1}: **IF** $z_i(k) < 5$, **THEN** $\tilde{z}_i(k) = z_i(k)$.
R_{b2}: **IF** $z_i(k) \in [5,30]$, **THEN** $\tilde{z}_i(k) = z_i(k) + \Delta z_i(k)$.
R_{b3}: **IF** $z_i(k) \in (30, 85]$, **THEN** $\tilde{z}_i(k) = z_i(k) + \Delta z_i(k)/K_b$.
R_{b4}: **IF** $z_i(k) > 85$, **THEN** $\tilde{z}_i(k) = z_i(k)$.

$z_i(k)$ and $\tilde{z}_i(k)$ are the valve positions before and after nonlinear compensation, respectively, and the gain, K_b, in R_{b3} depends on the characteristics of the specific butterfly valve.

5. Expert Multirate Sampling Strategy

The gas collection process is influenced by the changes in the environment as well as many disturbances. Expert operators deal with them skillfully to achieve satisfactory control performance. Close observation and analysis of their manual control techniques reveal that they generally use a variable sampling period to achieve good control performance, with the period ranging from 4 to 16 s. This is understandable. For instance, the pressures of the gas collectors fluctuate wildly when the suction power before the primary coolers and the pressure of the outlet gas change. If the change occurs just after a sample has been taken, then no control action to suppress the change can be taken until the next control period. That is, there is a delay in handling it, which degrades the control performance. So, a short sampling period should be used when the change in pressure is large, and a long sampling period should be used when the change is relatively small to avoid chatter caused by too many adjustments.

This scheme is implemented with a multirate sampling strategy [49] and an expert high-pass filter (Fig. 5.4). In the figure, the blocks labeled "Feedforward controller" and "Fuzzy expert feedback controller" are the same as those in Fig. 5.3. Except for P_4, all the variables in the figure are three-dimensional vectors, e.g., $P = [P_1,\ P_2,\ P_3]^T$, and $\tilde{z} = [\tilde{z}_1,\ \tilde{z}_2,\ \tilde{z}_3]^T$. Basically, P_4, P_r, and P are sampled by samplers S_{T_P} with a long sampling period, T_P (10 s in this study). These signals are sent to the feedforward and feedback controllers, and the control input produced by the controllers, $\tilde{z}_P$, using the control methods described above is sent to the process through a zero-order hold, H_{T_P}. On the other hand, a short sampling period, $T_{\delta P} = T_P/m$ (m is an integer that was set to 5 in this study), is used for sampling the pressure. For $k = 0, 1, \cdots$, let the pressure sampled at sampling time kT_P be $P(k)$ and the pressure sampled at $kT_P + jT_{\delta P}$ be $P(k, j)$ $(j = 0, 1, \ldots, m-1)$. Furthermore, let $\delta P(k, j) = P(k, j) - P(k)$. The following expert rules were designed to capture pressure fluctuations ($i = 1, 2, 3$):

$R_{\delta Pi}$: **IF** $|\delta P_i(k, j)| > 40$ Pa,
THEN $\delta \tilde{P}_i = \delta P_i(k, j)$,
ELSE $\delta \tilde{P}_i = 0$,

where $\delta \tilde{P}_i$ $(i = 1, 2, 3)$ is the output of the expert rules. These expert rules function as a high-pass filter for extracting large, rapid changes in pressure. The changes are

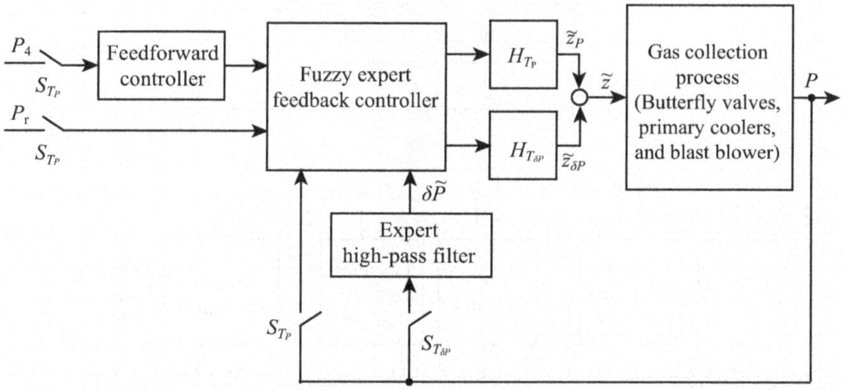

Fig. 5.4 Expert multirate sampling strategy

sent to the fuzzy expert feedback controller, which produces a correction and sends it out through a zero-order hold, $H_{T_{\delta P}}$, as shown in Fig. 5.4. In this way, pressure fluctuations are quickly suppressed. Actually, $\tilde{z}$ is constructed as follows:

$$\tilde{z}(k, j) = \tilde{z}_P(k) + \tilde{z}_{\delta P}(k, j), \quad k = 0, 1, \ldots; \; j = 0, 1, \ldots, m - 1. \tag{5.31}$$

It contains two parts: a slow main part, $\tilde{z}_P$, and a fast modification, $\tilde{z}_{\delta P}$. The expert rules guarantee that the fast-sampling path improves only the high-frequency characteristics of the control system and leaves the low- and middle-frequency characteristics unaffected. As a result, for a suitably designed feedback controller, this strategy does not destroy the stability of the whole control system.

5.3 System Implementation and Results of Actual Runs for Gas Collection Process

The intelligent decoupling control method explained earlier has been implemented in the control system of the gas collection process of an iron and steel company. It carries out real-time control of the pressure of the mixed gas. The results of actual runs demonstrate the validity of the method.

5.3.1 System Implementation

The hardware architecture and software functions of the intelligent decoupling control system are shown in Figs. 5.5 and 5.6, respectively. The system was implemented on an industrial computer. Control is handled by a Honeywell series DCS (Honeywell PlantScape R400). Programs were written in the Visual C++ language.

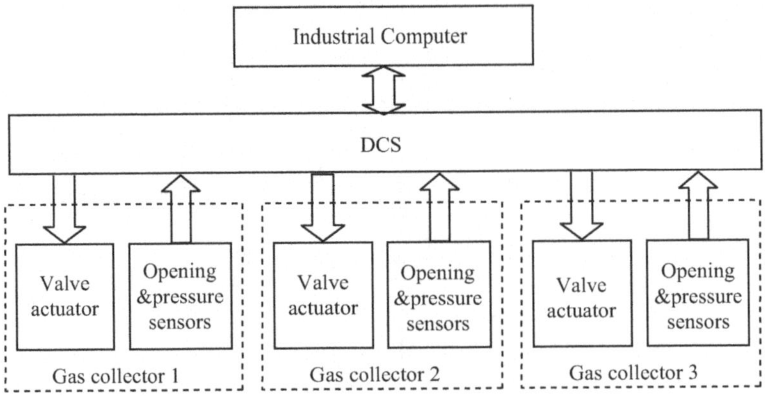

Fig. 5.5 Hardware architecture of control system

The software package contains a HMI module, an algorithm module, and an OPC client. The HMI module displays process data on the industrial computer, manages the security of the system (user registration and verification, monitoring of operations, etc.), sets the reference inputs (P_r), and issues warning alarms. The control methods explained in Sect. 5.2.3 were implemented in the algorithm module. The OPC client handles the communication of information between the industrial computer and the DCS.

The DCS contains an OPC server, an OPC configuration software, and a DCS module to carry out process control. The intelligent decoupling control system generates control inputs for the butterfly valves of the three gas collectors. These signals are sent to the DCS modules through OPC data communication technology. More specifically, the OPC server transfers data between the OPC client and the configuration software. The DCS modules convert those data into analog form and send them to the valve actuators. At the same time, the actual valve positions are measured by sensors, and the results are sent back to the intelligent decoupling controller. The configuration software controls and monitors the gas collection process, collects data in a real-time fashion, analyzes historical curves, and produces reports and tables.

The information and data flow in the control system is illustrated in Fig. 5.7.

In spite of the large uncertainties in the mathematical model of the process, the intelligent decoupling control system stabilizes the system and keeps the pressures of the three gas collectors in the ranges specified by the technological requirements.

5.3.2 *Results of Actual Runs*

Before work began on the intelligent decoupling control system, an attempt was made to use a PID control system. Suitable parameter sets of the PID controller were determined by exploring the relationship between the control performance and the

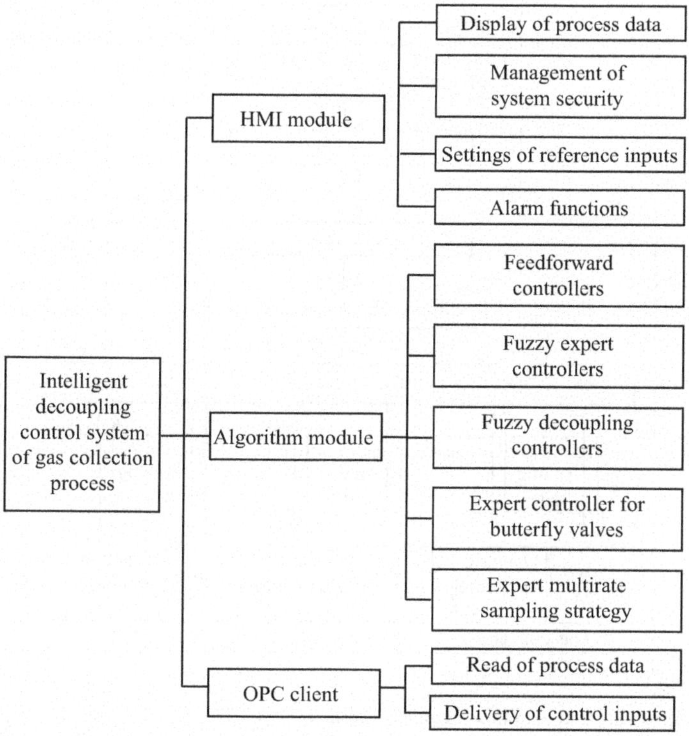

Fig. 5.6 Software package for control system

parameters over the three-parameter space using a finite-element search. For example, we first set the integral gain, and then examined the relationship between the settling time and proportional and derivative gains over the parameter space. However, the best control results obtained were a pressure change of ±50 Pa and a settling time of about 5 min. These values do not satisfy the technological requirements, which are a maximum pressure change of ±20 Pa and a settling time of less than 1 min.

The intelligent decoupling control system has been in operation in the coking plant of an iron and steel company. The control algorithms that were developed are simple, reliable, easy to run, and easy to extend. Figures 5.8, 5.9, and 5.10 show typical results of actual runs, for which the reference pressures for Gas collectors 1, 2, and 3 were 100 Pa, 100 Pa, and 30 Pa, respectively. The new control system better stabilizes the pressures of the gas collectors and more effectively suppresses pressure fluctuations than manual control, which was used until our method was implemented. Large disturbances caused by coaling, discharging coke, changes in gas consumption, etc. are stabilized within 30 s (e.g., as shown in Fig. 5.8c), which is much faster than the several minutes required by manual control.

Table 5.4 shows the success rate for keeping the total pressure within a given range for two ranges (±10 Pa and ±20 Pa) of the reference pressure. The pressures were

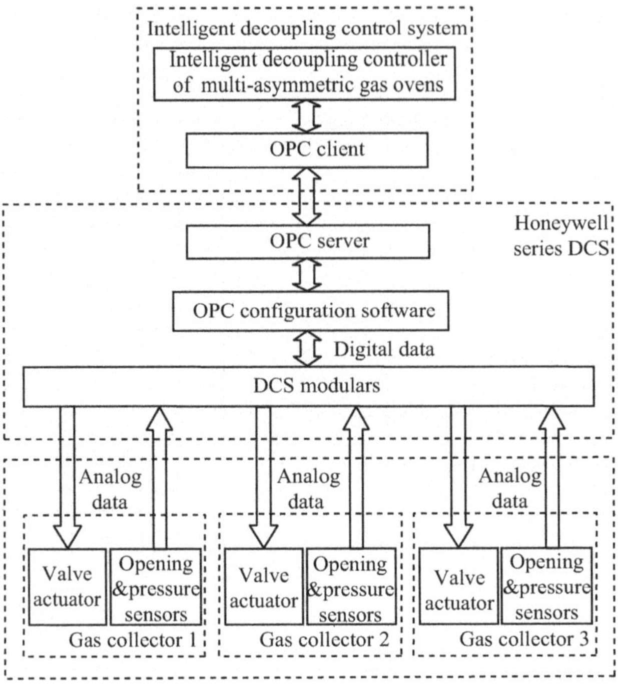

Fig. 5.7 Information and data flow in control system

within the ±20 Pa range more than 94% of the time. This satisfies the technological requirements, which is that the pressure be within ±20 Pa of the reference pressure except during coaling, discharging coke, and other such operations.

5.4 Intelligent Decoupling Control of Gas Mixing-and-Pressurization Process

The control inputs for gas mixing and pressurization are the positions of the valves in the pipelines from the coke-oven and blast furnace and the commands to the converters of the pressurization machines. The control outputs are the calorific value and pressure of the mixed gas. So, it is an MIMO process. Both calorific value and pressure are affected by many factors, such as the pressures of the source gases, the demand for mixed gas, the production policy of the company, etc. [50], and most of them change frequently and unpredictably. Consequently, large fluctuations in calorific value and pressure arise. This makes it very difficult to establish a precise mathematical model for the gas mixing-and-pressurization process. Furthermore, the calorific value and pressure are strongly coupled. That is, any adjustment of the calorific value destroys the pressure equilibrium, and vice versa. So, the crucial prob-

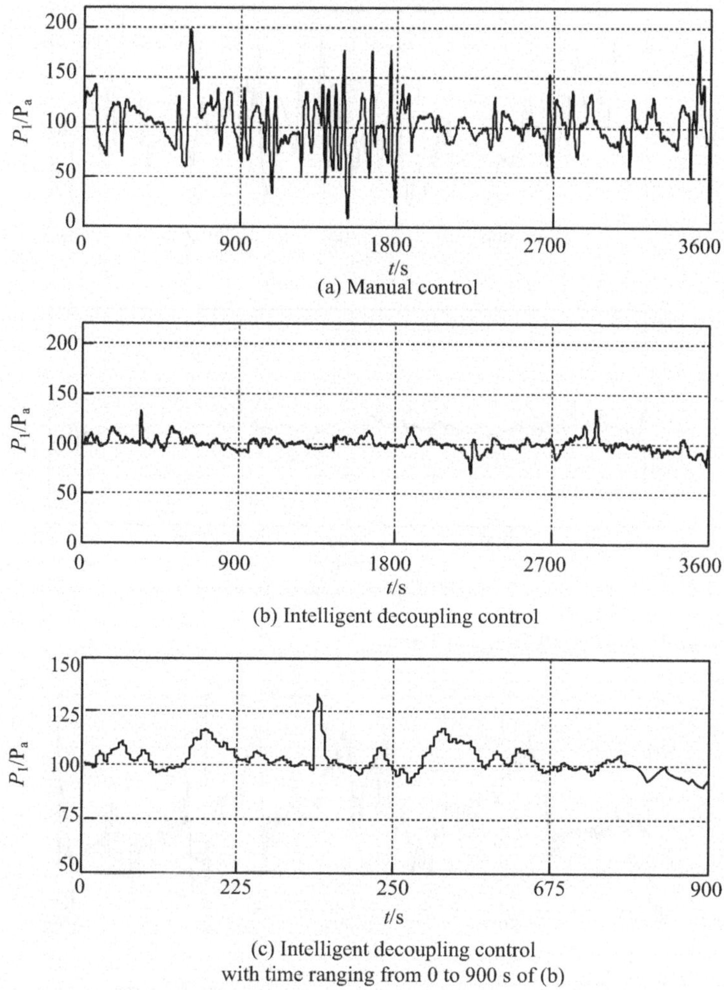

Fig. 5.8 Control results on P_1 for $P_{r1} = 100$ Pa

lems in gas mixing and pressurization are how to decouple the dynamic interaction between calorific value and pressure, and how to adjust the amounts that the valves are open and the frequencies of the converters of the pressurization machines, based on the settings for calorific value and pressure.

A gas mixing-and-pressurization station with two butterfly valves in each of two feed pipelines is a common piece of apparatus in a steel production process. This section presents an integrated intelligent control method for this configuration.

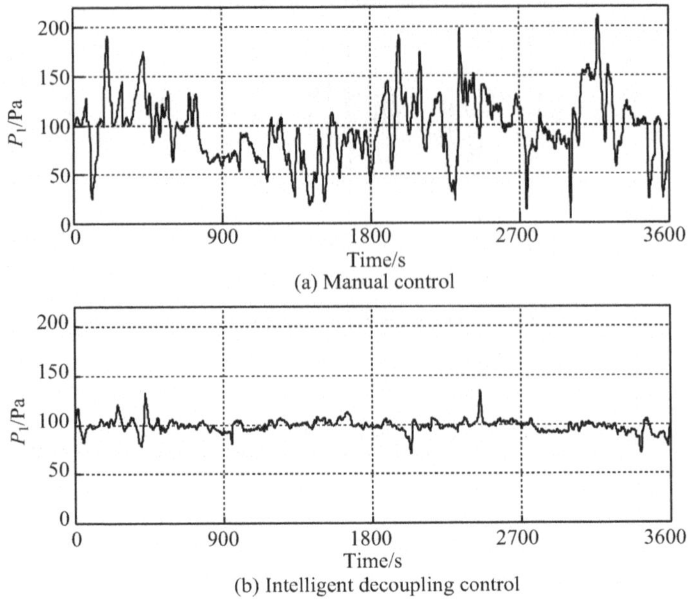

Fig. 5.9 Control results on P_2 for $P_{r2} = 100$ Pa

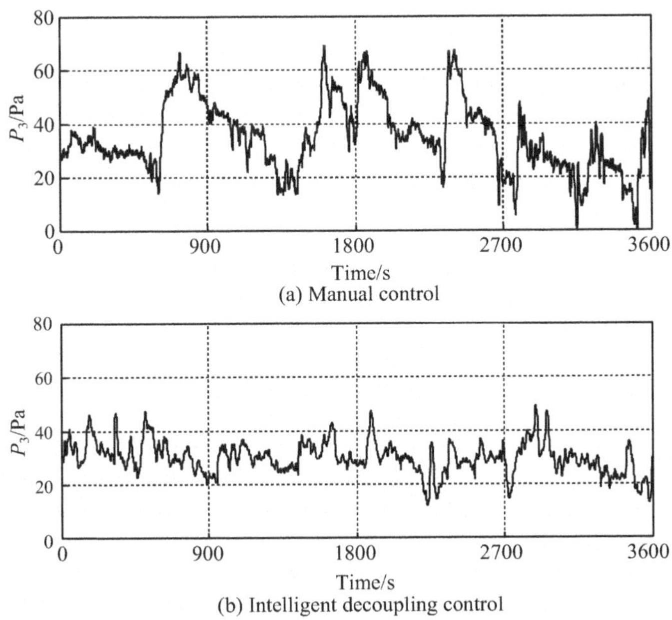

Fig. 5.10 Control results on P_3 for $P_{r3} = 30$ Pa

Table 5.4 Statistics on intelligent decoupling control

Gas collector	Variation in pressure	
	Within ±10 Pa (%)	Within ±20 Pa (%)
1	85	98
2	87	94
3	100	100

5.4.1 Configuration of Gas Mixing-and-Pressurization Control System

Based on an analysis of the physical layout of the equipment and the regulation of the gas mixing-and-pressurization process, we constructed a control system with two control subsystems: a calorific-value and pressure decoupling (CVPD) control subsystem and a pressurization control subsystem. The concept of relative gain matrix is also used. The CVPD control subsystem employs a fuzzy and expert control strategy to adjust the amounts that the butterfly valves are open, with the primary goal being to control the calorific value and the secondary goal being to keep the pressure of the pre-pressurized gas in a certain range. The pressurization control subsystem uses a TDF expert control strategy to regulate the converters of the boosters so as to ensure that the pressure of the boost-pressurized mixed gas tracks a specified value.

The reasons for the selection of the control objectives of the CVPD control subsystem are as follows:

(1) If the CVPD control subsystem focused only on the control of calorific value, then the valves might open or close by only very small amounts, which could produce a very low or even negative pressure in the mixed gas before pressurization.

(2) For the mixed gas, the pressure before pressurization very strongly affects the pressure after pressurization, and large fluctuations in the former may make it difficult to properly set the latter. So, stabilizing the pressure of the pre-pressurized mixed gas helps to prevent large, disruptive changes in the mixing-and-pressurization process.

(3) Using the CVPD control subsystem to make the pressure of the pre-pressurized mixed gas as high as possible allows the boosters to run at a lower frequency, which saves energy.

The strong interaction between the calorific value and pressure of the mixed gas before pressurization makes control of the pressure after pressurization difficult. The intelligent decoupling control strategy employed in the CVPD control subsystem solves the problem by decoupling this dynamic interaction. In addition, the pressurization control subsystem employs TDF expert control to suppress pressure variations in the post-pressurized mixed gas arising from changes in the pressure of the pre-pressurized mixed gas by using the measured pressures of the pre- and post-pressurized mixed gas as feedforward and feedback signals. These two subsystems are integrated into the gas mixing-and-pressurization process, as shown in Fig. 5.11.

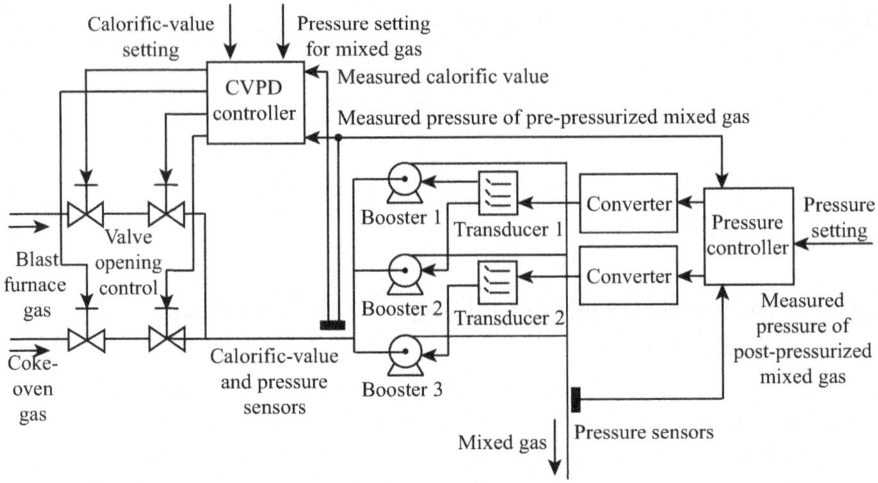

Fig. 5.11 Illustration of gas mixing-and-pressurization process and its control system

5.4.2 Design of Calorific-Value and Pressure Decoupling Control Subsystem

The calorific values and pressures of the gases from the blast furnace and coke-oven, and their mixing ratio, greatly influence the calorific value and pressure of the pre-pressurized mixed gas. The gas mixing-and-pressurization process is a highly nonlinear and strongly coupled MIMO system; so, an input–output decoupling controller was designed to obtain satisfactory control. Conventional decoupling methods usually require a precise mathematical model of the four butterfly valves. However, in an actual production process, some parameters (coefficient of friction of butterfly valves, density of gas, etc.) cannot be measured during operation, and the widely used static decoupling method cannot deal with real-time changes in the process. So, we need a new method to solve the decoupling problem. Note that on-site operators have accumulated a great deal of experience in decoupling control, and their experience should be exploited in devising a practical decoupling control strategy.

Fuzzy control has now become an important approach to designing nonlinear controllers due to its simplicity and the ease with which it can be designed and implemented using commercially available programming tools [51]. The control of knowledge-based systems using linguistic variables, which do not have precise values, is very practical because it allows us to use previously developed heuristic procedures and experience in designing systems [25]. Much experience cannot be expressed in numbers, but some of it can be translated into a fuzzy characterization involving conventional numerical probabilities [26]. Fuzzy control relies more on information from experts and less on the mathematical modeling of a physical system than conventional control approaches. It is also especially useful when low cost and easy operation are goals. The advantages of fuzzy control over other applicable techniques have been investigated by many researchers, e.g., [29, 30].

An expert system is also a very practical approach in the field of artificial intelligence and has been used in process control. Expert control can handle all kinds of quantitative and qualitative, precise and fuzzy information, and it can employ an open- or closed-loop control strategy, depending on the time-variability and uncertainty of the plant and disturbances [16]. It has become a powerful technique for the control of complex processes with nonlinearities and uncertainties due to the flexibility it provides with regard to control method and knowledge expression [17–19]. On the other hand, modern control theory was used to implement decoupling control in [34]. Recently, control methods employing adaptive decoupling [35, 36], fuzzy compensation decoupling [37], and neural network decoupling [38] and their application have been reported. In this study, the integration of operator experience with fuzzy control, expert control, and decoupling control allowed us to build a control system that adapts to a complex process and provides good control performance.

Conventionally, a fuzzy decoupling control method carries out fuzzy inference in one leap. In this study, fuzzy inference employs four inputs: the tracking errors for calorific value and pressure and the rates of change of those errors. And it produces four outputs: the amounts that the four valves are open. If we had designed our decoupling control law using a conventional fuzzy decoupling control method, it would have resulted in an enormous number of rules. This would have made the fuzzy reasoning very complex, which, in turn, would have led to a heavy workload and great computational expense [52]. So, we employed multilayer decoupling instead of one-leap decoupling, and designed some intermediate controllers. The structure of the resulting control system is shown in Fig. 5.12. This section explains each control block.

There are two kinds of fuzzy controllers in the CVPD control subsystem: one for control and the other for decoupling. Regarding the first kind, there are two fuzzy controllers: one controls the calorific value and the other controls the pressure of the pre-pressurized mixed gas. They have the same structure, and the inputs of each are the tracking error and its rate of change. In addition, each one has an auxiliary variable that reflects the tendency of the control error. These two fuzzy controllers are described below.

1. Design of Fuzzy Controller

The fuzzy controllers have a proportional–derivative control configuration [53, 54]. The values in Table 5.5 were selected because they provide the required control precision.

Regarding the membership functions for linguistic values, we used trapezoidal functions for e_P and $\dot{e}_P$ because they are relatively insensitive to linguistic values, and triangular functions for e_C and $\dot{e}_C$ because they are relatively sensitive to linguistic values [55]. Tables 5.6 and 5.7 are the corresponding rule tables.

The abbreviations for the values of the membership functions are NB—negative big, NM—negative medium, NS—negative small, ZO—zero, PS—positive small, PM—positive medium, and PB—positive big.

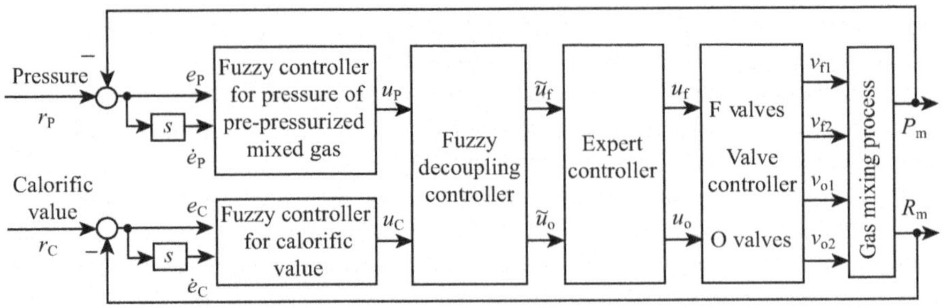

r_P (r_C): reference input for pressure (calorific value)
e_P (e_C): control error for pressure (calorific value)
$\dot{e}_P$ ($\dot{e}_C$): rate of change of control error for pressure (calorific value)
u_P(u_C): auxiliary variables
$\tilde{u}_f$, $\tilde{u}_o$: intermediate variables
u_f(u_o): variable specifying amount that F (O) valves are open

Fig. 5.12 Configuration of CVPD control system

Table 5.5 Design of fuzzy controllers

Variable	Number of linguistic values	
	Pressure controller	Calorific-value controller
Control error (e_P, e_C)	5	7
Rate of change of control error ($\dot{e}_P$, $\dot{e}_C$)	3	5
Auxiliary variable (u_P, u_C)	5	7

Table 5.6 Rule table for pressure control of pre-pressurized mixed gas

e_P	$\dot{e}_P$		
	NB	ZO	PB
NB	PB	PB	PM
NM	PB	PM	ZO
ZO	PM	ZO	NM
PM	ZO	NM	NB
PB	NM	NB	NB

2. Design of Fuzzy Decoupling Controller

u_f and u_o are variables used to specify how much the valves in the blast furnace pipeline and coke-oven pipeline, respectively, are open; they are determined by u_P and u_C. Since there is strong interaction between them, we use fuzzy decoupling control for them. That is, a new MIMO fuzzy approximate decoupling controller is constructed based on operator experience and the relative gain matrix of the parallel double valves. The design is explained below.

Table 5.7 Rule table for control of calorific value

e_C	$\dot{e}_C$				
	NB	NS	ZO	PS	PB
NB	PB	PB	PB	PM	PS
NM	PB	PB	PM	PS	ZO
NS	PB	PM	PS	ZO	NS
ZO	PM	PS	ZO	NS	NM
PS	PS	ZO	NS	NM	NB
PM	ZO	NS	NM	NB	NB
PB	NS	NM	NB	NB	NB

As previously mentioned, for an MIMO control system with n inputs and n outputs, the relative gain between the output y_i and the input u_j is

$$\lambda_{ij} = \frac{(\partial y_i/\partial u_j)|_{\mathcal{U}_j}}{(\partial y_i/\partial u_j)|_{\mathcal{Y}_i}}, \tag{5.32}$$

where $(\partial y_i/\partial u_j)|_{\mathcal{U}_j}$ is the gain of the channel from u_j to y_i when $\mathcal{U}_j = [u_1 \cdots u_{j-1}\ u_{j+1} \cdots u_n]^{\mathrm{T}}$ is invariable, and $(\partial y_i/\partial u_j)|_{\mathcal{Y}_i}$ is the gain of the channel from u_j to y_i when $\mathcal{Y}_j = [y_1 \cdots y_{i-1}\ y_{i+1} \cdots y_n]^{\mathrm{T}}$ is invariable.

For the parallel double valves in Fig. 5.1, the relative gain matrix is calculated as follows:

First, the process is described by

$$\begin{cases} R_{\mathrm{m}} = \dfrac{R_{\mathrm{o}}Q_{\mathrm{o}} + R_{\mathrm{f}}Q_{\mathrm{f}}}{Q_{\mathrm{o}} + Q_{\mathrm{f}}}, \\ Q_{\mathrm{m}} = Q_{\mathrm{o}} + Q_{\mathrm{f}}. \end{cases} \tag{5.33}$$

Let the control inputs be

$$u_1 = Q_{\mathrm{o}}, \quad u_2 = Q_{\mathrm{f}}, \tag{5.34}$$

and the control outputs be

$$y_1 = R_{\mathrm{m}}, \quad y_2 = Q_{\mathrm{m}}. \tag{5.35}$$

According to the definition of relative gain, we have

$$\begin{aligned} N_{\mathrm{rq}11} &= \frac{\partial}{\partial u_1}\left(\frac{R_{\mathrm{o}}u_1 + R_{\mathrm{f}}u_2}{u_1 + u_2}\right)\bigg|_{\mathcal{U}_2} = \frac{(R_{\mathrm{o}} - R_{\mathrm{f}})u_2}{(u_1 + u_2)^2} \\ &= \frac{R_{\mathrm{o}} - R_{\mathrm{m}}}{Q_{\mathrm{m}}}, \end{aligned}$$

$$D_{\mathrm{rq}11} = \frac{\partial}{\partial u_1}\left(\frac{R_\mathrm{o}u_1 + R_\mathrm{f}(y_2 - u_1)}{y_2}\right)\Bigg|_{y_2} = \frac{R_\mathrm{o} - R_\mathrm{f}}{Q_\mathrm{m}}.$$

Therefore,

$$\lambda_{\mathrm{rq}11} = \frac{N_{\mathrm{rq}11}}{D_{\mathrm{rq}11}} = \frac{R_\mathrm{o} - R_\mathrm{m}}{R_\mathrm{o} - R_\mathrm{f}}. \tag{5.36}$$

In the same manner, we obtain $\lambda_{\mathrm{rq}12}$, $\lambda_{\mathrm{rq}21}$, and $\lambda_{\mathrm{rq}22}$. So, the relative gain matrix is

$$\Lambda_{\mathrm{rq}} = \begin{bmatrix} \lambda_{\mathrm{rq}11} & \lambda_{\mathrm{rq}12} \\ \lambda_{\mathrm{rq}21} & \lambda_{\mathrm{rq}22} \end{bmatrix} = \frac{1}{R_\mathrm{o} - R_\mathrm{f}} \begin{bmatrix} R_\mathrm{o} - R_\mathrm{m} & R_\mathrm{m} - R_\mathrm{f} \\ R_\mathrm{m} - R_\mathrm{f} & R_\mathrm{o} - R_\mathrm{m} \end{bmatrix}. \tag{5.37}$$

As a result, we obtain the relationship between $[\Delta Q_\mathrm{o}\ \Delta Q_\mathrm{f}]^\mathrm{T}$ and $[\Delta R_\mathrm{m}\ \Delta Q_\mathrm{m}]^\mathrm{T}$ around an equilibrium point, where $\Delta x = x - \bar{x}$ is the change in x from its equilibrium point $\bar{x}$:

$$\begin{bmatrix} \Delta R_\mathrm{m} \\ \Delta Q_\mathrm{m} \end{bmatrix} = \Lambda_{\mathrm{rq}} \begin{bmatrix} \Delta Q_\mathrm{o} \\ \Delta Q_\mathrm{f} \end{bmatrix}. \tag{5.38}$$

In the production process considered in this paper, the calorific value of the coke-oven gas, R_o, is around 17000 $\mathrm{kJ/m^3}$; that of the blast furnace gas, R_f, is in the neighborhood of 3000 $\mathrm{kJ/m^3}$; and that of the mixed gas, R_m, is between 3000 and 17000 $\mathrm{kJ/m^3}$. Now, we consider the problem of controlling the calorific value based on Eq. (5.38).

There are two ways of changing the calorific value: adjust the coke-oven valves or adjust the blast furnace valves. When the calorific value of the mixed gas is low (for example, R_m is only 4000 $\mathrm{kJ/m^3}$), a simple calculation using Eq. (5.37) yields

$$\Lambda_{\mathrm{rq}} = \begin{bmatrix} \lambda_{\mathrm{rq}11} & \lambda_{\mathrm{rq}12} \\ \lambda_{\mathrm{rq}21} & \lambda_{\mathrm{rq}22} \end{bmatrix} = \begin{bmatrix} \frac{13}{14} & \frac{1}{14} \\ \frac{1}{14} & \frac{13}{14} \end{bmatrix}. \tag{5.39}$$

So,

$$\begin{cases} \Delta R_\mathrm{m} = \dfrac{13}{14} \times \Delta Q_\mathrm{o}, \\ \Delta Q_\mathrm{m} = \dfrac{1}{14} \times \Delta Q_\mathrm{o}. \end{cases} \tag{5.40}$$

From Eq. (5.40), we see that adjusting the coke-oven valves provides effective control of the calorific value while only slightly influencing the pressure. So, for this situation, this method almost decouples the interaction between calorific value and pressure.

On the other hand, if we adjust the blast furnace valves, then

$$\begin{cases} \Delta R_{\mathrm{m}} = \dfrac{1}{14} \times \Delta Q_{\mathrm{f}}, \\ \Delta Q_{\mathrm{m}} = \dfrac{13}{14} \times \Delta Q_{\mathrm{f}}. \end{cases} \tag{5.41}$$

For this situation, adjusting the blast furnace valves has very little effect on the calorific value, but it has a considerable influence on the pressure. So, when the calorific value of the mixed gas is low, Eqs. (5.40) and (5.41) tell us that we have to adjust the coke-oven valves, but not the blast furnace valves, to control the calorific value while maintaining the stability of the system.

Note that the non-diagonal elements of the relative gain matrix in Eq. (5.37) are not zero and are strongly dependent on the calorific values of the blast furnace gas, the coke-oven gas, and the mixed gas; they change in a real-time fashion. Generally speaking, just adjusting valves based on a relative gain matrix for the special calorific values of those gases cannot provide good decoupling control. In this study, we designed a fuzzy decoupling controller based on the characteristics of the parallel double valves and the experience of experts. The inputs of the controller are u_{P} and u_{C}, and the outputs are $\tilde{u}_{\mathrm{f}}$ and $\tilde{u}_{\mathrm{o}}$. The fuzzy decoupling rules have the form of IF-THEN statements. For example,

IF (u_{P} = PB) AND (u_{C} = PM),
THEN ($\tilde{u}_{\mathrm{f}}$ = PS) AND ($\tilde{u}_{\mathrm{o}}$ = PB).

This rule says that, if u_{P} is positively big and u_{C} is positively medium, then open the coke-oven valves by a large amount and the blast furnace valves by a medium amount. Table 5.8 shows the 35 fuzzy decoupling rules that were established.

Remark 5.1 The decoupling method presented here provides only approximate decoupling, not complete decoupling, which is almost impossible to achieve owing to the complexity of the process. However, actual runs showed that our decoupling method provided good enough decoupling control for a production process.

3. Design of Expert Controller

The complexity of the process makes it difficult to use fuzzy reasoning to control the calorific value and pressure under abnormal conditions. So, an expert controller is designed to handle these cases. The expert rules are divided into two classes, one for calorific value and one for pressure. They have the form of IF-THEN statements, with the ith rule being indexed as R_i. Let $\delta\tilde{u}_{\mathrm{f}}$ and $\delta\tilde{u}_{\mathrm{o}}$ be the changes in $\tilde{u}_{\mathrm{f}}$ and $\tilde{u}_{\mathrm{o}}$, respectively, in a control operation. Under normal conditions, the outputs of the expert controller are

$$\delta u_{\mathrm{f}} = \delta\tilde{u}_{\mathrm{f}}, \quad \delta u_{\mathrm{o}} = \delta\tilde{u}_{\mathrm{o}}. \tag{5.42}$$

Some typical rules are given below for abnormal conditions. In those rules, $P_{\min}$ is the minimum of P_{m}, and $\delta\tilde{u}_{\mathrm{fM}}$ and $\delta\tilde{u}_{\mathrm{vM}}$ are the maxima of $\delta\tilde{u}_{\mathrm{f}}$ and $\delta\tilde{u}_{\mathrm{o}}$, respectively.

Table 5.8 Rule table for fuzzy decoupling control

u_C	u_P				
	NB	NM	ZO	PM	PB
NB	$\tilde{u}_f = \text{ZO}$	$\tilde{u}_f = \text{PS}$	$\tilde{u}_f = \text{PM}$	$\tilde{u}_f = \text{PB}$	$\tilde{u}_f = \text{PB}$
	$\tilde{u}_o = \text{NB}$	$\tilde{u}_o = \text{NB}$	$\tilde{u}_o = \text{NM}$	$\tilde{u}_o = \text{NS}$	$\tilde{u}_o = \text{ZO}$
NM	$\tilde{u}_f = \text{NS}$	$\tilde{u}_f = \text{ZO}$	$\tilde{u}_f = \text{PS}$	$\tilde{u}_f = \text{PM}$	$\tilde{u}_f = \text{PB}$
	$\tilde{u}_o = \text{NB}$	$\tilde{u}_o = \text{NM}$	$\tilde{u}_o = \text{NS}$	$\tilde{u}_o = \text{ZO}$	$\tilde{u}_o = \text{PS}$
NS	$\tilde{u}_f = \text{NM}$	$\tilde{u}_f = \text{NS}$	$\tilde{u}_f = \text{ZO}$	$\tilde{u}_f = \text{PM}$	$\tilde{u}_f = \text{PB}$
	$\tilde{u}_o = \text{NB}$	$\tilde{u}_o = \text{NM}$	$\tilde{u}_o = \text{NS}$	$\tilde{u}_o = \text{PS}$	$\tilde{u}_o = \text{PM}$
ZO	$\tilde{u}_f = \text{NM}$	$\tilde{u}_f = \text{NS}$	$\tilde{u}_f = \text{ZO}$	$\tilde{u}_f = \text{PS}$	$\tilde{u}_f = \text{PM}$
	$\tilde{u}_o = \text{NM}$	$\tilde{u}_o = \text{NS}$	$\tilde{u}_o = \text{ZO}$	$\tilde{u}_o = \text{PS}$	$\tilde{u}_o = \text{PM}$
PS	$\tilde{u}_f = \text{NB}$	$\tilde{u}_f = \text{NM}$	$\tilde{u}_f = \text{ZO}$	$\tilde{u}_f = \text{PS}$	$\tilde{u}_f = \text{PM}$
	$\tilde{u}_o = \text{NM}$	$\tilde{u}_o = \text{NS}$	$\tilde{u}_o = \text{PS}$	$\tilde{u}_o = \text{PM}$	$\tilde{u}_o = \text{PB}$
PM	$\tilde{u}_f = \text{NB}$	$\tilde{u}_f = \text{NM}$	$\tilde{u}_f = \text{NS}$	$\tilde{u}_f = \text{ZO}$	$\tilde{u}_f = \text{PS}$
	$\tilde{u}_o = \text{NS}$	$\tilde{u}_o = \text{ZO}$	$\tilde{u}_o = \text{PS}$	$\tilde{u}_o = \text{PM}$	$\tilde{u}_o = \text{PB}$
PB	$\tilde{u}_f = \text{NB}$	$\tilde{u}_f = \text{NB}$	$\tilde{u}_f = \text{NM}$	$\tilde{u}_f = \text{NS}$	$\tilde{u}_f = \text{ZO}$
	$\tilde{u}_o = \text{ZO}$	$\tilde{u}_o = \text{PS}$	$\tilde{u}_o = \text{PM}$	$\tilde{u}_o = \text{PB}$	$\tilde{u}_o = \text{PB}$

R_{pc1}: **IF** $P_m < P_{\min}$, **THEN** $(\delta\tilde{u}_f = \delta\tilde{u}_{fM})$ AND $(\delta\tilde{u}_o = \delta\tilde{u}_{vM})$.

Explanation When the pressure of the mixed gas is below the minimum acceptable level, the booster becomes unsafe and the pressure should be increased quickly. So, both the coke-oven and blast furnace valves should be fully opened.

R_{pc2}: **IF** $P_m > P_f$, **THEN** $(\delta\tilde{u}_f = -\delta\tilde{u}_{fM})$ AND $(\delta\tilde{u}_o = -\delta\tilde{u}_{vM})$.

Explanation When the pressure of the mixed gas is higher than the pressure of the blast furnace gas, it means that the pressure and calorific value of the coke-oven gas are both too high. So, both the coke-oven and blast furnace valves should be completely closed.

R_{pc3}: **IF** $P_m > P_o$, **THEN** $(\delta\tilde{u}_f = -\delta\tilde{u}_{fM})$ AND $(\delta\tilde{u}_o = \delta\tilde{u}_{vM})$.

Explanation If the pressure of the mixed gas is higher than that of the coke-oven gas, then the pressure of the blast furnace gas is very high and the calorific value is very low. So, the coke-oven valves should be fully opened, and the blast furnace valves should be completely closed.

R_{pc4}: **IF** $R_m < 4500\ \text{kJ/m}^3$, **THEN** $(\delta\tilde{u}_f = -\delta\tilde{u}_{fM})$ AND $(\delta\tilde{u}_o = \delta\tilde{u}_{vM})$.

Explanation When the calorific value of the mixed gas is lower than $4500\ \text{kJ/m}^3$, the calorific value should be increased quickly by fully opening the coke-oven valves and completely closing the blast furnace valves.

R_{pc5}: **IF** $(P_o > 8.5\ \text{kPa})$ AND $(\delta\tilde{u}_o < 0)$, **THEN** $\delta\tilde{u}_o = 0$.

Explanation If the pressure of the coke-oven gas is higher than 8.5 kPa and the coke-oven valves are closed, then those valves must be opened to ensure the safety of the coke-oven. This is because the coke-oven gas cannot be vented into the air and a very high pressure makes the coke-oven unsafe.

4. Design of Valve Controller

There are two butterfly valves in the pipeline from the blast furnace and two in the one from the coke-oven. So, u_P (u_C) must be used to specify the amounts that valves F_1 and F_2 (O_1 and O_2) are open. Furthermore, the butterfly valves are highly nonlinear, and the nonlinearity has to be compensated for to achieve satisfactory control performance. Since the basic principle of specifying the inputs is the same for valves F_1 and F_2 as for valves O_1 and O_2, in the rest of this subsection, only the control law for the pair of coke-oven valves, O_1 and O_2, is explained.

As can be seen in Fig. 5.1, the relationships among Q_o, P_o, P_{oi}, and v_{oi} $(i = 1, 2)$ are

$$Q_o = v_{o1}(P_o - P_{o1}) = v_{o2}(P_{o1} - P_{o2}) = \frac{v_{o1}v_{o2}(P_o - P_{o2})}{v_{o1} + v_{o2}}. \tag{5.43}$$

That is,

$$\begin{cases} Q_o = \dfrac{v_{o1}v_{o2}}{v_{o1} + v_{o2}}(P_o - P_{o2}), \\ P_1 = \dfrac{v_{o1}P_o + v_{o2}P_{o2}}{v_{o1} + v_{o2}}. \end{cases} \tag{5.44}$$

And the relative gain matrix from the amounts that the valves are open to the pressures and the flow rate is

$$\begin{bmatrix} \Delta Q_o \\ \Delta P_{o1} \end{bmatrix} = \Lambda_p \begin{bmatrix} \Delta v_{o1} \\ \Delta v_{o2} \end{bmatrix}, \tag{5.45}$$

$$\begin{aligned} \Lambda_p &= \begin{bmatrix} \dfrac{(\partial Q_o)/(\partial v_{o1})|_{v_{o2}=ct}}{(\partial Q_o)/(\partial v_{o1})|_{P_{o1}=ct}} & \dfrac{(\partial Q_o)/(\partial v_{o2})|_{v_{o1}=ct}}{(\partial Q_o)/(\partial v_{o2})|_{P_{o1}=ct}} \\ \dfrac{(\partial P_{o1})/(\partial v_{o1})|_{v_{o2}=ct}}{(\partial P_{o1})/(\partial v_{o1})|_{Q_o=ct}} & \dfrac{(\partial P_{o1})/(\partial v_{o2})|_{v_{o2}=ct}}{(\partial P_{o1})/(\partial v_{o2})|_{Q_o=ct}} \end{bmatrix} \\ &= \frac{1}{P_o - P_{o2}} \begin{bmatrix} P_o - P_{o1} & P_{o1} - P_{o2} \\ P_{o1} - P_{o2} & P_o - P_{o1} \end{bmatrix}, \end{aligned} \tag{5.46}$$

where *ct* stands for "constant". Analysis of the above gain matrix, Λ_p, yields the following control strategy:

(1) If the difference between P_{o1} and P_{o2} is small, then the flow rate Q_o should be adjusted using valve O_1, and the pressure P_{o1} should be adjusted using O_2.

(2) If the difference between P_o and P_{o1} is small, then Q_o should be adjusted using valve O_2, and P_{o1} should be adjusted using O_1.

(3) If the difference between P_{o1} and $(P_o + P_{o2})/2$ is small, then both Q_o and P_{o1} should be adjusted using O_1 and O_2.

The relationship between the amount that a butterfly valve is open and flow rate is strongly nonlinear [56]:

(1) When the amount is in the range 0–5%, the flow rate is zero.

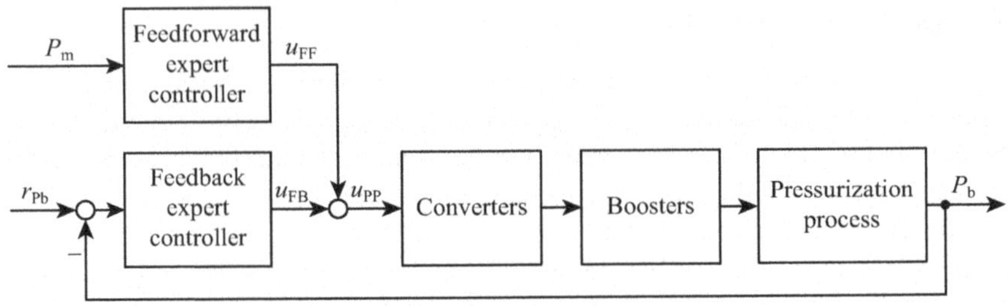

Fig. 5.13 Configuration of TDF expert pressurization control subsystem

(2) When the amount is in the range 5–30%, the flow rate changes linearly with the amount.

(3) When the amount is in the range 30–85%, a small change in the amount results in a large change in the flow rate.

(4) When the amount is in the range 85–100%, the flow rate is at a maximum.

So, depending on the current position of a valve, opening or closing it by a particular amount may produce different changes in the flow rate. We use expert rules to compensate for the nonlinearity, some of which are shown below. Note that μ is the measured amount that a valve is open, K_p is a gain, χ is the control input for the valve, and $\delta\chi$ is the calculated change in χ.

R_{p1}: **IF** $\mu \in [5, 30]$, **THEN** $\chi = \chi + \delta\chi$.
R_{p2}: **IF** $\mu \in (30, 85]$, **THEN** $\chi = \chi + \delta\chi / K_p$.
R_{p3}: **IF** $(\mu > 85)$ AND $(\delta\chi \geqslant 0)$, **THEN** $\chi = \chi + 0$.
R_{p4}: **IF** $(\mu < 5)$ AND $(\delta\chi \leq 0)$, **THEN** $\chi = \chi + 0$.

5.4.3 Design of Pressurization Control Subsystem

The pressurization control subsystem keeps the pressure of the post-pressurized mixed gas at a specified value. This is achieved by adjusting the frequencies of the converters, which govern the rotational speeds of the boosters. The TDF expert control configuration in Fig. 5.13 suppresses the influence of fluctuations in P_m on P_b, thus guaranteeing precise control.

The control system employs expert proportional control in both feedforward and feedback controllers. The proportional gains are tuned in a real-time fashion. Below are some typical expert control laws for the proportional gain of the P feedback controller, K_{FB},

R_{fb1}: **IF** $0.5\ \text{kPa} \leq |\Delta P_b| < 1\ \text{kPa}$, **THEN** $K_{FB} = 6$.
R_{fb2}: **IF** $1\ \text{kPa} \leq |\Delta P_b| < 2\ \text{kPa}$, **THEN** $K_{FB} = 8$.
R_{fb3}: **IF** $2\ \text{kPa} \leq |\Delta P_b|$, **THEN** $K_{FB} = 10$.

In these rules, ΔP_b is the difference between the actual pressure and the pressure setting.

P_m and P_b strongly affect each other. When the rotational speeds of the boosters are constant, a higher P_m results in a higher P_b, and vice versa. We can treat P_m as the sum of a constant value, $\bar{P}_m$, and a change, ΔP_m. So, using a simple P feedforward controller to send the measured P_m to the control system effectively suppresses the influence of the change in P_m. This feedforward control technique also provides the system with greater flexibility and precision. For the feedforward controller, $\bar{P}_m$ is chosen to be 3.5 kPa, and the gain of the feedforward controller, K_{FF}, is tuned based on $\Delta P_m = P_b - \bar{P}_m$.

Finally, the control input, u_{PP}, of the converters is obtained by combining the feedforward control input, u_{FF}, and the feedback control input, u_{FB}:

$$u_{PP} = u_{FF} + u_{FB}. \tag{5.47}$$

5.5 System Implementation and Results of Actual Runs for Gas Mixing-and-Pressurization Process

The intelligent decoupling control system was implemented in the gas mixing-and-pressurization process of a steel mill to carry out real-time control of the calorific value and pressure of the mixed gas. The results of actual runs have demonstrated the validity of this strategy.

5.5.1 System Framework

Considering of the practical operation and characteristics of the gas pressurization station, a two-layer distributed control structure including direct digital control and process monitoring is adopted for the control system. The host computer carries out the process monitoring and the DCS runs the direct digital control. The host computer not only runs the configuration software to complete scheduling and monitoring of the entire production process, and carries out the statistics and analysis of real-time data and historical curve to generate the corresponding report forms, but also runs intelligent decoupling control software to process the real-time data and completes the decoupling control. The DCS gathers the parameters of production process and the states of the operation equipment online and then controlled all the related equipments in the production site and operated the control system according to the control signal of the actuators.

The logical structure of the control system is shown in Fig. 5.14. The control objective is divided into two parts. One part is to control the calorific value, and the other part is to control the pressure. Since there are mutual coupling effects among

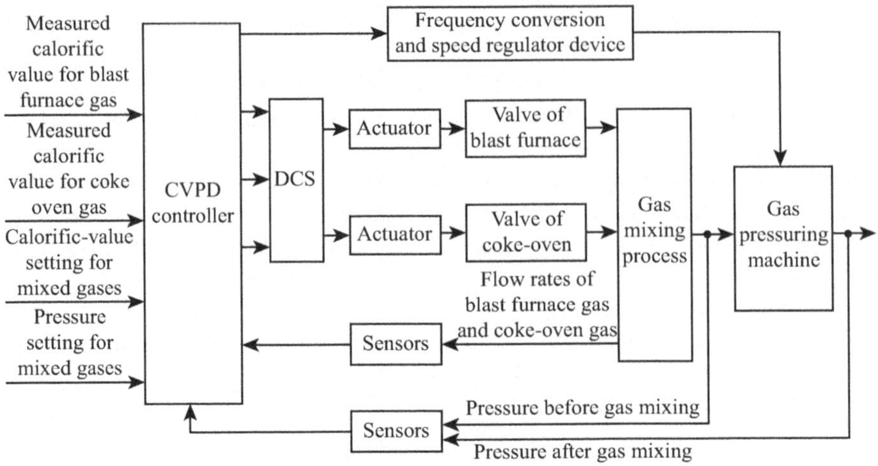

Fig. 5.14 Logical structure of intelligent decoupling control system

the adjustment of four butterfly valves and the original control system does not conduct the CVPD control, it easily leads to calorific-value fluctuations, the pressure instability of the manifold output of the mixed gas, and other issues. The CVPD controller introduced into the control system, which is the core of the system, enables to decouple the calorific-value control and the pressure control to eliminate the mutual coupling effects of the adjustment of r butterfly valves and adjust the valves more accurately. Meantime, the frequency conversion speed regulator is embedded into the control system to achieve the automatic control of the pressure of the mixed gases.

As we can see from Fig. 5.14, there are two loops in the control system. The inner loop is of the feedback control of the flow rate, which is mainly to stabilize the calorific value of the mixed gases by mixing sufficiently the blast furnace and coke-oven gases. The outer loop is of the feedback control of the calorific value and pressure, which is mainly to stabilize the mixed gas pressure and also plays a certain role in stabilizing the fluctuations of the calorific value. Eventually, the automatic control system based on the feedback control of mixed gas calorific value and pressure is constituted.

In Fig. 5.14, first the CVPD controller obtains the set points of the positions of the valves of the blast furnace and coke-oven according to the set points of the calorific value and pressure for the mixed gases, and the measured calorific values and flow rates of the blast furnace gas and the coke-oven gas. The control frequency by the frequency conversion speed regulator is also obtained. By the DDS module, the rotational speed of the booster is controlled by the frequency conversion speed regulator, so as to make the pressure of the mixed gas meet the requirement. Note that the flow rates of the blast furnace gas and the coke-oven gas, together with the pressure of mixed gases before and after gas mixing, are measured online and then are fed back to the CVPD controller to form the closed control loop.

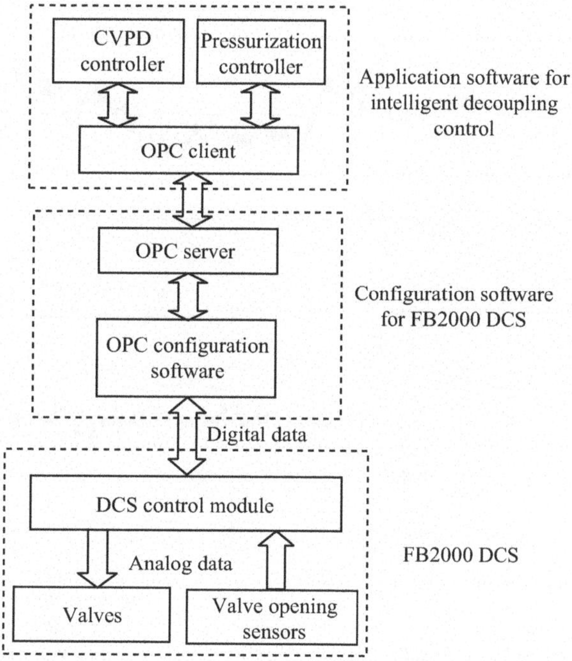

Fig. 5.15 Flow of information and data in the control system

5.5.2 *System Implementation*

The intelligent decoupling control system was implemented using application software for intelligent decoupling control developed especially for this study, an FB2000 DCS, and configuration software. The application software, which was written in Visual C++, carries out the control algorithm for the decoupling control of calorific value and pressure. The configuration software performs the following functions:

(1) Real-time monitoring and management of the whole process.
(2) Fault diagnosis, emergency alarms, and accommodation to faults.
(3) Statistics and analysis of real-time data.
(4) Establishment of material balance sheets.
(5) Display of dynamic graphs and process data in a real-time fashion.
(6) Real-time recording of calorific values and pressures once per second.

The application software package contains a parameter identification program and an integrated intelligent control program. The system uses OPC data communication technology, which is based on the ATL package, to carry out the real-time transmission of information and data in the DCS. The flow of information and data in the control system is shown in Fig. 5.15. The control period was 10 s for the valves and 15 s for the converters.

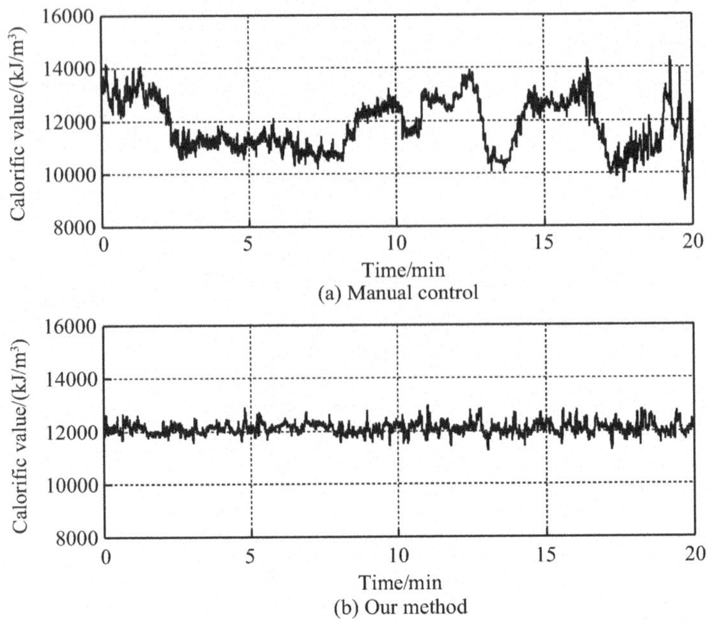

Fig. 5.16 Control results on calorific value of mixed gas for $r_C = 12000\ \text{kJ/m}^3$

5.5.3 Results of Actual Runs

Figures 5.16 and 5.17 show some typical control results on the calorific value and pressure of the mixed gas obtained with the integrated intelligent control method presented in this chapter. For comparison, the results for manual control, which was used until our method was implemented, are also shown.

For manual control, the standard deviation and variance of the tracking error for calorific value (Fig. 5.16) were 65.12 kJ/m^3 and 6472.21 kJ/m^3, respectively. Our method reduced those values to 26.04 and 282.19 kJ/m^3. Similarly, our method reduced the standard deviation and variance of the pressure (Fig. 5.17) from 0.59 kPa and 1.57 kPa, respectively, to 0.04 kPa and 0.26 kPa. For our method, both tracking errors are much smaller than the required value of 5%.

Use of the integrated intelligent control system provides three main benefits:

(1) The calorific value and pressure of the mixed gas are stabilized, and meet the requirements of the process. As a result, the time needed to heat the reheating furnace to the requisite temperate is shorter, and the lifetime of the reheating furnace is extended. Furthermore, statistical data show that the fuel consumption of the reheating furnace is reduced by about 3%.

(2) The utilization rate of the blast furnace gas is improved. This allows the use of coke-oven gas to be reduced by about 2%. This has a beneficial effect on the environment.

(3) Control of the converters reduces electricity consumption by about 20%.

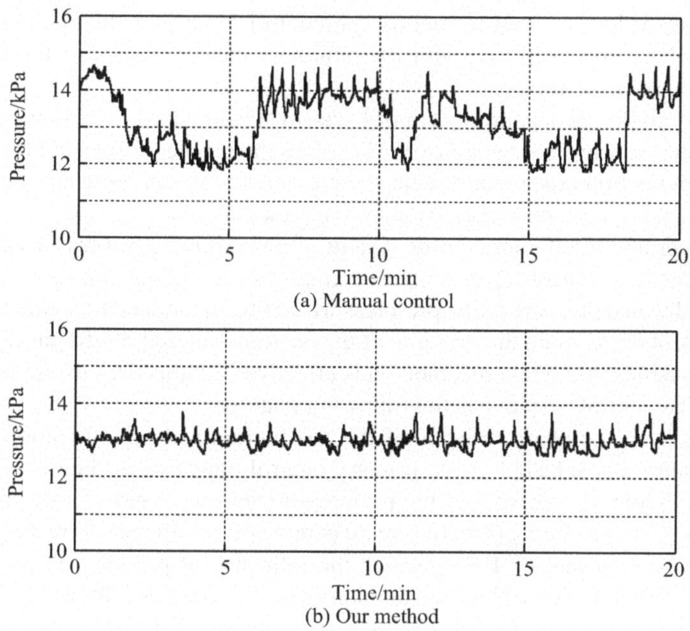

Fig. 5.17 Control results on pressure of post-pressurized mixed gas for $r_P = 13$ kPa

5.6 Conclusion

An intelligent decoupling control system has been designed for gas collection from multiple asymmetric coke-ovens and an integrated intelligent control method for a gas mixing-and-pressurization process with the objective of controlling the calorific value and pressure of the mixed gas has been presented. The main points of the intelligent decoupling control system of three asymmetric coke-ovens of an iron and steel company are as follows:

(1) Based on an analysis of the coupling characteristics of the gas collection process, an intelligent grouping-and-decoupling strategy was devised to solve the problem of decoupling multiple asymmetric coke-ovens. The intelligent decoupling control system contains fuzzy expert controllers, feedforward controllers, intragroup and intergroup fuzzy decoupling controllers, and an expert butterfly valve controller.

(2) The fuzzy expert controllers determine the operating state and select appropriate fuzzy controllers to carry out fast or fine adjustment so as to balance the requirements of quick response and stability, and to yield satisfactory disturbance rejection performance.

(3) Fuzzy theory is used to carry out decoupling on the basis of the coupling relationships within and between groups.

(4) An expert multirate sampling strategy quickly suppresses large fluctuations and avoids chatter.

(5) Expert rules are used to modify the control inputs for the positions of the butterfly valves in accordance with the nonlinear characteristics of the valves to improve control performance.

(6) This intelligent decoupling control system is being used for actual gas collection from three asymmetric coke-ovens. It solves the problem of decoupling control of the gas collection process and keeps the pressures of the gas collectors within the technological requirements, thereby stabilizing the process.

The integrated intelligent control system of gas mixing-and-pressurization process contains two control subsystems: one employs decoupling control to adjust the calorific value and pressure of the pre-pressurized mixed gas, and the other uses TDF expert control to maintain the pressure of the post-pressurized mixed gas. Results of actual runs demonstrate that this method is effective and improves the quality of the product. The control system has the following features:

(1) Based on the characteristics of the gas mixing-and-pressurization process, calorific value was selected as the primary control objective of the CVPD control subsystem, while the pressure of the pre-pressurized mixed gas is a secondary objective. CVPD control integrates fuzzy and expert control, fuzzy decoupling control, and a relative gain matrix. It suppresses the influence of pressure on the calorific value, and rationally determines the amounts that the four butterfly valves are open.

(2) The pressurization control subsystem incorporates TDF, fuzzy, and expert control. The TDF control structure effectively suppresses the influence of the pressure of the pre-pressurized mixed gas on the pressure after pressurization. The pressurization control subsystem properly adjusts the frequencies of the converters of the pressurization machines.

(3) ATL-based OPC data communication technology enables the real-time transmission of information and data in the DCS. This allows the application software to have a simple framework and good portability.

(4) Actual runs of the system show that our control strategy is effective, and that fluctuations in calorific value and pressure are kept at a low level. The system provides significant economic and social benefits.

Investigation of the stability of the whole control system is of great importance and can provide a theoretical guarantee of the control performance of the designed system. A considerable number of studies have been devoted to this topic, e.g., [57–60] and the references therein. An analysis of intelligent decoupling control for a gas collection process will be carried out in the future.

References

1. Sadaki J, Tanaka K, Naganuma Y (1993) Automatic coking control system. In: Proceedings of IEEE Conference on Control Applications, pp 531–538
2. Loison R, Foch P, Boyer A (1989) Coke: Quality and Production. Butterworths, London
3. Jenkins DR (2001) Plastic layer permeability estimation using a model of gas pressure in a coke oven. Fuel 80(14):2057–2065

4. Gray RJ (1990) Coke Oven Wall Pressures: Measurement, Cause and Effect. Iron and Steel Society Inc, Pittsburgh
5. Wu M, Yan J, She JH, Cao WH (2009) Intelligent decoupling control of gas collection process of multiple asymmetric coke ovens. IEEE Trans Ind Electron 56(7):2782–2792
6. Gamrat S, Poraj J, Bodys J, Smolka J, Adamczyk W (2016) Influence of external flue gas recirculation on gas combustion in a coke oven heating system. Fuel Process Technol 152:430–437
7. Weng Y, Gao X (2017) Data-driven robust output tracking control for gas collector pressure system of coke ovens. IEEE Trans Ind Electron 64(5):4187–4198
8. Coking Experts Group (1978) Coking technologies. Metallurgical Industry Press, Beijing (In Chinese)
9. Wu M, Cao WH, He CY, She JH (2009) Integrated intelligent control of gas mixing-and-pressurization process. IEEE Trans Control Syst Technol 17(1):68–77
10. Yuan C, Cao WH, Yuan Y, Wu M, Zhang Y (2014) Integrated modeling for gas mixing process based on mechanism and subspace identification. In: The 26th Chinese control and decision conference, pp 1912–1917
11. Li Z (1998) Discussion on methods of stabilizing caloric value of mixed gas used for a heating furnace. Metall Power 2:14–17 (In Chinese)
12. Niemueller T, Zug S, Schneider S, Karras U (2016) Knowledge-based instrumentation and control for competitive industry-inspired robotic domains. Künstliche Intelligenz 30(3–4):1–11
13. Liu J (2018) Intelligent control design and Matlab simulation. Springer, New York
14. Zuhrie MS, Munoto, Hariadi E, Muslim S (2018) The design of artificial intelligence robot based on fuzzy logic controller algorithm. IOP Conf Ser: Mater Sci Eng 336(1):012020-1-012020-9
15. Xie S, Xie Y, Yang C, Gui W, Wang Y (2018) Distributed parameter modeling and optimal control of the oxidation rate in the iron removal process. J Process Control 61:47–57
16. Cai ZX (1997) Intelligent control: principles, techniques and applications. World Scientific Publishing Co., Ltd., Singapore
17. Åström KJ, Anton JJ, Årzén KE (1986) Expert control. Automatica 22(3):277–286
18. Wang H, Sheng C, Lu X (2017) Knowledge-based control and optimization of blast furnace gas system in steel industry. IEEE Access 5(99):25034–25045
19. Albusac J, Vallejo D, Castro-Schez JJ, Gzlez-Morcillo C (2018) An expert fuzzy system for improving safety on pedestrian crossings by means of visual feedback. Control Eng Pract 75:38–54
20. Hou LL, Yang HP, Zhao YJ (1995) Microcomputer control system design for coke oven gas collecting pressure process. Coal Convers 18(2):91–95 (In Chinese)
21. Yang CH, Wu M, Shen DY, Deconinck G (2001) Hybrid intelligent control of gas collectors of coke ovens. Control Eng Pract 9(7):725–733
22. Mahmoud MS (2018) Fuzzy control, estimation and diagnosis. Springer, New York
23. Su X, Xia F, Liu J, Wu L (2018) Event-triggered fuzzy control of nonlinear systems with its application to inverted pendulum systems. Automatica 94:236–248
24. Precup RE, David RC, Petriu EM (2016) Grey wolf optimizer algorithm-based tuning of fuzzy control systems with reduced parametric sensitivity. IEEE Trans Ind Electron 64(1):527–534
25. Ahmed MS, Bhatti UL, Al-Sunni FM, El-Shafei M (2001) Design of a fuzzy servo-controller. Fuzzy Sets Syst 124(2):231–147
26. Yuvapriya T, Lakshmi P (2017) Design of fuzzy logic controller for reduction of vibration in full car model using active suspension system. Asian J Res Soc Sci Hum 7(3):302–313
27. Osorio R, Alonso JM, Vazquez N, Pinto SE, Sorcia-Vazquez FDJ, Martnez M, Barrera LM (2018) Fuzzy logic control with an improved algorithm for integrated led drivers. IEEE Trans Ind Electron 65(9):6994–7003
28. Hu X, Xu B, Hu C (2018) Robust adaptive fuzzy control for HFV with parameter uncertainty and unmodelled dynamics. IEEE Trans Ind Electron 65(11):8851–8860

29. Feng G, Lai C, Kar N (2016) A closed-loop fuzzy logic based current controller for PMSM torque ripple minimization using the magnitude of speed harmonic as the feedback control signal. IEEE Trans Ind Electron 64(4):2642–2653
30. Eker İ, Torun Y (2006) Fuzzy logic control to be conventional method. Energy Convers Manag 47(4):377–394
31. Li CH, Liu CS, Li HJ, Zheng ZM (1997) Development and application of a pressure self-optimizing fuzzy control system for gas collector of coke oven. Metall Ind Autom 5:32–34 (In Chinese)
32. Zhou B, Li WY (2003) Application of fuzzy theory on gas collector control. In: International conference on machine learning and cybernetics, pp 2516–2519
33. Fang KL, Zhou HJ, Huang WH (2003) Fuzzy decoupling adjustment for the pressure of coke oven collecting main. In: International conference on machine learning and cybernetics, pp 2605–2608
34. Li H, Yang S, Ren H (2016) Dynamic decoupling control of DGCMG gimbal system via state feedback linearization. Mechatronics 36:127–135
35. Fang J, Yin R, Lei X (2015) An adaptive decoupling control for three-axis gyro stabilized platform based on neural networks. Mechatronics 27:38–46
36. Wang Y, Chai T, Fu J, Sun J, Wang H (2013) Adaptive decoupling switching control of the forced-circulation evaporation system using neural networks. IEEE Trans Control Syst Technol 21(3):964–974
37. Zhu HL, Meng Z, Lu W, Zhu ZN, Zhang D (2015) Control system design for lattice distortion modification test equipment based on fuzzy compensated decoupling. Appl Mech Mater 709:308–311
38. Gao SZ, Yang J, Wang JS (2014) D-FNN based modeling and BP neural network decoupling control of PVC stripping process. Math Probl Eng 835–892
39. Wang C, Zhao W, Luan Z, Gao Q, Deng K (2018) Decoupling control of vehicle chassis system based on neural network inverse system. Mech Syst Signal Process 106:176–197
40. Sun W, Yang DP, Yue D, Cheng YH (2000) Intelligent control for pressure system of gas collecting pipe of coke oven. J China Univ Min Technol 29(5):503–505 (In Chinese)
41. Qin B, Wang X, Wu M (2005) Application of multi-agent system to coke ovens. Iron Steel 40(10):25–28 (In Chinese)
42. Liu XQ, Hao R, Tan DJ, Sun W (2000) Application of compensated decoupling arithmetic in discharge header pressure control system. J China Univ Min Technol 29(2):2115–218 (In Chinese)
43. Cao WH, Wu M, Hou SY (2006) Implementation of intelligent decoupling control method for gas mixing and pressurization. J Central South Univ 37(4):780–785 (In Chinese)
44. Lin B (2003) The computer control on gas calorific value. Control Instrum Chem Ind 30(1):75–77 (In Chinese)
45. Qin L (2002) Control for mixed gas calorific value with decoupling and Smith compensator. Control Eng China 9(4):73–75 (In Chinese)
46. Ren HL, Li SY (2004) Fuzzy decoupling control of pressure and heating power for mixed gas. Process Autom Instrum 25(3):65–67 (In Chinese)
47. Jing YH, Fang CZ (1993) Process control. Tsinghua University Press, Beijing (In Chinese)
48. Bristol EH (1966) On a new measure of interaction for multivariable process control. IEEE Trans Autom Control 11(1):133–134
49. Mizuochi M, Tsuji T, Ohnishi K (2007) Multirate sampling method for acceleration control system. IEEE Trans Ind Electron 54(3):1462–1471
50. Li CT (2003) Review for commissioning of gas mixing and pressuring station. Metall Power 18(6):31–32 (In Chinese)
51. Bartos FJ (1996) Fuzzy logic reaches adulthood. Control Eng 43(10):50–56
52. Lee SJ, Kwon YA (2007) Study on fuzzy reasoning application for sensory evaluation of sausages. Food Control 18(7):811–816
53. Meng JE, Mandal S (2016) A survey of adaptive fuzzy controllers: nonlinearities and classifications. IEEE Trans Fuzzy Syst 24(5):1095–1107

54. Wiktorowicz K (2017) Design of state feedback adaptive fuzzy controllers for second-order systems using a frequency stability criterion. IEEE Trans Fuzzy Syst 25(3):499–510
55. Wu HC (2007) Using fuzzy sets theory and Black-Scholes formula to generate pricing boundaries of European options. Appl Math Comput 185(1):136–146
56. Sha TJ (2003) The capability and application of control valves group. China Instrum 20(1):16–19 (In Chinese)
57. Hu Y, Liu G (2016) Intelligent decoupling of gas collector pressure based on internal model control. In: Chinese control and decision conference, pp 4302–4305
58. Wang JS, Xie C (2006) Redial basis function neural network-based robotic joint intelligent decoupling control. In: Wavelet active media technology and information processing, pp 86–91
59. Fu Y, Chai T (2009) Intelligent decoupling control of nonlinear multivariable systems and its application to a wind tunnel system. IEEE Trans Control Syst Technol 17(6):1376–1384
60. Jantzen J (2007) Foundations of fuzzy control. Wiley, Chichester

Chapter 6
Intelligent Optimization and Control for Reheating Furnaces

A reheating furnace is not only a crucial apparatus but also a principal source of power consumption in a tandem hot-rolling steel mill [1]. Regenerative pusher-type reheating furnaces and compact strip production (CSP) soaking furnaces are common reheating furnaces for hot-rolled production lines of billets. The combustion processes in the reheating furnaces have the features of strong coupling among zones, large inertia, strong nonlinearities, as well as time-varying and distributed parameters. The control objective is to minimize the energy consumption and the amount of scale on the surface of the billets under the condition that the billets are uniformly reheated to the specified temperature for rolling.

Since the temperature control performance and the temperature settings for a rolling mill not only directly affect the quantity and quality of rolled steel, but also significantly influence the fuel consumption and the lifetime of the rollers, there has been increasing interest in the modeling, the optimal control, and the temperature settings of the combustion process in recent years [2–11]. Wang et al. used a decoupling technique and the Smith compensation method in [8] to solve the problems of the strong coupling between variables and a time delay in the reheating furnace. Hu et al. employed a zone method based model with a self-adapting predictive control scheme to dynamically respond to the changes that occurred in the furnace operation. These control methods are all based on a linear model of the process; but since the reheating furnace is highly nonlinear, they do not provide satisfactory control performance. Wang et al. made some efforts in [9] to simulate the thermo-fluid mechanical phenomenon inside the reheating furnace. Kim et al. combined principal component analysis with a hierarchy of expert networks in [10] to extract an inverse model for calculating three zone temperature settings for a given target temperature of a billet at the exit. But the thickness, the type, the initial temperature, or the rolling speed of billets were not taken into account in this model, even though these factors have a significant influence on the zone temperatures. In addition, the processing time

M. Wu et al., *Intelligent Optimization and Control of Complex Metallurgical Processes*, Engineering Applications of Computational Methods 3,
https://doi.org/10.1007/978-981-15-1145-5_6

was very long (1 min/generation). Sibarani et al. employed H_∞ control theory to minimize the temperature gradient within a billet to optimize the zone temperature settings of the reheating furnace [11], and demonstrated the validity of this method through simulations. However, due to the complexity and the large uncertainties of an actual reheating furnace, it is difficult to use it in practice.

This chapter analyzes the combustion processes of a regenerative pusher-type reheating furnace and a CSP soaking furnace, and explains the requirements of the control of production processes. A prediction model of furnace temperature is established based on the mechanism analysis. An optimization and control system for the regenerative heating furnace and the CSP soaking furnace is designed for different control requirements, and the operation effects of the system are tested in an actual production line.

6.1 Process Description and Control Requirements

The reheating furnace being considered includes a regenerative pusher-type reheating furnace and a CSP soaking furnace. In this section, the combustion processes for those furnaces are described and the control requirements are presented.

6.1.1 Combustion Process and Control Requirements for the Regenerative Pusher-Type Reheating Furnace

The combustion process of the regenerative pusher-type reheating furnace is first described, and then its control requirements are presented.

1. Process Description

The regenerative pusher-type reheating furnace considered here has three zones (preheating, heating, soaking), the temperatures of which are coupled. The fuel is a gas–air mixture. Figure 6.1 shows a cross section of the furnace. The key components of a zone are a pair of regenerative burners, one on each side of the furnace, two gate valves for gas, one for each burner, and a reversal valve, which enables the pipes of the burners to switch from supplying air to supplying exhausting flue gas.

A regenerative burner has two functions: burning gas and venting flue gas. When the burner on one side of the furnace is burning gas, the one on the other side is venting flue gas, and their functions are switched about every 2.5 min by the reversal valve. More specifically, regarding the burning (Regenerative burner B in Fig. 6.1), the gate valve allows the gas that has been preheated by the recuperator in the flue into the burner, and a blower blows the air into the regenerator of the burner, where it passes around hot ceramic pellets to preheat it. The gas–air mixture is ejected through nozzles and burned. At the same time, the flue gas is drawn into the regenerator (Regenerative burner A in Fig. 6.1) on the other side of the furnace, where it heats up the ceramic pellets, which constitute the heat exchange medium for preheating

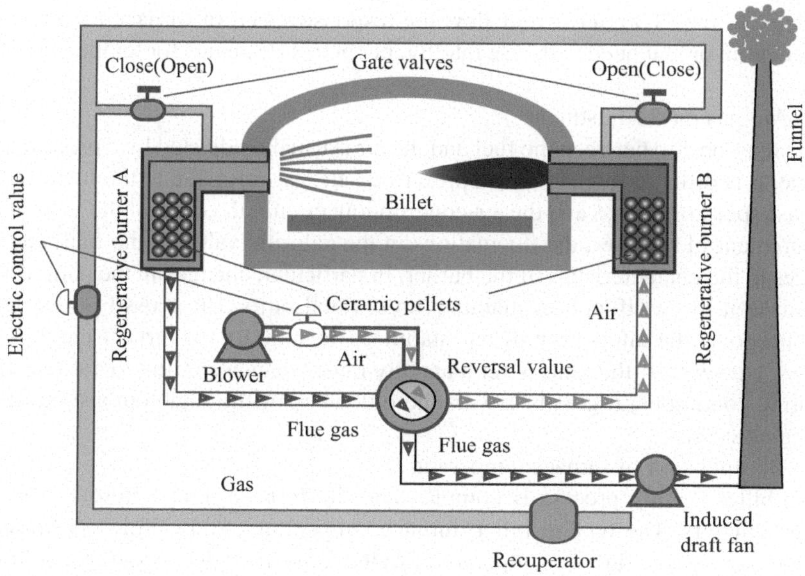

Fig. 6.1 Cross section of zone of regenerative pusher-type reheating furnace

the air during a burning operation. It is discharged through the reversal valve. In addition, two electric control valves regulate the gas and air flows to adjust the zone temperature.

2. Control Requirements

The aim of the reheating furnace combustion process is to control the furnace temperature, the air and gas flows, the air–gas ratio, the air–fuel pressure, the flue gas residual oxygen concentration, and the furnace pressure according to the parameters of a billet, the operating conditions, and the process indicators, thereby reducing the fuel consumption and the oxidation of burning. So, in the actual production process, the advanced control methods are used to reduce the energy consumption and to improve the control accuracy and product quality.

There are four main control requirements in the reheating furnace combustion process with respect to furnace temperature automatic control, air–gas ratio adjustment, furnace temperature optimization settings, and accessory circuit adjustment and parameters correction.

(1) Automatic control of furnace temperature

Dual closed-loop PID control is usually used for the control of temperature and flows. The control of the furnace temperature, and the control of the fuel flow and the air flow compose a cascade control configuration. The set value of the flow control circuit is given by the output of the furnace temperature regulator. In an actual industrial production process, large fluctuations in operating conditions, especially big changes in the gas calorific value and the billet temperature, result in a large overshoot of the PID control and slow the response speed of furnace temperature. So there is an urgent need to devise intelligent control strategies for temperature and flows.

(2) Air–gas ratio adjustment

The relationship between the fuel and the air is usually adjusted by a proportional scheme. It is difficult to guarantee a prescribed air–gas ratio due to the inconsistent response speed of the gas and the air-conditioning circuit, the error of the volumetric measurement of the flow, the fluctuations in the calorific value of the fuel, and the changes in the characteristics of the burner. In particular, changes in the combustion load make it more difficult to maintain an optimal ratio. The cross-limiting technique improves the safety running and stabilizes the ratio for the furnace combustion process. However, if the ratio is set manually rather than according to the real-time operating conditions, it greatly lowers the efficiency of the fuel combustion in the heating process.

(3) Optimization of furnace temperature

The billet heating process is complex and the furnace temperature is affected by many factors. The type of billet, furnace temperature, and rolling rhythm have direct impacts on the furnace temperature. At the same time, the operating conditions are complex and change easily, the changes of gas heat value, gas pressure, and air pressure affect the optimal temperature. The optimization of furnace temperature is the basis to achieve the combustion process control for the reheating furnace.

(4) Circuit adjustment and parameters correction

The key to ensuring an optimal combustion condition includes the regulation of the pressures of the furnace chamber, the main air pipes, and the main gas pipes, and the corrections of the temperature, the pressures, and the calorific value. Those factors are mainly controlled using the PI or PID method. In the combustion control of the reheating furnace, the induced draft fans and the blowers, which are a principal source of energy consumption, usually run at full loads. The amount of air and exhausting gas are controlled by adjusting the baffles of the main air pipe and the main exhausting gas pipe, and the openings of the corresponding air and exhausting gas valves. Thus, it is important to reduce the energy consumption of the reheating furnace by performing an intelligent variable-voltage variable-frequency control of the draft fans and blowers, as well as adjusting the related circuits.

6.1.2 Combustion Process of and Control Requirements for Compact Strip Production Soaking Furnace

This subsection describes the combustion process of the CSP soaking furnace and explains its control requirements.

1. Process Description

The CSP soaking furnace uses a high-efficiency thermal combustion technology. It uses a mixed gas, which is the mixture of two kinds of gases delivered separately

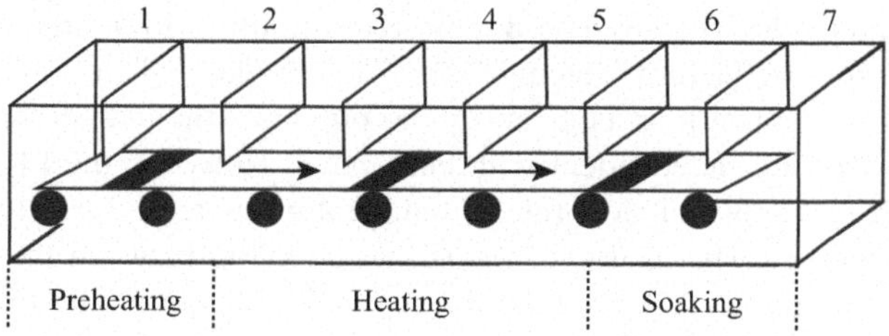

Fig. 6.2 Process of CSP soaking furnace

from a coke-oven and a blast furnace, as the main fuel to heat billets. According to the different heating effects, the soaking furnace is divided into preheating section, heating section, and soaking section. There are no clear boundaries among the three sections.

The billet heating process in the three sections of the soaking furnace is basically the same as that in the regenerative pusher-type reheating furnace. But in accordance with the requirements of the process, the soaking furnace is divided into seven temperature control parts with a big gas valve in each part (Fig. 6.2). The body of the furnace has two sides: south and north. A string of nozzles is arranged at each side. When the mixed gas is ejected into the furnace through the nozzles and burned on one side, the flue gas is drawn into the regenerative burner on the other side. Three reversal valves (one for each section) switch the functions periodically.

When billets enter the heating furnace, the billet temperature gradually reaches the target temperature by preheating, heating, and soaking. The temperature in the preheating section should be maintained in [850, 1100] °C. A thermal stress occurs in a billet at the early stage of heating due to the large change in temperature. So, we need to control the heating rate.

Billets enter the heating section after the preheating section, and the heating section is the most important stage of temperature control in the reheating furnace. The heating degree of a billet determines whether or not the billet is heated thoroughly and whether or not it can be properly discharged from the furnace. The temperature should be maintained in [1150, 1220] °C.

A billet is uniformly heated to 1200–1300 °C in the soaking section. A very high temperature in the soaking section causes the steel slippage phenomenon. On the other hand, since a very low temperature in the soaking section results in that a billet cannot be discharged from the furnace because it will damage a rolling mill.

The temperatures of the preheating, the heating, and the soaking sections are coupled with and affected each other. The furnace temperature distribution in these three sections determines the temperature of a billet, which is an important factor affecting the quality of the product.

2. Control Requirements

Since the combustion process in the soaking furnace is complex, nonlinear, and time-varying, and has a large time delay, it is difficult to establish a model of it using conventional methods. The cascade double cross-limiting PID strategy is usually used for the control of the furnace temperature, and the gas and air flows. The setting value of the flow control loop is given by the output of the temperature regulator, and the PID loops are independent with each other. However, the PID controllers exhibit large overshoots and slow rate of temperature increase for actual industrial production processes. This is due to the strong coupling between the temperature control circuits for different parts. So, this method has a poor ability to suppress the changes in the calorific value and gas pressure.

These problems affect the quality of the billet heating and result in serious billet oxidation and a waste of gas resource. It is important to improve the basic automation control level and to optimize the control of the combustion process of the soaking furnace so as to reduce workers' labor intensity and ensure normal operation in sheet production lines.

In order to solve these problems and to ensure that the heating process of the soaking furnace is desirable, the optimization and control of the furnace temperature and the air–gas ratio is carried out by establishing a relationship model between the gas calorific value and the air–gas ratio, and a model between the changes in furnace temperature and the gas flow. At the same time, the control of the caloric valve is introduced to enhance the control effect, and the furnace temperature control system is also improved. The fuzzy expert control method is used to design the air–gas ratio optimization controller and the gas flow fuzzy expert controller.

The intelligent optimization and control for the CSP soaking furnace is devised to meet the actual production requirements by providing the following countermeasures:

(1) Carry out the closed-loop control of the furnace temperature by establishing a model for the temperature optimization and control. This ensures uniform heating in the soaking furnace. Thus, it improves the accuracy of plate rolling. So, this provides a solid foundation for the improvement of product quality.

(2) Establish a model for the optimization of the air–gas ratio so as to produce appropriate flows of gas and air. This avoids burning through caused by too high temperature of heat exchangers. This also avoids an excess of air and reduces oxidation burning loss, energy consumption, and sparks and smoke.

(3) Establish a model for valve control so as to reduce workers' labor intensity, improve the control accuracy of valves, and extend the service lives of valves.

The optimization and control of the reheating system of the soaking furnace is carried out from three aspects: optimization of the furnace temperature, optimization of the air–gas ratio, and optimal tracking control of the furnace temperature. At the same time, the monitoring screens display the temperature of the soaking furnace, the air–gas ratio, and the valve openings. This simplifies the operation of operators, reduces the labor intensity of workers, improves the control precision of the furnace temperature, implements the optimal setting of the soaking furnace temperature, ensures the balanced heating of each area in the soaking furnace, and finally reduces the energy consumption.

6.2 Temperature Prediction Models

The industrial processes considered here are highly complex systems with nonlinearities, large time delays, uncertainties, and multi-levels. The main features of the systems are the roughness and the imperfectness of measurement information, the randomness of disturbances, dynamic behavior, high orders, high nonlinearities, the uncertainties of dynamic models, the hybridity of continuous and discrete levels, the diversity of levels, the distributed characteristic of state variables, and strong coupling between subsystems. It requires that industrial process control have the ability to learn and recognize the dynamics models of plants, and the ability in the adaptation and robustness to the changes in environment and disturbances. This accelerates the investigation of thinking-simulation-based modeling methods.

Even though the design of an intelligent control system does not need an accurate model, the analysis of control effect and the stability of the system requires a process model. The modeling method based on thinking simulation is expected to solve the problems of the lack of information, incomplete information, an ill structure, computational complexities, real-time requirements, and difficulties in the description of nonlinearities.

AI provides us an effective way to solve some complex problems, and to deal with noise, incomplete data, and nonlinearities. It has been widely used in modeling and forecasting of combustion processes.

In this section, we first analyze dynamic NNs and their learning algorithms that are widely used in industrial processes. Then, we establish recurrent-neural-network (RNN) models to predict the furnace and the steel temperatures. Finally, we combine these NN models with predictive models built based on the heat transfer mechanism and establish an integrated model for the billet temperature prediction.

6.2.1 Recurrent-Neural-Network Model

The thermal process is nonlinear, slowly time-varying, and uncertain, and has a large time delay. So, it is difficult to establish an accurate mathematical model for the process. Conventional linear control methods may be difficult to deal with a big change in the operating condition. Not only does it cause the degradation of control quality, but also it may affect the normal operation. Thus, the establishment of an accurate mathematical model for the thermal process is the basis of improving the performance of the control system. Compared with a feedforward NN, a dynamic NN is more suitable for the description of a dynamic system because of its dynamic memory function.

NNs are primarily classified into two types: feedforward and recurrent. While a feedforward NN is essentially a static model and has the ability to describe a continuous nonlinear function with arbitrary precision, an RNN is basically a dynamic model and can precisely describe the behavior of a dynamic physical system [12].

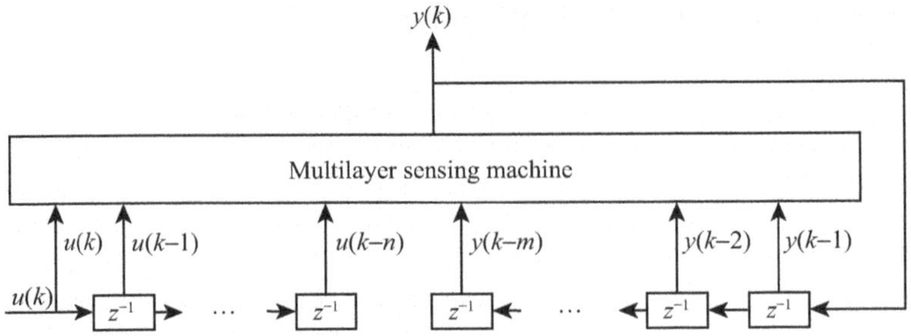

Fig. 6.3 Narenda dynamic neural network

NNs and fuzzy logic (FL) are very useful for the control of a complex system with a large uncertainty [13–20]. A genetic algorithm (GA) is employed to optimize the weights of a neural network classifier [21]. A crucial problem in designing an intelligent controller is how to automatically derive its structure and parameters so as to achieve an acceptable trade-off between accuracy and interpretability. One possible solution is to use a radial-basis-function NN (RBF-NN), which is equivalent to a fuzzy system [22] and is capable of nonlinear mapping [23]. So, the combination of the knowledge representation of fuzzy inference and the nonlinear mapping ability of an RBF-NN can provide a solution to the trade-off problem [24], and it can also yield satisfactory control performance for an uncertain plant.

Dynamic NNs are divided into time-delay NNs (TDNNs) and RNNs. RNNs are also known as feedback NNs, and they are further divided into output feedback networks and state-feedback networks according to the feedback signals [25].

(1) Output Feedback Recurrent Neural Network

Narenda added output feedback on the basis of a TDNN to construct an RNN in 1990 (Fig. 6.3). The network is described as

$$y(k) = f[u(k), u(k-1), \ldots, u(k-n), y(k-1), y(k-2), \ldots, y(k-m)], \tag{6.1}$$

where m is the number of the lag steps in the output vector.

(2) Time-Delay Neural Network

A TDNN is described by a difference equation

$$y(k) = f[u(k), u(k-1), \ldots, u(k-n)], \tag{6.2}$$

where $u(k)$ is the input vector of the neural network at k, $y(k)$ is the output vector of the neural network at k, and n is the number of the lag steps in the input vector. No feedback is used in this network.

(3) State-Feedback Recurrent Neural Network

The structure of a state-feedback dynamic NN, Elman network, is shown in Fig. 6.4. The recurrent layer of the hidden layer is described as

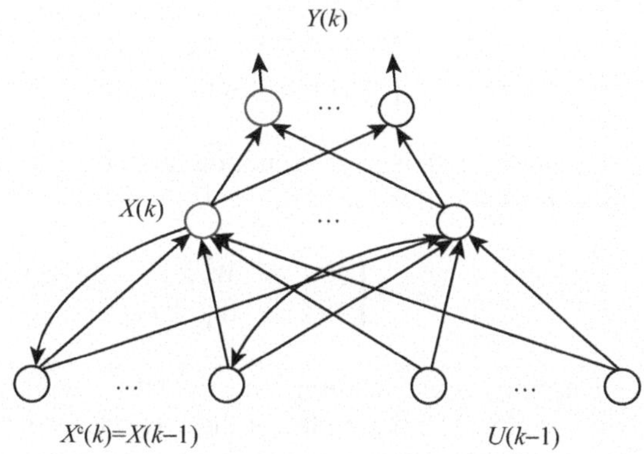

Fig. 6.4 Elman recurrent neural network

$$X^{c}(k) = X(k-1). \tag{6.3}$$

A multi-input, multi-output Elman network is represented by a state-space model

$$\begin{cases} x_i(k) = f\left(\sum_{j=1}^{N} w_{i,j}^{x} x_j(k-1) + \sum_{m=1}^{M} w_{i,m}^{u} u_m(k-1)\right), & i = 1, 2, \ldots, N, \\ y_l(k) = \sum_{i=1}^{N} w_{i,l}^{y} x_i(k), & l = 1, 2, \ldots, L, \end{cases} \tag{6.4}$$

where $w_{i,j}^{x}$, $w_{i,m}^{u}$, and $w_{i,l}^{y}$ is the weights from the jth node in the recurrent layer and from the mth node in the input layer to the ith node in the hidden layer, respectively, and the weight from the ith node in the hidden layer to the lth node in the output layer; $f(\cdot)$ is the excitation function of the hidden neurons; and N, M, and L are the numbers of the hidden neurons, the input neurons, and the output neurons, respectively. In general, a monotonic sigmoid function or a hyperbolic tangent function is chosen to be the nonlinear hidden layer excitation function to realize the modeling and the control of the nonlinear system, for example, $f(x) = 1/(1 + \mathrm{e}^{-x})$.

The three weights are

$$\begin{cases} W_x = [w_{i,j}^{x}]_{N \times N}, \\ W_u = [w_{i,m}^{u}]_{N \times M}, \\ W_y = [w_{i,l}^{y}]_{N \times L}. \end{cases} \tag{6.5}$$

The vector forms are

$$\begin{cases} X(k) = f(W_x X(k-1) + W_u U(k-1)), \\ Y(k) = W_y{}^{\mathrm{T}} X(k), \end{cases} \tag{6.6}$$

where

$$\begin{cases} Y(k) = [y_1(k), \ldots, y_L(k)]^{\mathrm{T}}, \\ U(k-1) = [u_1(k-1), \ldots, u_M(k-1)]^{\mathrm{T}}, \\ X(k) = [x_1(k), \ldots, x_N(k)]^{\mathrm{T}}. \end{cases} \tag{6.7}$$

If a linear hidden layer excitation function is used, then Eq. (6.6) becomes a linear Elman network. If we let $f(x) = x$, then

$$\begin{cases} X(k) = W_x X(k-1) + W_u U(k-1), \\ Y(k) = {W_y}^{\mathrm{T}} X(k). \end{cases} \tag{6.8}$$

The fixed point learning and trajectory learning methods can be applied to a dynamic NN. The basic idea of these two learning algorithms is similar to the back propagation algorithm.

6.2.2 *Estimation of Zone Temperature*

Since the combustion process is dynamic and complex, it is almost impossible to build a precise model of a combustion process based on an analysis of the mechanism(s) involved. Our solution is an RBF-based recurrent NN (RBF-RNN) model for predicting zone temperatures. The model contains a delayed feedback loop and can describe the dynamics of the process with a prescribed precision. In addition, a sequential learning (SL) algorithm is employed to enable the model to adapt to changes in the process in a real-time fashion.

Considering the phase lag and dynamic characteristics of the combustion process and that the air–gas ratio is fixed at a prescribed value, we select the delayed gas flow and the furnace temperature distribution in each furnace temperature control zone as the input in the feedforward NN of the RBF-RNN, and the predicted furnace temperature distribution to be the output of the RBF-RNN. The structure of the RBF-RNN model is given by

$$\hat{T}_{\mathrm{f}}(k) = f(T_{\mathrm{f}}(k-1), \ldots, T_{\mathrm{f}}(k-m), q(k-1), \ldots, q(k-n)), \tag{6.9}$$

where $f(\cdot)$ is the mapping of the RBF-RNN; m and n are the delay steps of the output and input, respectively; q is the gas flow vector that takes the gas flow of each furnace temperature control area as its elements; T_{f} is the furnace temperature distribution vector that takes the furnace temperature of each control zone as its elements; and $\hat{T}_{\mathrm{f}}$ is the predicted output of the furnace temperature. The model plays an important role in the prediction of the furnace temperature at the next step.

The structure is shown in Fig. 6.5. The RBF-RNN is trained using the SL algorithm that combines growing and pruning with an extended Kalman filter (GAP-EKF). The SL algorithm using the GAP-EKF starts with no hidden neuron. It adds hidden neurons according to the significance of the present observation or regulates the network parameters by means of the EKF and determines whether or not a hidden

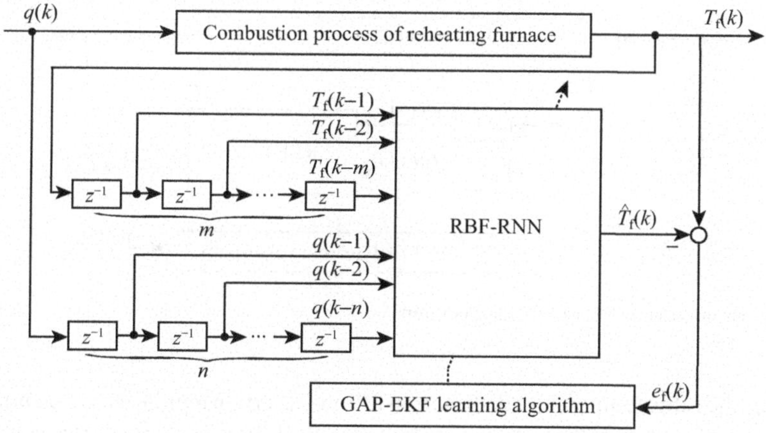

Fig. 6.5 Estimation model of zone temperature

neuron should be removed based on the pruning criterion. This results in a compact RNN. SL is faster than batch learning in online applications because it does not require retraining [26, 27]; and an EKF converges more quickly than a BP algorithm in the calculation of the gradient of the cost function [28]. Thus, the convergence speed of a sequential GAP-EKF is higher than that of a dynamic BP algorithm, which is usually used to train an NN. The selection of the parameters for the RBF-RNN is discussed in [29].

6.2.3 Estimation of Billet Temperature

According to the mechanism analysis of the billet heating process, the temperature distribution of a billet in the furnace at a time point (or a position) depends on the temperature distributions of the furnace and the billet at the previous time point, and its input and output relationship is given by

$$\hat{T}_b(t) = G(T_b(t-1), \tilde{T}_f(t-1)), \quad 0 \leqslant t \leqslant t_r, \tag{6.10}$$

where $T_b(t) = [T_b(0,t), \ldots, T_b(h,t)]^T$ is the billet temperature distribution, h is the billet thickness, $\tilde{T}_f(t)$ is the furnace temperature at the place where the billet is, and t_r is the total time for heating the billet. The discretization of (6.10) yields

$$\hat{T}_b(k) = G(T_b(k-1), \tilde{T}_f(k-1)). \tag{6.11}$$

The structure is shown in Fig. 6.6, where $\hat{T}_b(k)$ is the predicted output.

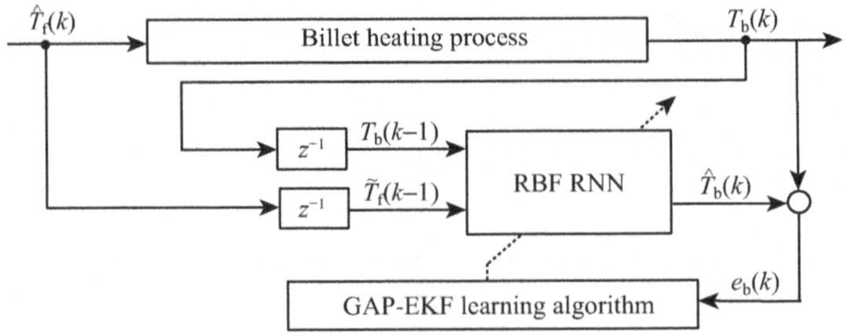

Fig. 6.6 Estimation model of billet temperature

Since the billet temperature distribution follows a certain regularity, we do not need to observer the whole temperature distribution of the billets, but only the temperatures at two points: one on the surface and one at the center of the billet. This gives

$$T_b(k) = \left[T_b(0, k), T_b\left(\frac{h}{2}, k\right) \right]^T. \tag{6.12}$$

The RNN for prediction of the billet temperature distribution requires online learning, which imposes a strict requirement for the real-time adjustment of the model. And we can choose the SL algorithm because of its rapid learning speed and high learning precision.

6.2.4 Integrated Model of Billet Temperature Prediction

The key to the optimization and setting of the furnace temperature is to achieve a precise online forecast of the billet temperature in the heating process. The conventional modeling method for the billet temperature prediction is based on the mechanism of the energy conservation. A model is usually described by partial differential equations of heat transfer, and an NN can also be used to establish a nonlinear model of the process. Note that an RNN is suitable for the online prediction of the billet temperature and self-adjustment but is not suitable for the offline investigation of the optimization and setting of the furnace temperature because no new sample data can be obtained. Thus, it is possible to build a temperature prediction model for a billet based on the heat transfer mechanism. And then, incorporating it with the RNN model performs the prediction of the billet temperature with high precision.

1. Heat Transfer Model

A billet moves at a certain speed after entering the furnace. Its surface is heated by radiation from the atmosphere and the walls of the furnace, and the heat is transferred from the surface of the billet to the inside by convective heat transfer so as to achieve required average temperature and sectional temperature difference of the billet. So, the billet temperature prediction model has two parts: an equation of heat conduction and the related boundary condition.

A heat transfer model is used to predict the temperature distribution and mean temperature of a billet based on the following assumptions:

(1) The furnace temperature is a linear function of time and position along the length of the furnace.

(2) Heat transfer can be ignored between a billet and fixed beams.

(3) The total heat absorptance in one furnace section is constant.

The billet temperature prediction model is a model that tracks the heating process of a billet and describes the temperature variation of the entire heating process of the billet. The position of a billet in the furnace changes with time. And a moving coordinate system, which moves synchronously with the billet, is usually used so that the boundary condition of the billet is transformed into a time-dependent temperature field. The heating of a billet in the heating furnace features that heating flow in the lateral direction is basically the same, and the amount of transferred heat in the longitudinal direction in a billet is much lower than the amount of heat transferred between the surface of a billet and the atmosphere in the furnace. So, the heat transfer differential equation can be simplified to be

$$\frac{\partial T(x,t)}{\partial t} = \frac{1}{c\rho}\frac{\partial}{\partial x}\left[\lambda_{\mathrm{r}}\frac{\partial T(x,t)}{\partial x}\right], \quad 0 \leqslant t \leqslant t_{\mathrm{r}}, \ 0 \leqslant x \leqslant h, \tag{6.13}$$

where $T(x,t)$ is a specified billet temperature distribution, x is the coordinate in the direction of the thickness of the billet, λ_{r} is the coefficient of heat transfer of the billet, t_{r} is the billet heating time in the furnace, h is the thickness of the billet, c is the specific heat capacity of the billet, and ρ is the density of the billet.

The position of a billet in the furnace, y, is determined by the moving speed of the billet v:

$$y(t) = \int_0^t v(\tau)\mathrm{d}\tau, \quad 0 \leqslant t \leqslant t_{\mathrm{r}}. \tag{6.14}$$

The boundary condition of the tracking model of a billet is

$$\lambda_{\mathrm{r}}\frac{\partial T(x,t)}{\partial x} = \begin{cases} -Q(y,t), & x = 0, \\ 0, & x = h, \end{cases} \tag{6.15}$$

where $Q(y,t)$ is the density of heating flow on the billet surface.

Except for the temperature, many other factors also affect the heat transfer from atmosphere to the billet surface, but the mechanisms are complex. According to Stefan–Boltzmann law, heat is mainly exchanged by radiation at high temperatures. Thus, only a small percentage of heat transfer is convected heat. In order to simplify the mathematical description of the heat transfer process on the surface of a billet, we consider the convected heat and the heat transfer caused by other factors in the radiation heat transfer process, and give the mathematical expression of overall heat transfer to the surface of a billet in the furnace as

$$Q(y,t) = \Phi(y)\sigma\left[T_f^4(y,t) - T^4(0,t)\right], \tag{6.16}$$

where $\Phi(y)$ is the overall heat absorptance of the surface, $\sigma = 5.67 \times 10^{-8}\,\mathrm{W/(m^2 \cdot K^4)}$ is the Stefan–Boltzmann constant, and $T_f(y,t)$ is the furnace temperature.

For the heating of a cold billet, the initial distribution of the billet temperature distribution can be set at room temperature, T_0, that is,

$$T(x,0) = T_0. \tag{6.17}$$

The temperature of a billet at time t, $T_m(t)$, is

$$T_m(t) = \frac{1}{h}\int_0^h T(x,t)\mathrm{d}x. \tag{6.18}$$

Combining Eqs. (6.13), (6.15), (6.16), and (6.18) yields

$$\frac{\partial}{\partial t}T_m(t) = \frac{\Phi(y)\sigma}{hc\rho}\left[T_f{}^4(y,t) - T^4(0,t)\right]. \tag{6.19}$$

2. Integrated Model

Generalization ability is an important index to measure the result of NN learning. The larger the number of the test samples that can be correctly classified using a trained NN is, the higher the generalization ability of the NN is. While the RNN model has a high learning speed, it has high prediction accuracy only near the current sample, and its generalization ability is poor. On the other hand, the heat transfer model is obtained by fitting the whole billet heating process, and it is in good agreement with all data of the billet heating process. Therefore, it is desirable to perform the soft measurement of the steel temperature by taking full advantage of the above two kinds of steel temperature prediction models.

Figure 6.7 shows the structure of the integrated model for the steel temperature prediction. $T_{b,1}(k)$, $T_{b,2}(k)$, and $T_b^o(k)$ are the predicted steel temperatures given by the RNN model and the heat transfer model, and the steel temperature target value at time k, respectively.

Assume $e_1(k) = \left\|T_{b,1}(k) - T_b^o(k)\right\|$ and $e_2(k) = \left\|T_{b,2}(k) - T_b^o(k)\right\|$. Then, the model selection rule of the coordination algorithm is

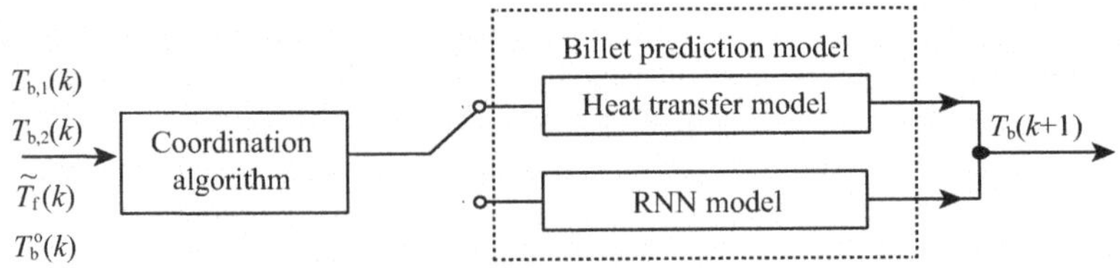

Fig. 6.7 Integrated model of billet temperature prediction

R_1: **IF** $(e_1(k) \leqslant e_2(k))$, **THEN** output the steel temperature prediction result of the RNN model at time $k+1$.

R_2: **IF** $(e_1(k) > e_2(k))$, **THEN** output the steel temperature prediction result of the heat transfer model at time $k+1$.

6.3 Optimization and Control for Regenerative Pusher-Type Reheating Furnace

PID decoupling is a conventional control method for multivariable systems [30]. It requires precise values for the parameters of a plant to carry out satisfactory decoupling. To obtain good control performances for the combustion process, a zone temperature control system is established based on a fuzzy NN decoupling controller (FNNDC), for which the structure is determined by fuzzy c-means (FCM) clustering and the weight parameters are adjusted using a hybrid particle swarm optimization (HPSO) algorithm. In addition, the HPSO is also used to determine the optimal zone temperature settings for the furnace.

6.3.1 Configuration of Optimization and Control System

The temperature distribution of the furnace is influenced by the flow and calorific value of the gas, the rolling speed, the initial billet temperature, periodic changes in the gas flow caused by the opening and the closing of the gate valves, etc. Furthermore, these factors interact with each other. Due to the strong nonlinearity of the process, conventional linear control methods cannot meet practical technical requirements. This subsection presents an HPSO-based adaptive fuzzy NN decoupling control (FNNDC) method that solves the temperature control problem. The configuration of the control system is shown in Fig. 6.8. The output of the system is $T_z(k) = [T_p(k)\ T_h(k)\ T_s(k)]^T$, and the reference input is $r(k) = [r_p(k)\ r_h(k) r_s(k)]^T$.

The control system contains two feedback loops. The outer FNNDC loop carries out zone temperature decoupling control to produce appropriate settings for the gas and air flows. The inner PID control loop adjusts those flows. It employs the PID control with cross magnitude limiting to guarantee that the flows quickly track given values and are within a reasonable range. The parameters of the PID controllers are automatically tuned by the NN modules of the DCS, and the control loops for different zones are independent of each other. Since the time constants of the flows are much smaller than those of the zone temperatures, this two-control-loop strategy easily provides the good zone temperature control. Note that, in the inner PID control loop, an insufficient supply of the air results in incomplete combustion and leads to air pollution and reduced thermal efficiency, while excess oxygen increases the amount of scale. So, it is important to determine the optimal air–gas ratio, L_0, under ideal

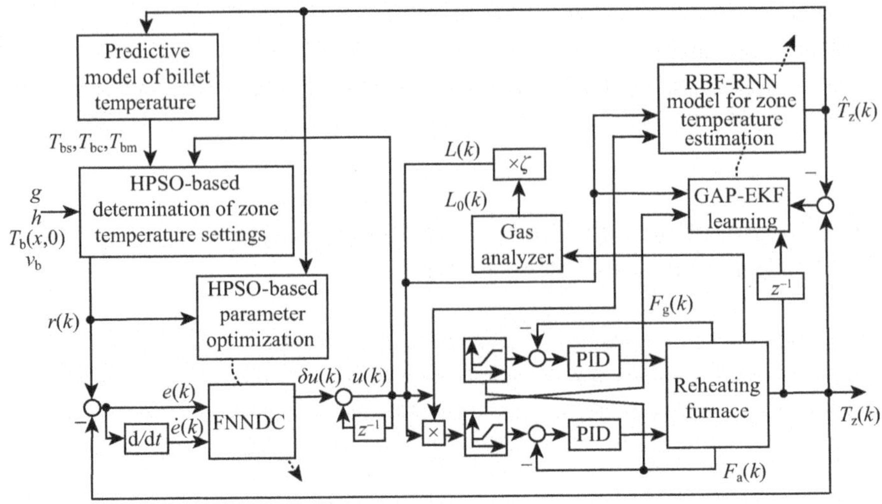

T_{bs}/T_{bc} : surface/center temperature of billet at exit

T_{bm} : mean temperature of billet at exit

$T_z(k)/\hat{T}_z(k)$: actual/predicted zone temperature vector

$T_b(x,0)$: initial temperature of billet

$r(k)$: target zone temperature vector

$e(k)$: error vector of zone temperature

$L_0(k)/L(k)$: ideal/actual air-gas ratio

$F_g(k)/F_a(k)$: detected gas/air flow

$\dot{e}(k)$: rate of change of $e(k)$

$u(k)$: gas flow vector

$\delta u(k)$: change in gas flow vector

T_b : rolling speed

g : billet type

h : billet thickness

ζ : air surplus coefficient

Fig. 6.8 Configuration of zone temperature FNNDC system

conditions. A gas analyzer directly shows the minimum amount of the air needed. However, since the gas–air mixture is inhomogeneous, the actual air–gas ratio, L, needed to achieve complete combustion of the gas is usually larger than L_0. It is given by

$$L = \zeta L_0, \tag{6.20}$$

where ζ is an air surplus coefficient with a value in the range [1.05, 1.20] for an actual production process.

An analysis of actual runs showed that the combustion process was bounded-input, bounded-output (BIBO) stable. Restricting the control inputs produced by the FNNDC for the three zone temperatures and the PID controllers for the gas and air flows within prescribed ranges is a very practical strategy to ensure the BIBO stability of the control system.

6.3.2 Decoupling Control Based on Fuzzy Neural Network

A heat transfer analysis shows that the general direction of the heat flow in the furnace is from the exit toward the entrance (hotter to cooler), so the temperature of one zone strongly influences that of the zone(s) before it. We exploit this fact to solve the coupling problem by employing a controller with the structure in Fig. 6.9 and by adjusting the weights connecting the hidden neurons of the heating zone to the output neuron of the preheating zone, and those of the soaking zone to that of the heating zone. The inputs of the FNNDC are the vector of the temperature error, $e(k) = [e_{\mathrm{p}}(k), e_{\mathrm{h}}(k), e_{\mathrm{s}}(k)]^{\mathrm{T}}$, and the vector of the rate of change in the error, $\dot{e}(k) = [\dot{e}_{\mathrm{p}}(k), \dot{e}_{\mathrm{h}}(k), \dot{e}_{\mathrm{s}}(k)]^{\mathrm{T}}$. The output is the increase of control input $\delta u(k) = [\delta u_{\mathrm{p}}(k), \delta u_{\mathrm{h}}(k), \delta u_{\mathrm{s}}(k)]^{\mathrm{T}}$. The temperature error is given by

$$e(k) = r(k) - T_{\mathrm{z}}(k). \tag{6.21}$$

The decoupling control law extracted from the knowledge of experts is simple and static. It can easily be represented by an FNN. Fuzzy logic (FL) is a model-free method that depends on the experience of experts and heuristic evaluation, but it lacks

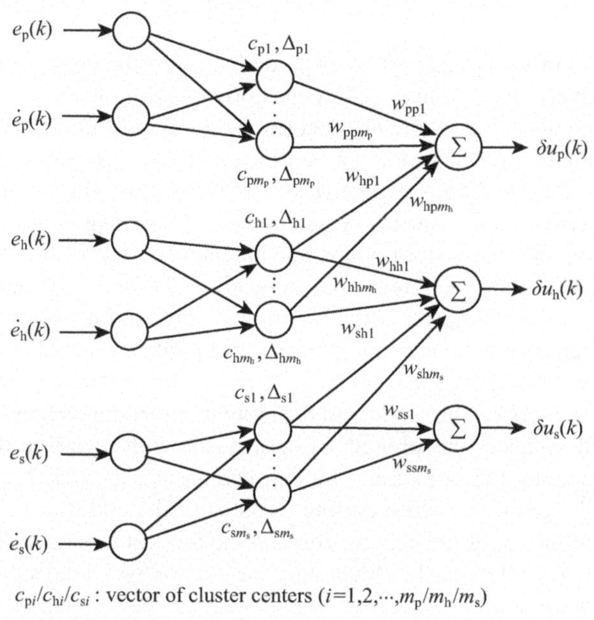

Fig. 6.9 Configuration of FNNDC

$c_{pi}/c_{hi}/c_{si}$: vector of cluster centers (i=1,2,⋯,$m_p/m_h/m_s$)

$\Delta_{pi}/\Delta_{hi}/\Delta_{si}$: width matrix (i=1,2,⋯,$m_p/m_h/m_s$)

$w_{ppi}/w_{hhi}/w_{ssi}$: output weight (i=1,2,⋯,$m_p/m_h/m_s$)

w_{hpi}/w_{shi} : decoupling weight (i=1,2,⋯,m_h/m_s)

$m_p/m_h/m_s$: number of neurons

the capabilities of online self-learning and self-adjustment. By fusing the knowledge representation of the FL with the self-adaptation of an NN, we devised an FNNDC to implement zone temperature control. It adapts to environmental changes by means of parameter adjustments made by an HPSO. It employs an FNN to make expert forward inferences and consists of three FNN inference machines (FNNIMs), one for each zone, and a defuzzifier with a decoupling function. Since the three FNNIMs all have the same structure, we used the one for the preheating zone as an example and explain it in detail. It is described by

$$\begin{aligned} R_{\mathrm{p}i} : \textbf{IF } (e_{\mathrm{p}}(k) \in A_{\mathrm{ep}i}) \text{ AND } (\dot{e}_{\mathrm{p}}(k) \in A_{\dot{\mathrm{e}}\mathrm{p}i}), \textbf{ THEN} \\ \delta u_{\mathrm{p}}(k) \text{ is chosen from } B_{\mathrm{p}i}, \quad i = 1, 2, \ldots, m_{\mathrm{p}}. \end{aligned} \tag{6.22}$$

The hidden layer clusters the inputs of the network and employs the following Gaussian function to calculate the activations of the neurons in the first layer

$$\phi_{\mathrm{p}i}(k) = \exp\left\{-[x_{\mathrm{p}}(k) - c_{\mathrm{p}i}]^{\mathrm{T}} \Delta_{\mathrm{p}i}^{-2} [x_{\mathrm{p}}(k) - c_{\mathrm{p}i}]\right\}, \quad i = 1, 2, \ldots, m_{\mathrm{p}}. \tag{6.23}$$

And the vector of cluster centers $c_{\mathrm{p}i}$ and the width matrix $\Delta_{\mathrm{p}i}$ are given by

$$c_{\mathrm{p}i} = \begin{bmatrix} c_{\mathrm{ep}i} \\ c_{\dot{\mathrm{e}}\mathrm{p}i} \end{bmatrix}, \quad \Delta_{\mathrm{p}i} = \begin{bmatrix} \sigma_{\mathrm{ep}i} & 0 \\ 0 & \sigma_{\dot{\mathrm{e}}\mathrm{p}i} \end{bmatrix}. \tag{6.24}$$

In Eqs. (6.22)–(6.24), $c_{\mathrm{ep}i}$ and $c_{\dot{\mathrm{e}}\mathrm{p}i}$ are the cluster centers of e_{p} and $\dot{e}_{\mathrm{p}}$, respectively, and $\sigma_{\mathrm{ep}i}$ and $\sigma_{\dot{\mathrm{e}}\mathrm{p}i}$ are the corresponding widths, $x_{\mathrm{p}}(k) = [e_{\mathrm{p}}(k),\ \dot{e}_{\mathrm{p}}(k)]^{\mathrm{T}}$ is the input vector, where m_{p} denotes the number of the fuzzy rules, which is equal to the number of the m hidden neurons, and $A_{\mathrm{ep}i}$, $A_{\dot{\mathrm{e}}\mathrm{p}i}$, and $B_{\mathrm{p}i}$ are the ith fuzzy sets of e_{p}, $\dot{e}_{\mathrm{p}}$, and δu_{p}, respectively. The FCM clustering algorithm relies on training patterns to determine the cluster centers. Analyzing the clustering results and taking the requirements of simplicity and precision into account, we determined the numbers of clusters for the three zones to be $m_{\mathrm{p}} = 21$, $m_{\mathrm{h}} = 26$, and $m_{\mathrm{s}} = 27$. The width of the corresponding Gaussian function is equal to the average distance between a cluster center and the training patterns, and it takes values in the ranges [39, 100] for $e_{\mathrm{p}}(k)$ and [40, 105] for $\dot{e}_{\mathrm{p}}(k)$.

$\phi_{\mathrm{p}i}(k)$ is the product of the input membership values in the ith rule of the FNNIM. It indicates the strength of the ith rule with regard to the inputs. The weights connecting the output and hidden layers, $w_{\mathrm{pp}i}$ $(i = 1, 2, \ldots, m_{\mathrm{p}})$, are the changes in the gas flow corresponding to the ith rule, and $w_{\mathrm{hp}i}$ $(i = 1, 2, \ldots, m_{\mathrm{h}})$ cancels the influence of the heating zone on the preheating zone. The output layer defuzzifies a fuzzy inference by calculating the linearly weighted sum of the outputs of the hidden neurons

$$\delta u_{\mathrm{p}}(k) = \frac{\sum_{i=1}^{m_{\mathrm{p}}} w_{\mathrm{pp}i}\phi_{\mathrm{p}i}(k) + \sum_{i=1}^{m_{\mathrm{h}}} w_{\mathrm{hp}i}\phi_{\mathrm{h}i}(k)}{\sum_{i=1}^{m_{\mathrm{p}}} \phi_{\mathrm{p}i}(k) + \sum_{i=1}^{m_{\mathrm{h}}} \phi_{\mathrm{h}i}(k)}. \tag{6.25}$$

The call interval of the HPSO is much longer than the control period of the FNNDC because it is necessary to optimize the FNNDC only when the performance deteriorates owing to a change in status, such as a change in the billet type. The performance of the FNNDC is evaluated using

$$J_{\mathrm{c}} = \sum_{k=7}^{N_{\mathrm{r}}} \|M[r(k) - \hat{T}_{\mathrm{z}}(k)]\|^2, \tag{6.26}$$

where $M = \mathrm{diag}\{0.2, 0.3, 0.5\}$ represents the weights for the preheating, the heating, and the soaking zones, and N_{r} is the prescribed number of the evaluation steps. In this chapter, N_{r} was set to 20, $w_{\mathrm{pp}i}$ and $w_{\mathrm{hp}i}$ are adjusted by the HPSO algorithm explained below. Since we focused on the steady state, we ignored the transient response ($k = 1, 2, \ldots, 6$).

6.3.3 Optimization for Temperature

Based on a metaphor for social behavior, Kennedy and Eberhart developed PSO as a parallel evolutionary computation method in 2002 [31]. The PSO, which searches for an optimal solution on the basis of past information and interactions among particles, is a population-based optimization technique that provides a quick search of the solution space [32, 33]. It has been used to solve the optimization problem for an FNN and has been used in many applications in control engineering, e.g., controlling a brushless DC Motor Drives [34] and a voice-controlled robot [35]. But the phenomenon of population regression due to information sharing may produce local minima, and there is no satisfactory refined local search capability [36]. To solve this problem, Lin et al. improved the PSO by introducing a local approximation, a multi-elite strategy, and recursive singular value decomposition [37]. Javidrad et al. combined PSO with simulated annealing together, which integrated the merits of good exploration capability of PSO and good local search properties of simulated annealing [38]. These improved types of PSOs are called HPSOs.

1. Optimization Scheme

Determining the optimal zone temperature settings is essential to the temperature control of the combustion process. For a billet at the exit, we let the difference between the mean temperature, T_{bm}, and the target temperature, T_{bm}^*, be

$$\delta T_{\mathrm{bm}} = T^*_{\mathrm{bm}} - T_{\mathrm{bm}}(r), \tag{6.27}$$

and the difference between the temperature of the surface, T_{bs}, and that of the center, T_{bc}, be

$$\delta T_{\mathrm{bsc}} = T_{\mathrm{bs}}(r) - T_{\mathrm{bc}}(r). \tag{6.28}$$

The following performance index is used to formulate the decision problem:

$$J_{\mathrm{s}}(r) = \frac{1}{2}k_1 \delta T^2_{\mathrm{bm}}(r) + \frac{1}{2}k_2 \delta T^2_{\mathrm{bsc}}(r) + \frac{1}{2}k_3 \int_0^{t_{\mathrm{s}}} \|u(r,t)\| \mathrm{d}t, \tag{6.29}$$

where k_1, k_2, and k_3 ($k_1, k_2 \gg k_3$) are weighting coefficients, and t_{s} is the evaluation time for the FNNDC ($t_{\mathrm{s}} < t_{\mathrm{r}}$). The values of k_1, k_2, and k_3 should be selected so that the three items in Eq. (6.29) have the same order of magnitude.

The optimization problem is to find an optimal temperature vector, namely, $r = [r_{\mathrm{p}},\ r_{\mathrm{h}},\ r_{\mathrm{s}}]^{\mathrm{T}}$, such that

$$\min J_{\mathrm{s}}(r) \tag{6.30}$$

subject to the physical constraints

$$r_{\mathrm{L}} \leqslant r \leqslant r_{\mathrm{H}} \tag{6.31}$$

and technical constraints, such as

$$|\ \delta T_{\mathrm{bm}}\ | \leqslant 30\,^{\circ}\mathrm{C}, \quad |\ \delta T_{\mathrm{bsc}}\ | \leqslant 30\,^{\circ}\mathrm{C}. \tag{6.32}$$

In Eq. (6.31), r_{L} and r_{H} are the permissible lower and upper limit vectors, respectively, and the optimal temperature vector, r, is determined by the HPSO algorithm described below. To reduce the thickness of the iron scale on a billet caused by surface melting and carbon loss during the reheating process, the surface temperature of the billet should not be very high, and the furnace temperature cannot exceed the permissible upper limit for the fireproof materials. Moreover, the temperature gradient of the billet must be kept under a certain level, and this along with the strong interaction between zones necessitates that the gradient of the furnace temperature in the lengthwise direction also be under a certain level.

2. Optimum Setting of Furnace Temperature

According to the process analysis, the furnace temperature can be described as a quadratic function of the longitudinal direction of the furnace

$$u(s) = a + bs + cs^2, \tag{6.33}$$

where s is the longitudinal position. Equation (6.33) has to satisfy the following constraints:

$$u_{mi} \leqslant a + bs + cs^2 \leqslant u_{Mi}, \quad s = s_i, \tag{6.34}$$

where u_{mi} and u_{Mi} are the lower and upper limits of the furnace temperature setting for the furnace at the ith section, respectively.

Taking into account the requirements on the heating quality of billets and the minimization of energy consumption, the optimization problem is set to be

$$\min J = \min \left\{ \frac{1}{2} P[T_m(s_e) - T_m^*(s_e)]^2 + \frac{1}{2} Q[T_s(s_e) - T_c(s_e)]^2 + \frac{1}{2} R \int_0^{t_e} u^2(s)dt \right\}, \tag{6.35}$$

where $T_m(s_e)$, $T_s(s_e)$, $T_c(s_e)$, and $T_m^*(s_e)$ are the predicted average temperature and the surface temperature of the billet, the center temperature of the billet, and the average expected temperature of the billet, respectively; $u(s)$ is the furnace temperature; s is the longitudinal position of the billet in the furnace; t_e is the billet heating time; and P and Q are the weights that satisfy that $P, Q \gg R$.

Optimal a^*, b^*, and c^* of the coefficients a, b, and c are obtained using the prediction model for different types of billets. Then, optimal temperature settings are calculated based on those estimates for the three sections.

For example, the coefficients of the temperature prediction model for HR345-type billets are obtained by using the internal penalty function method to solve the aforementioned multi-constrained nonlinear optimal problem:

$$a^* = 935.23, \quad b^* = 14.88, \quad c^* = -0.17.$$

The temperature profile of the corresponding furnace along the furnace length is shown in Fig. 6.10. According to the optimal distribution curve of the furnace temperature, the optimum temperature settings of the three-section furnace temperature are

$$\begin{cases} \text{Preheating section: } u^*(s_1) = a^* + b^* s_1 + c^* {s_1}^2 = 950, \\ \text{Reheating section: } u^*(s_2) = a^* + b^* s_2 + c^* {s_2}^2 = 1200, \\ \text{Soaking section: } u^*(s_3) = a^* + b^* s_3 + c^* {s_3}^2 = 1250. \end{cases}$$

The optimal temperature settings for different types of billets are obtained using this method (Table 6.1).

3. Particle Swarm Optimization

PSO can solve many nonlinear optimization problems with a multimode structure. It starts with a population of random candidates, which are conceptualized as particles. Each particle has two attributes: position and velocity. The velocity (speed and direction) of particles is adjusted so that each particle flies toward a global best based on its own experience and the social information of the particle swarm, i.e., population. The search procedure, or in other words, the flight of a particle, seeks through the solution space. Let N_p be the population size of the particle swarm, $p_j(n)$ be the

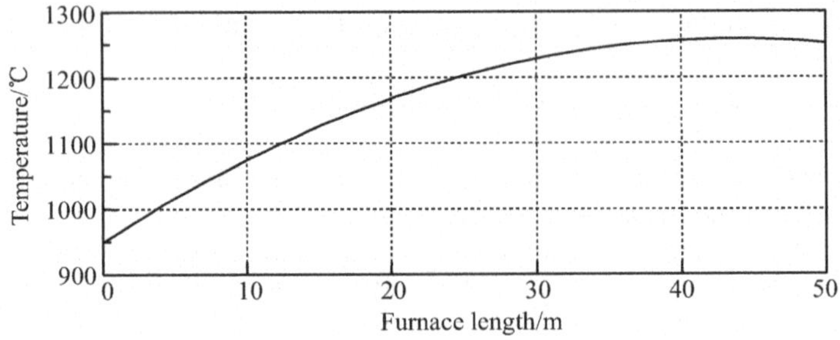

Fig. 6.10 Optimization curve of billet temperature

Table 6.1 Optimal temperature settings of furnace for different types of billets

Billet type	Optimal furnace temperature setting/°C		
	Preheating section	Reheating section	Soaking section
LZ06	700	1000	1200
45	820	1150	1250
50	850	1200	1250
50AA	716	1104	1218
55	800	1150	1250
60	665	1120	1200
Q345A	800	1050	1200
HR345	950	1200	1250
23MnV	800	1100	1220
40Mn	700	1100	1250
Q195	900	1120	1250
Q235B	780	1100	1200
LZ08	900	1115	1220

best position that the jth particle has achieved so far for $j = 1, 2, \ldots, N_p$, $p_g(n)$ be the global best position that the whole population has achieved so far, and $z_j(n)$ and $v_j(n)$ be the position and velocity of the jth particle at the nth iteration, respectively. Then, $z_j(n)$ and $v_j(n)$ are adjusted using the following kinematics equations:

$$\begin{cases} v_j(n+1) = \gamma(n)v_j(n) + \alpha_1\psi_1[p_j(n) - z_j(n)] + \alpha_2\psi_2[p_g(n) - z_j(n)], \\ z_j(n+1) = z_j(n) + v_j(n+1), \end{cases} \tag{6.36}$$

where $\gamma(n)$ is an inertial constant, and α_1 and α_2 are constants representing a cognitive and a social component, respectively, that reflect what degree a particle is directed toward a good position. The stochastic exploration capability of the particle swarm is provided by introducing two random constants, ψ_1 and ψ_2.

Regarding the problem of the optimal control, the particle position, $z_j(n)$, consists of the connecting weights of the FNNDC in Fig. 6.9 when those weights are being determined, and it consists of the three zone temperature settings when those settings are being determined.

The vector $v_j(n)$ indicates the velocity of the search for an optimal solution. During flight, the jth particle is attracted toward the best position, which is determined by its own previous best position, $p_j(n)$, and the global best position, $p_g(n)$, of the whole population. The strength of attraction is determined by two products: α_1 times ψ_1 and α_2 times ψ_2. Without loss of generality, set

$$\alpha_1 = \alpha_2. \tag{6.37}$$

ψ_1 and ψ_2 are generated in the range (0, 1), with all numbers having the same probability of being selected. A suitable $\gamma(n)$ prevents a particle from changing directions too frequently. During the optimization process, it decreases linearly

$$\gamma(n) = \gamma_{\max} - \frac{\gamma_{\max} - \gamma_{\min}}{n_m} n, \tag{6.38}$$

where n is the iteration number; n_m denotes the maximum number of iterations; and $\gamma_{\max}$ and $\gamma_{\min}$ are the maximum and minimum of $\gamma(n)$, respectively.

The fitness of the jth particle is given by

$$F(z_j) = \frac{1}{a + J(z_j)}, \tag{6.39}$$

where $J(z_j)$ is the performance evaluation function Eq. (6.26) or Eq. (6.29), and a is a small positive number that guarantees that the denominator is larger than zero. In this study, the parameters were chosen to be

$$\begin{cases} N_p \in [20, 50], & \alpha_1, \alpha_2 \in [1.6, 2.4], \\ \gamma_{\min} = 0.1, & \gamma_{\max} = 2.2, \quad a = 0.1. \end{cases} \tag{6.40}$$

Even though the PSO quickly produces an optimal solution, the problems of population regression, low accuracy, and local minima need to be dealt with.

4. Hybrid Particle Swarm Optimization

The drawbacks of the PSO can be overcome by incorporating the density-based selection and clonal expansion [39] of the immune system approach into the PSO. For example, the search precision can be improved by applying a clonal operator to the global best position, $p_g(n)$. Cloning creates new particles in the neighborhood of $p_g(n)$, which is defined to be

$$\mathrm{SN}(p_g(n)) = \{z : \|z - p_g(n)\| \leqslant R, z \in \Omega, R > 0\}, \tag{6.41}$$

where Ω is the space of candidate solutions, and R is the radius of extension. The selection of R involves a trade-off: a large R reduces the convergence speed of the optimization, and a small R reduces the search precision. Thus, it is important to choose an appropriate value. In this study, it was in the range [0.05, 0.1].

In the evolution stage, the particles with a high fitness usually survive. However, if the density of high-fitness particles is too high, particles with a low fitness but good evolutionary tendencies will be discarded. This reduces the diversity of the population, which can lead to the local minima. Introducing density-based selection into PSO suppresses the regressive tendency and the occurrence of the local minima, thus enhancing the global-search capability. The density of particles in the neighborhood of the jth one is given by

$$D(z_j) = \frac{1}{\sum_{l=1}^{N} |F(z_l) - F(z_j)|}, \quad j = 1, 2, \ldots, N, \tag{6.42}$$

where $N \geqslant N_p, N = N_p + m_1 + m_2$ is the size of the intermediate population created by performing one step of a PSO, cloning, and mutation operation. m_1 and m_2 are the numbers of particles created by cloning and mutation, respectively. The choice of m_1 is critical because it strongly influences the control performance. While a larger m_1 provides a more accurate local solution, it also results in lower diversity in the population, which could lead to the local minimum. In this study, a suitable value of m_1 was determined from an analysis of the sensitivity of the performance index to m_1.

The probability that the jth particle will be selected is

$$P(z_j) = \frac{\dfrac{\sqrt[q]{F(z_j)}}{D(z_j)}}{\sum_{l=1}^{N} \dfrac{\sqrt[q]{F(z_l)}}{D(z_l)}}, \quad j = 1, 2, \ldots, N. \tag{6.43}$$

In this study, the constant q in Eq. (6.43) was selected so as to maximize the range of $\sqrt[q]{F}$ for the given domain of F. It is clear from Eq. (6.43) that the greater the fitness of the jth particle is, the higher the probability that it will be selected is. In addition, the more particles there are that are similar to the jth particle, the less chance there is that the jth particle will be selected to be one of the new parents. This gives particles with a low fitness a chance to evolve.

HPSO, which integrates clonal expansion and a density-based selection with PSO, solves the problems of determining the optimal weights of the FNNDC and the optimal zone temperature settings for the reheating furnace. It involves the following steps:

Step 1: Determine the parameters for optimization. Select the population size, N_p; the same value for α_1 and α_2 in Eq. (6.36); γ_{min}, γ_{max}, and n_m in Eq. (6.38); R in Eq. (6.41); and q in Eq. (6.43).

Step 2: Set $n = 1$ and construct the initial population. The position of a particle is determined as follows:

(1) To optimize the zone temperature settings, form a particle swarm by first using a lookup table and interpolation to construct a particle that encodes the zone temperatures, and then creating other particles, $z_j(n)$, based on a logistic mapping.

(2) To optimize the FNNDC weights, place particle j at a random position based on the following rules so as to accelerate the convergence of the HPSO (Note: The preheating zone is used as an example).

$$\begin{aligned} &\textbf{IF } (c_{\mathrm{ep}i} > 0) \text{ AND } (\dot{c}_{\mathrm{ep}i} \geqslant 0), \textbf{ THEN } w_{\mathrm{pp}i} > 0, \\ &\textbf{IF } (c_{\mathrm{ep}i} < 0) \text{ AND } (\dot{c}_{\mathrm{ep}i} \leqslant 0), \textbf{ THEN } w_{\mathrm{pp}i} < 0, \\ &\qquad i = 1, 2, \ldots, m_{\mathrm{p}}. \end{aligned} \tag{6.44}$$

The initial velocity, $v_j(n)$, is set to 0, and the previous best position of the particle is set to $p_j(n) = z_j(n)$, where $j = 1, 2, \ldots, N_{\mathrm{p}}$.

Step 3: Determine the global best position, $p_{\mathrm{g}}(n)$, for the whole population, $A(n)$, by calculating the fitnesses Eq. (6.39) of all the particles. If the current population contains an optimal particle, then output the results and stop; otherwise, go to Step 4.

Step 4: Create new particles by the following three methods:

(1) PSO: Produce N_{p} new particles using Eq. (6.36); and then find their best positions, $p_j(n)$ and $p_{\mathrm{g}}(n)$, based on their fitness.

(2) Cloning: Create m_1 new particles with random positions in the neighborhood of $p_{\mathrm{g}}(n)$ (see Eq. (6.41)).

(3) Mutation: Create m_2 new particles either with random positions for the zone temperature optimization, or based on IF-THEN rules (for example, Eq. (6.44)) for the weights of the FNNDC in the solution space.

The velocities of the particles produced by cloning and mutation are set to zero.

Step 5: Select particles on the basis of density. Calculate the selection probability of the $N_{\mathrm{p}} + m_1 + m_2$ particles produced in Step 4, and then select N_{p} particles from them to form a temporary population, $B(n)$, based on Eq. (6.43).

Step 6: Renew the particle swarm. Change the position of the particle with the lowest fitness in $B(n)$ to the global best position, $p_{\mathrm{g}}(n)$, set its velocity to zero, and form the next population, $A(n + 1)$. Then, let $n = n + 1$ and go to Step 3.

6.3.4 *Verification and Discussion*

This subsection presents the results of simulations on the modeling and the optimal control of a regenerative pusher-type reheating furnace and the results of actual runs of the optimization and control system described above. The length and the width of the furnace were 26.3 m and 6.4 m, respectively. The furnace was divided into three zones (preheating, heating, and soaking). Gas with a high calorific value from a coke-oven and gas with a low calorific value from a blast furnace were mixed to form the fuel for the furnace. After being loaded into the furnace, a billet was propelled through the three heating zones by a stock pusher and then sent to a mill.

First, real-time GAP-EKF sequential learning was utilized to train the RBF-RNN so as to establish a predictive model for zone temperatures. And a model for estimating billet temperatures was extracted using an analysis based on the theory of heat transfer. The effectiveness of the models was verified through simulations. Then, FCM clustering was employed to extract the cluster centers of the FNNDC, and an HPSO strategy was used to optimize the connecting weights, and also to search for the optimal zone temperature settings.

1. Prediction of Zone and Billet Temperatures

(1) Prediction of Zone Temperatures

Training and testing based on 2000 sets of sampling data from actual furnace operations yielded an RBF-RNN with 7 input neurons, 559 hidden neurons, and 3 output neurons. Figure 6.11 shows typical results on the predicted temperature of the soaking zone. The standard deviations of the predicted temperatures of the preheating, heating, and soaking zones are 29.6 °C, 24.9 °C, and 16.8 °C, respectively. Clearly, this model satisfies the requirement that the standard deviation must be less than 50 °C.

(2) Prediction of Billet Temperatures

The required mean temperature of billets at the furnace exit, which depends on the type of billet, was in the range [1130, 1250] °C. The absolute value of the prediction error in the billet temperature must be less than 50 °C. Figure 6.12 shows typical results on the predicted and actual temperatures of a billet of Type 45 steel during the reheating process. The billet was divided into six layers in the thickness direction. At the furnace entrance, the billet was at the ambient temperature (20 °C). The maximum absolute values of the prediction errors for the surface and center temperatures were 27.8 °C and 29.8 °C, respectively. They are both less than 50 °C, thereby satisfying the technical requirements.

2. Optimization of Fuzzy Neural Network Decoupling Control System

(1) Optimal Weights of FNNDC

The HPSO algorithm adjusts the weights of the FNNDC in accordance with the predictive RBF-RNN model. The required precision for temperature control is 100 °C for preheating, 50 °C for heating, and 30 °C for soaking. Based on these requirements and the clustering results for the temperature error and its rate of change obtained from 3000 data sets, repeated adjustment yielded 21 hidden neurons in the FNNDC (Fig. 6.9) for the preheating zone, 26 for the heating zone, and 25 for the soaking zone. Thus, it resulted in 123 ($21 + 2 \times (26 + 25)$) connecting weights, which were optimized by HPSO algorithm. The weights are closely related to changes in the gas flow, which has to be in the range [$-3000\ \text{m}^3/\text{h}$, $3000\ \text{m}^3/\text{h}$]. Considering that $F \in [0.0125, 0.02]$, we chose q in Eq. (6.43) to be 4 to maximize the range of $\sqrt[q]{F}$. The other parameters were

$$\begin{cases} \alpha_1 = \alpha_2 = 2, \quad \gamma_{\min} = 0.1, \quad \gamma_{\max} = 2.2, \quad R = 0.05, \\ N_p = 25, \quad m_1 = 2, \quad m_2 = 2, \quad n_m = 30. \end{cases} \tag{6.45}$$

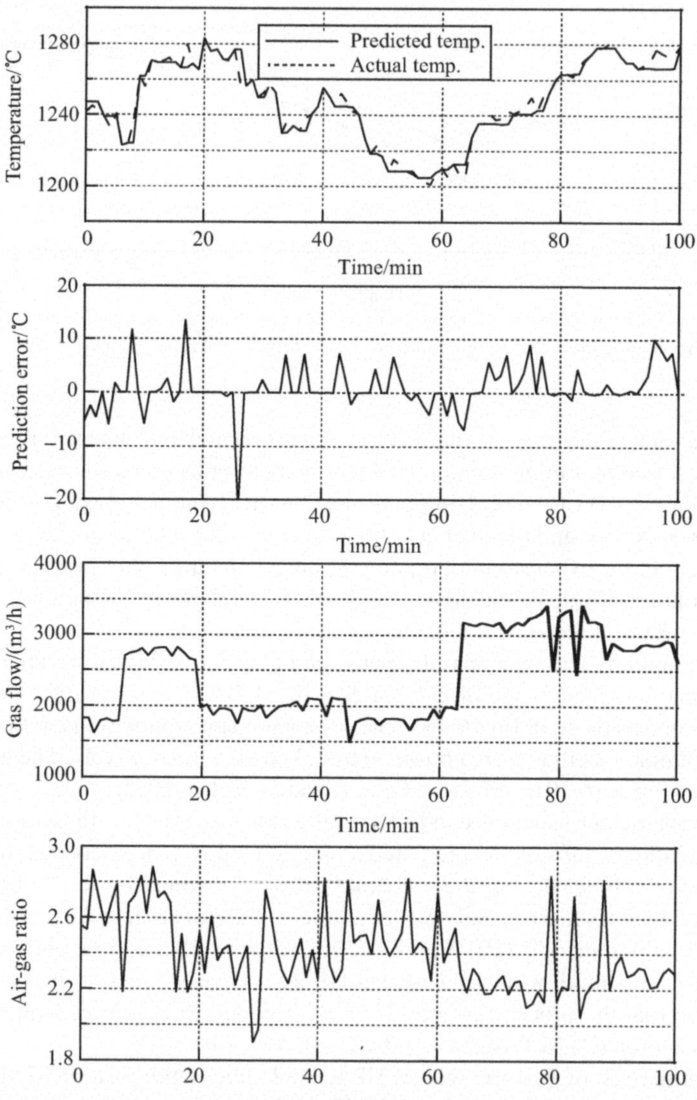

Fig. 6.11 Verification results of predictive model for soaking zone

(2) Optimal Zone Temperature Settings

For evaluation purposes, a simplex algorithm was also used to optimize the furnace temperatures [40]. This subsection compares the performance of the Nelder–Mead (N–M) simplex algorithm, which is an improved version, to that of HPSO.

A simplex algorithm constructs a regular simplex in an n-dimensional space from $n + 1$ mutually equidistant points in the space. The search begins with the setting-up of a regular simplex in a hyperplane of the space and the evaluation of the objective

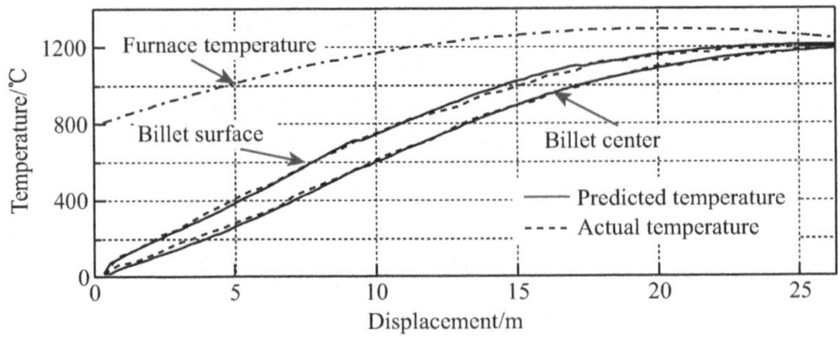

Fig. 6.12 Actual and predicted temperatures of billet during reheating process

function at each of the $n + 1$ vertices. The vertex for which the objective function has the largest value in a minimization problem is then replaced by its reflection point around the centroid of the remaining vertices in the hyperplane, which results in a new simplex. Nelder and Mead altered both the size and configuration of the simplex by introducing an expansion/contraction strategy. Iterating this process yields an optimal solution. In this subchapter, the vertices correspond to candidate particles related to a set of three zone temperatures.

For simulations and analysis, the parameters for the HPSO were chosen to be $q = 3$ to maximize the range of $\sqrt[q]{F}$ for $F \in [0.04, 0.05]$, and the other parameters were the same as those in Eq. (6.45). The expansion and contraction coefficients for the N–M simplex method were chosen to be 1.2 and 0.8, respectively. The weighting coefficients in Eq. (6.29) were $k_1 = k_2 = 1$, and $k_3 = 1 \times 10^{-4}$.

The target mean temperature of billets at the exit was 1160 °C. Figure 6.13 shows some simulation results on the predicted temperature difference between the surface and center of a billet during the reheating process obtained by the N–M simplex method and by HPSO. δT_{bm} and δT_{bsc} of a billet at the exit are listed in Table 6.2. For HPSO, the values are only 0.01% and 35.29%, respectively, of those obtained by the N–M simplex method, which means that HPSO is clearly superior. It is also worth mentioning that the processing time for HPSO is only 0.26 min/generation (CPU: Intel (R) Celeron (R) M Processor, 1.3 GHz; RAM: 240 MB).

(3) Comparison of Performance of HPSO and Other Optimization Methods

Table 6.3 shows some statistics on HPSO and other methods for the same performance index Eq. (6.29). Ten trials were carried out for each method; the table lists the averages of the values obtained. The target mean temperature of a billet was 1160 °C. The population size was 25 and the maximum number of generations was 30. Regarding the coefficients for the various methods, those in the previous subsection were used for the N–M simplex method; the crossover and mutation rates were 0.8 and 0.08 for GA; those in Eq. (6.45) were used for HPSO and PSO, and $q = 3$ in Eq. (6.43) was used for HPSO. The GA used real-coded individuals, the roulette wheel operator for selection, and a single crossover [41]. It employed the same methods to create an initial population and carry out mutations as those used in PSO and HPSO. Clearly, GA is better than N–M simplex; PSO is superior to GA;

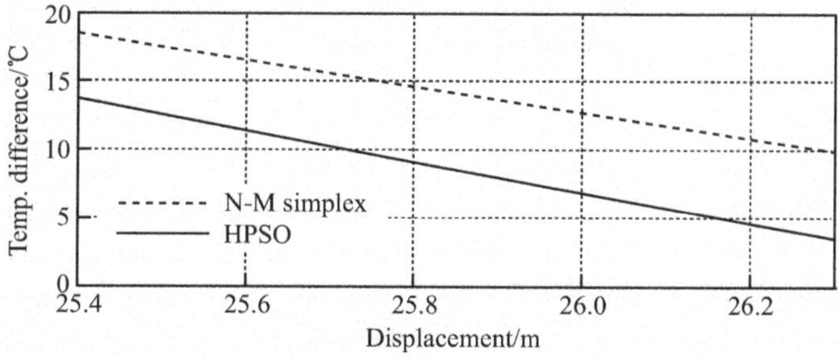

Fig. 6.13 Predicted temperature difference between surface and center of billet during reheating process obtained by N–M simplex method and by HPSO

Table 6.2 δT_{bm} and δT_{bsc} of billet at furnace exit

Method	δT_{bm}/°C	δT_{bsc}/°C
N–M simplex	1.17	9.89
HPSO	1.34×10^{-4}	3.49

Table 6.3 Comparison of performance index Eq. (6.29) for HPSO and other methods of optimizing zone temperature settings

Method	Maximum	Minimum	Average	Standard deviation
N–M simplex	36.9878	25.9682	29.6427	3.6923
GA	27.2258	25.1066	26.0889	0.8278
PSO	25.1610	23.2611	23.5439	0.5790
HPSO	23.2662	23.2595	23.2610	0.0020

and HPSO is superior to PSO. For $n_m = 100$, Fig. 6.14 shows that HPSO provides higher search accuracy because it overcomes the regression problem.

3. Sensitivity Analysis of Hybrid Particle Swarm Optimization

The relationships between the performance index Eq. (6.29) and the parameters of the HPSO were investigated through simulations. As expected, increasing the population size, N_p, improves the performance, but the rate of improvement drops appreciably when N_p is larger than 25. An interesting point is that the relationship between the performance index and the number of clones, m_1, is not monotonic. Figure 6.15 shows an example of the relationship for the parameters in Eq. (6.45). Roughly speaking, the performance index is low when $2 \leqslant m_1 \leqslant 5$ and high elsewhere. This can be explained as follows: The search precision is low when m_1 is small, and the diversity of the population decreases as m_1 increases.

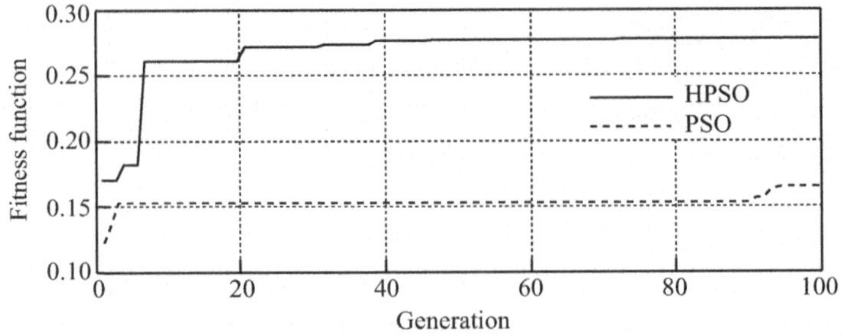

Fig. 6.14 Evolution curves for HPSO and PSO

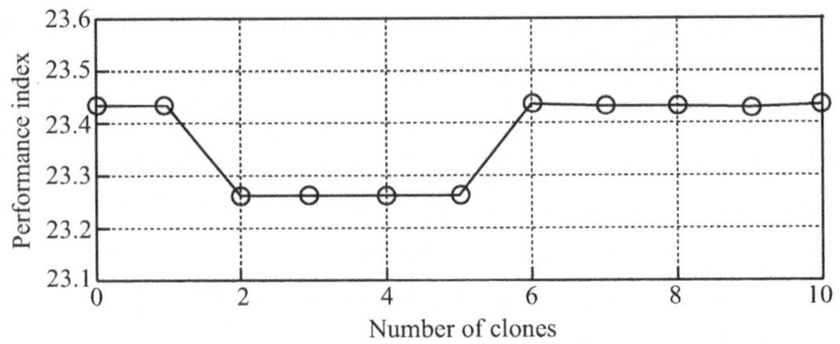

Fig. 6.15 Sensitivity of HPSO to clonal expansion parameter, m_1

6.3.5 Implementation and Results of Actual Runs

The control system has been used for the combustion process of a regenerative pusher-type reheating furnace. Application software was written in the Visual C++ 6.0 language, and the database was designed using SQL Server 7.0. The techniques of structured programming and multithread programming were used to improve the reliability, maintainability, and real-time implementation of the overall system. All programs were designed in accordance with the principle of high cohesion and low coupling. Each part of the system was connected to the global data area so as to share the same information.

1. Hardware Configuration

The structure of the distributed computer control system has two levels (Fig. 6.16). The first level consists of three C40 field controllers and six C31 controllers, ACS600 inverters, a data collector JTM350A, digital display meters TRM006, etc. It performed the distributed real-time monitoring and the control of the combustion process of the reheating furnace. The control of the preheating section, the heating section, the hot section gas, and the air valve openings were directly carried out so as to achieve desired control performance for the furnace temperature. The digital display meters TRM006 were used to display the temperatures of the preheating section, the heating section, the heat storage regenerator, the exhaust gas, and gas preheating before and

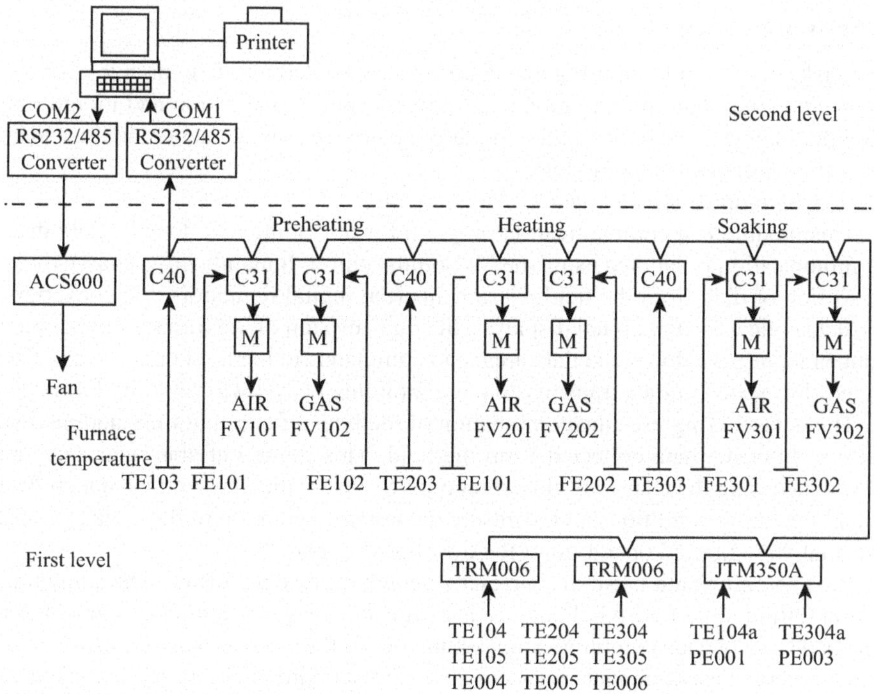

Fig. 6.16 Structure of distributed computer control system for reheating furnace

after heat exchange. The data collectors JTM350A collected the pressures of the furnace, the main gas pipe, and the main air pipe; the temperatures of the regenerator, and the preheating, heating, and soaking sections. The inverters ACS600 were used to implement the air flow control of the fans and to prevent the explosion caused by the inflow of the air into the gas pipelines. This ensured the safety of equipment and workers. The outputs of the inverters were controlled in the range [25%, 100%], and the openings of the fans were adjusted according to the operating conditions.

The second level contains an upper industrial control computer. It mainly functioned as real-time monitoring including the display of and inquiries about parameters and historical curves; the display of real-time curves; the modification and transmission of the furnace temperature, the setting values of the gas and air value openings; and the setting values of the air–gas ratio in the preheating, heating, and soaking sections.

Since the locations of the inverters were far from the monitoring room, and the communication lines were wired close to high-voltage lines, signals were strongly affected by electromagnetic interference. The RS485 system was used for the two communication lines between the upper industrial computer and the inverters, and between the upper industrial computer and the field instruments so as to suppress the influence on the communication controllers in the control room, the data collectors, the displays, and the industrial computers.

2. Software Design

The tasks of system monitoring and control were divided into five modules: a communication and data filtering module, a process state visualization and monitoring module, a control and optimization module, a database management module, and an information management module.

(1) Communication and data filtering module

This module is a communication program for uploading and downloading data. Communication is handled using the MsComm control. It mainly takes charge of the collection of data from the field instruments (the digital regulators, the converters, the data collectors, the digital displays), the transmission of the furnace temperature control inputs to a download data area, and commands to the field instruments. Due to the slow speed of data transmission, the sampling period was set to be 1 min.

The data filtering uses the combination of methods of clipping filter and median filter to filter the data collected from the field. This filters out abnormal data and flow data during the change of flow directions to ensure that the sampled data reflect actual operating conditions and to ensure the normal operation of the control loops, especially the normal operation of the flow tracking part.

Data communication uses a character-oriented frame data format. The communication setting unit of each field device is assigned a unique communication address. The baud rate of each communication setting unit is the same as that of the host computer, and the host computer can access the corresponding field device according to the address.

(2) Visualization and monitoring of process status: module

Animation, tables, and graphs are used for real-time monitoring of various parameters of the billet furnace combustion process. The animation uses process drawing as the background for the intuitive display of the process flow and the operating conditions such as the furnace temperature, the furnace pressure, the pressures of the gas and airlines, and the flow of the gas and air. All monitoring parameters are displayed in tables on one screen. In addition, the parameters for the animation and other parameters such as the exhaust gas temperature are also displayed. The graphs include the temperatures and pressures of the three stages in the furnace, the pressures of the main gas and air pipes.

(3) Control and optimization: module

This module implements the optimization and setting of furnace temperature, the multi-model control of the furnace temperature, and the frequency control of the inverters of the fans. Except for automatic switching between multi-models, manual switching is also available by choosing a model using a bar graph on the display.

There are two kinds of switching modes: manual and automatic. The manual switching mode allows us to directly set the references of the valve openings through a bar graph interface. This interface also has a function of blower frequency control. It allows us to directly set the speed of the blowers and displays the speed and the currents of the blowers.

(4) Database management: module

The system database contains a sampling and control database and an alarm database. The sampling and control database is used to store operating condition

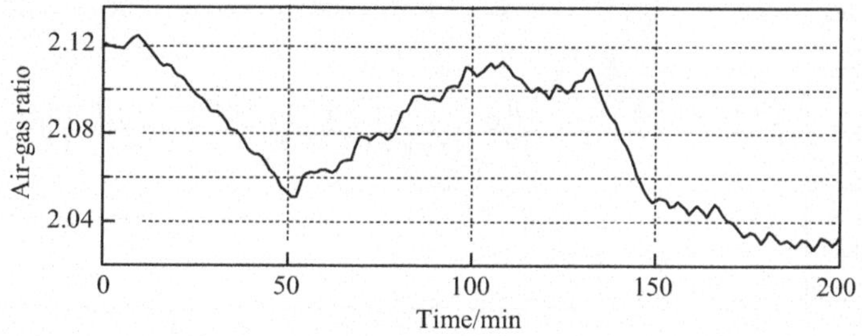

Fig. 6.17 Actual run results of air–gas ratio

data and control data. The alarm database stores the out-of-range information of the process parameters. Before sampled data are stored in the sampling and control database, the data of the variables that affect safe operation, for example, the pressures of the gas and air, are first checked if they are out of their allowable ranges. If they are, they are stored in the alarm database, and warning alarms are given off through a bell and lights depending on the order of priority. Up to five different alarm events can be issued simultaneously. To ensure that there is enough space to store new data, the module has a function of creating new tables and deleting old ones.

(5) Information management: module

It has functions of printing reports and alarm records, and querying data. Passwords at the levels of a system expert and an operator are set to ensure the security of the system operation. This module also provides a user detailed help files, operating instructions, exception handling strategies, etc.

3. Results of Actual Runs

Figures 6.17, 6.18, and 6.19 show some typical results of actual runs of the zone temperature FNNDC system, which was implemented in a regenerative pusher-type reheating furnace. In each zone, the gas flow fluctuates periodically due to the switching of the regenerative burners between burning and venting, which is needed to keep the temperature within a zone laterally uniform. This causes temperature fluctuations in the zones. It was found that our control and optimization method for the combustion process not only lightened the operators' workload, but also improved system performance: Fuel consumption was 12% lower than for conventional manual control. Thus, the method provides significant economic benefits.

6.4 Intelligent Control System for Soaking Furnace of Compact Strip Production

In this section, the structure of the intelligent control system for a soaking furnace of compact strip production (CSP) (hereafter referred to as CSP soaking furnace) is first devised according to the characteristics of its combustion process. Then, intelligent control methods are presented according to the control system structure. The meth-

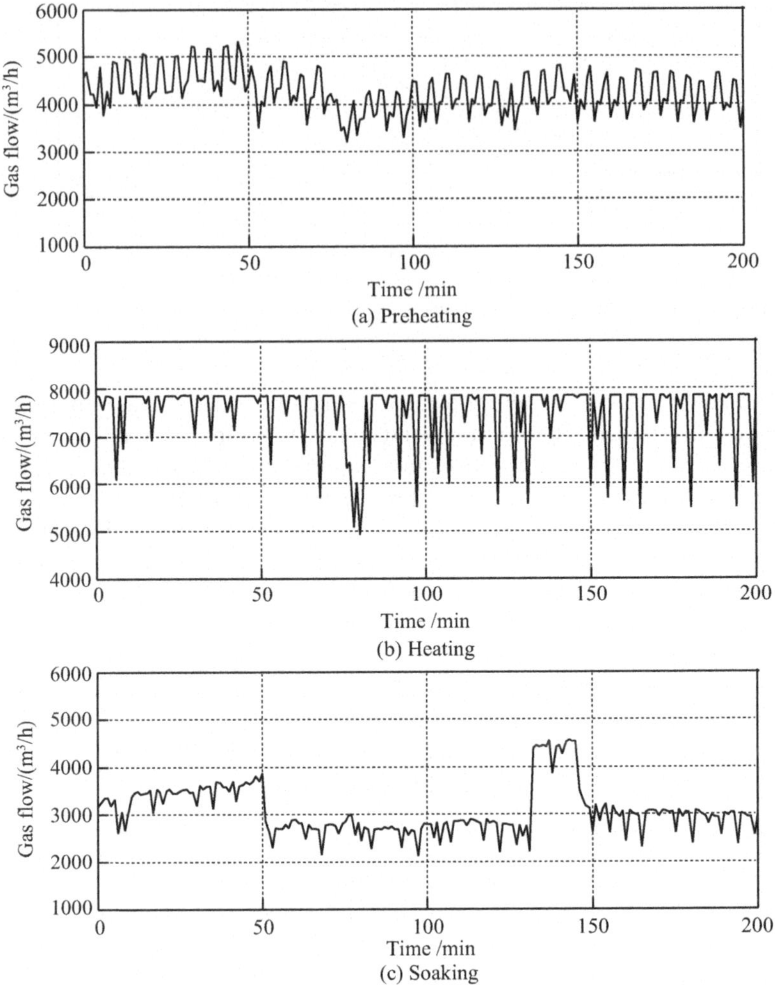

Fig. 6.18 Actual run results of gas flow

ods contain the temperature fuzzy expert control, the air–gas ratio self-optimization algorithm, and the valve linear compensation control. Finally, the implementation of the system is explained, and real run results are used to show the validity of the intelligent control system.

6.4.1 Configuration of Intelligent Control System

An intelligent control system (Fig. 6.20), which incorporates fuzzy expert control, is used to control the temperature to yield satisfactory control performance for combustion process of a CSP soaking furnace. The goal of the control is to achieve an

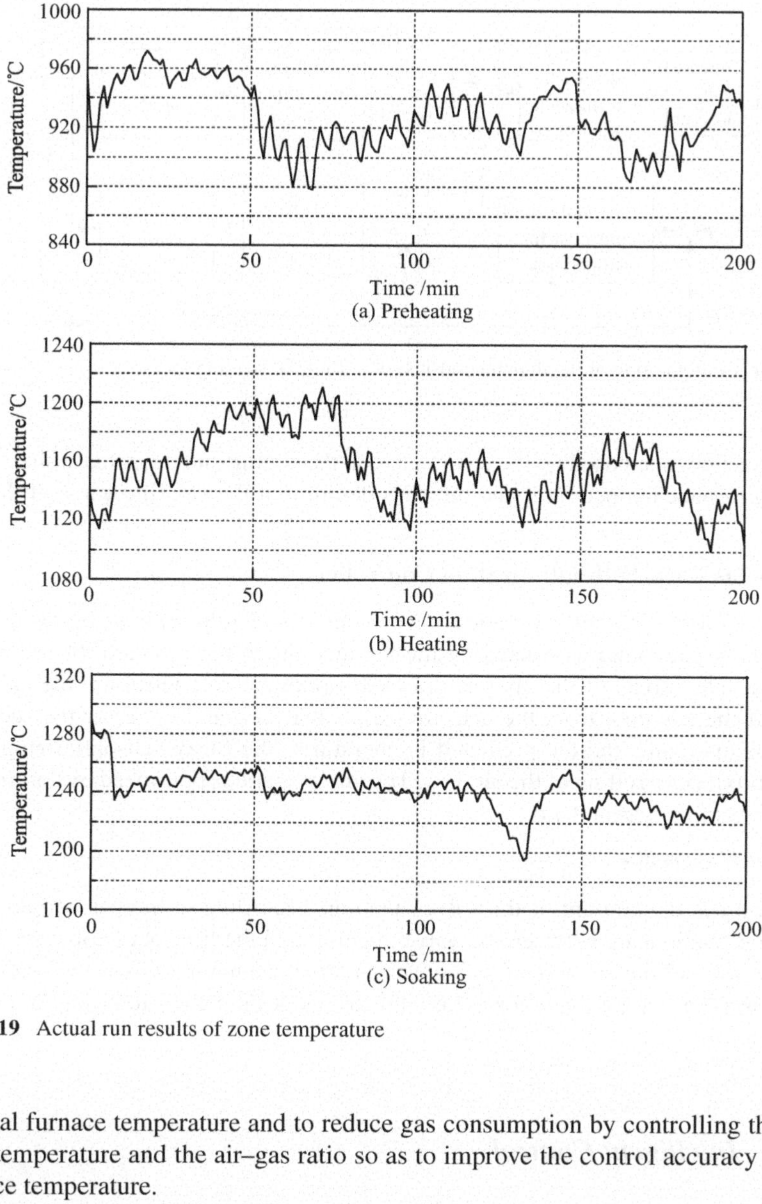

Fig. 6.19 Actual run results of zone temperature

optimal furnace temperature and to reduce gas consumption by controlling the furnace temperature and the air–gas ratio so as to improve the control accuracy of the furnace temperature.

The control system is divided into three parts according to the optimization control requirements and the block diagram of the system structure:

1. Fuzzy and Expert Controller for Temperature

It is the core of the control system; it not only ensures that the soaking furnace temperature is within a given range, but also stabilizes the gas flow and the calorific value. The fuzzy controller and the expert controller are switched to calculate the

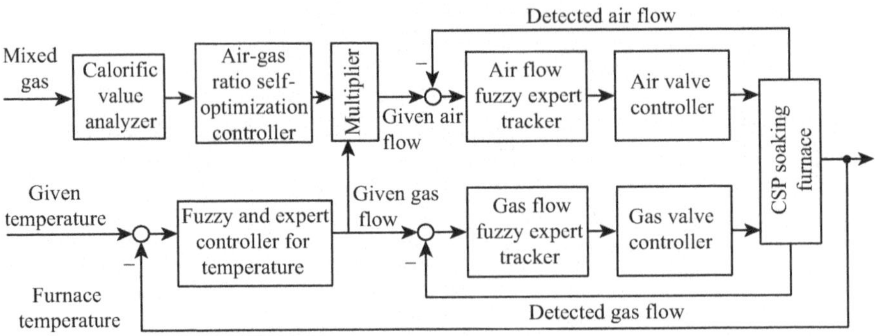

Fig. 6.20 Configuration of intelligent control system for CSP soaking furnace

target gas flow rate based on the error between the setting and the actual value of the soaking furnace temperature, and the consideration of the calorific value of the gas flow.

2. Air–Gas Ratio Self-optimization Controller

It provides a suitable air–gas ratio to avoid the loss of iron scale of billets as well as smoking phenomenon caused by the fluctuations in the gas calorific value and pressure. The inputs of the air–gas ratio self-optimization control are the calorific value of the gas flow from the calorific-value analyzer, and the error between the actual temperature and the predicted temperature. The fuzzy self-optimizing controller produces a setting of the air flow. The air–gas ratio is optimized by controlling the valves of the gas and air.

3. Valve Controller

It determines the amount of the valve openings according to the gas and air flow. First, the relationship between the valve opening and the flow is established. Then, the flow rates of the gas and air are converted into the amount of the valve openings. The controller compensates for the nonlinearities of the valves to ensure a quickly response.

6.4.2 *Intelligent Control*

The algorithm used in the previous control system for the soaking furnace combustion process was first improved to perform the intelligent control of the process. However, the required control performance could not be achieved due to the uncertainty of the gas supply from the soaking furnace. Then, we collected operators experience thoroughly and built a integrated system combining fuzzy control and expert control to carry out the automatic control of the soaking furnace.

1. Fuzzy and Expert Control for Temperature

The main purpose of controlling the soaking furnace temperature is to enhance the ability of suppressing the fluctuations in the coal calorific value and pressure. It ensures each area of the soaking furnace to be heated uniformly and improves the accuracy of the billet rolling and the product quality. When the calorific value and the gas pressure are stable, the furnace temperature is proportional to the gas flow rate. This means that the temperature of the furnace can be adjusted by regulating the gas flow rate. However, when the gas flow rate changes, the amount of air also needs to be regulated so as to ensure the full combustion of gas.

The amount of the increasement of heat is determined by expert rules according to temperature error between the setting and the measured ones, the gas calorific value, the pressure, and other factors. Then, the fuzzy expert controller produces a setting of the gas flow rate based on the amount of the calculated heat. Fuzzy control is used when the temperature error is within a prescribed range, and expert control is used instead to quickly suppress the error when the error is out of the range. The switching between fuzzy control and expert control ensures that the control system adapts the operating state of the process. Detail design methods of the two controllers are described below.

(1) Temperature Fuzzy Controller

The fuzzy-control-based optimization and control of the CSP soaking furnace combustion process is the core of the temperature control. The fuzzy controller has the structure of two inputs and single output.

The fuzzy control contains fuzzification, fuzzy reasoning, and defuzzification. The fuzzification is the process of converting the temperature error and the rate of the error, e and ec, into fuzzy quantities. Its output is the values of membership degrees corresponding to the inputs. The fuzzification of e and ec gives the corresponding fuzzy variables E and EC. A rule base is composed by control rules related to operation, in which control rules are determined according to operators' or experts' experience, and the table of control rules is built. This rule table collects information during the operation in an online fashion and is adjusted and improved by a self-learning expert controller. The fuzzy reasoning activates a rule corresponding to an input fuzzy variable and yields an output variable, U. Defuzzification converts U to a crisp control variable, u.

The fuzzy controller is designed by the following steps:

Step 1: Determine the structure of the fuzzy controller, that is, select the input and output variables according to the concrete system, and choose the domains and membership functions for the inputs and outputs.

Step 2: Design fuzzification methods for the input and output variables, and transform the precise quantities of inputs and outputs into fuzzy sets of corresponding language variables.

Step 3: Design a fuzzy reasoning and decision-making algorithm and produce the output of a fuzzy quantity using fuzzy reasoning based on fuzzy rules.

Table 6.4 Rules of fuzzy control for temperature

EC	E							
	$(-\infty, -50)$ (%)	$[-50, -20)$ (%)	$[-50, -20)$ (%)	$[-10, 0)$ (%)	$[0, 10)$ (%)	$[10, 20)$ (%)	$[20, 50)$ (%)	$[50, +\infty)$ (%)
$(-\infty, -7)$	0	0	1	2	3	5	8	10
$[-7, -5)$	−1	0	0	1	2	3	5	8
$[-5, -2)$	−2	−1	0	0	1	2	3	5
$[-2, 0)$	−3	−2	−1	0	0	1	2	3
$[0, 2)$	−5	−3	−2	−1	0	0	1	2
$[2, 5)$	−8	−5	−3	−2	−1	0	0	1
$[5, +\infty)$	−10	−8	−5	−3	−2	−1	0	0

Step 4: Design a defuzzification algorithm for the fuzzy quantities of the outputs, use the center of gravity method to perform defuzzification, and convert a fuzzy quantity to a crisp quantity.

The temperature fuzzy controller deduces the setting of the amount of the valve opening based on the fuzzy rule base for e and ec. The inputs of the controller are e and ec. The output variable is the change in the amount of the valve opening, u. The opening of the valve is increased when the temperature error is larger than 0, and it is decreased when the error is less than 0. In the controller, the domain of E is set to be $[-50, 50]$, that of EC is $[-7, 7]$, and that of U is $[-2, -2]$. The table of the fuzzy rules for the amount of the valve opening is shown in Table 6.4. It is a 8×7 incremental-type fuzzy control rule table.

(2) Temperature Expert Control

Expert control is an important tool for intelligent control. It has the characteristics of high reliability, long-running reliability, real-time control, excellent control performance, good disturbance rejection characteristic, flexibility of use, and convenience of maintenance. Expert control chooses different control strategies to adapt the system to changes in and uncertainties of processes, and random noises. Control results can meet the requirements of the industrial process.

Knowledge base is the core of an ES. It has two parts: one is data related to the selection of a proportional parameter for the temperature error, and the other one is the general knowledge and the knowledge of experts and experience for reasoning. Experts' control experience is collated and summarized in the form of **IF** ⟨condition⟩ **THEN** ⟨action⟩, and is stored in the knowledge base.

The temperature expert controller of the soaking furnace switches control laws between two control strategies. Fuzzy control is used when the error is between the temperature setting and that is fed from a PLC system is in a prescribed range. However, if the error is out of the range, fuzzy control is not used anymore. Instead, expert control is used to quickly reduce the error. Observation and experience show that the primary task of the controller is to reduce the error as soon as possible when the error is larger than 150 °C. The main rules are as follows:

R_1: **IF** ($e < -150\,°\text{C}$), **THEN** increase Δu_1 by 10%.

R_2: **IF** ($e \geqslant -150\,°\mathrm{C}$) AND ($e < -100\,°\mathrm{C}$), **THEN** increase u_1 by 7%.
R_3: **IF** ($e \geqslant -100\,°\mathrm{C}$) AND ($e < -70\,°\mathrm{C}$), **THEN** increase u_1 by 5%.
R_4: **IF** ($e > 70\,°\mathrm{C}$) AND ($e \leqslant 100\,°\mathrm{C}$), **THEN** decrease u_1 by 5%.
R_5: **IF** ($e > 100\,°\mathrm{C}$) AND ($e \leqslant 150\,°\mathrm{C}$), **THEN** decrease u_1 by 7%.
R_6: **IF** ($e > 150\,°\mathrm{C}$), **THEN** increase Δu_1 by 10%,

where e is the temperature error, and u_1 is the gas flow rate in percent of full-scale gas flow. The reasoning algorithm carries out reasoning using knowledge and the experience in the knowledge base and makes a conclusion.

2. Self-optimization Algorithm for Air–Gas Ratio

Finding an optimal air–gas ratio is an important task for the soaking furnace combustion control. A suitable air–gas ratio ensures a fast rise in the furnace temperature to the setting. When the gas flow rate is constant, the control of the air–gas ratio is the adjustment of the air flow rate. A large change in the air flow rate results in a large amount of exhaust gas. This not only loses a lot of heat, but also leads to a very high oxygen content of furnace gas, which causes serious burning loss of billets. On the other hand, excessive gas causes in incomplete combustion and low thermal efficiency. This results in a large amount of smoke polluting the air or flaming at the end of the furnace impacting on the safety of operators and the equipment.

It is difficult to establish an accurate mathematical model to describe the relationship between the temperature, the gas flow rate, and the air flow rate because the CSP soaking furnace combustion process is complicated. While the cross-limiting technique can improve the safety of the operation and ensure the air–gas ratio in a prescribed range, it cannot perform adjustment according to actual situation in a timely fashion. This affects the efficiency of the combustion process. Fuzzy expert control is used to optimize the air–gas ratio and to provide a suitable air–gas ratio for the combustion process. This method is also used to deal with abnormal air–gas ratio caused by fluctuations in the gas calorific value and equipment abnormality.

As explained in the combustion theory, whether or not the air–gas ratio is reasonable finally reflects on the furnace temperature. From this viewpoint, self-optimization control for the air–gas ratio is carried out based on the temperature error. It searches an optimal air–gas ratio by calculating the air flow rate using the fuzzy self-optimizing controller and performs optimal control of the amounts of valve openings.

The self-optimization algorithm for the air–gas ratio has two parts. First, the calorific value of the gas is analyzed by a calorimeter, and an appropriate air–gas ratio is obtained according to the expert control rules. Second, the setting of the air flow rate is given by the product of the current gas flow rate and the air–gas ratio. Since the coal gas calorific value changes frequently, the following expert rules are used to adjust it:

R_1: **IF** ($Q < a_1$), **THEN** $\alpha = b_1$.
R_2: **IF** ($Q \geqslant a_1$) AND ($Q < a_2$), **THEN** $\alpha = b_2$.
R_3: **IF** ($Q \geqslant a_2$) AND ($Q < a_3$), **THEN** $\alpha = b_3$.
R_4: **IF** ($Q \geqslant a_3$) AND ($Q < a_4$), **THEN** $\alpha = b_4$.
R_5: **IF** ($Q \geqslant a_4$) AND ($Q < a_5$), **THEN** $\alpha = b_5$.

R_6: **IF** ($Q \geqslant a_5$), **THEN** $\alpha = b_6$.

In the above rules, Q is the gas calorific value, and α is the air–gas ratio. a_1-a_5 represent the thresholds of the calorific value. b_1-b_6 represent the air–gas ratios for the corresponding calorific values. Both a_i and b_i $(i = 1, 2, \ldots, 6)$ can be updated online according to actual situation.

Note that the air–gas ratio obtained from the gas calorific value is only a preliminary result. The air–gas ratio is further optimized by the fuzzy self-optimizing controller when the furnace temperature is stabilized at the setting. This guarantees the optimality of the ratio.

It is difficult to perform optimal temperature control for the soaking furnace due to the difficulty in modeling. A possible way to solve this problem is the application of the fuzzy control theory. The fuzzy self-optimizing controller is designed by setting the furnace temperature as the control objective to carry out optimization.

This fuzzy self-optimizing controller also has functions of automatically changing search direction and step length. If the effect of a controller is observable, then the controller increases the step length on the same search direction. If the effect is not clear, then the controller decreases the step length on the same search direction. However, if the control law has a countereffect, then the controller changes the search direction.

The fuzzy self-optimizing controller avoids too much surplus air, rationalizes the air–gas ratio, ensures full combustion of gas, reduces the scale loss of bullets, suppresses energy consumption, and improves the production efficiency.

3. Fuzzy Expert Control for Valve

Actuators are the indispensable part of the control system. They behave precisely corresponding to the control input to complete a control task. The responding speed and accuracy directly affect control performance. Valves are used as actuators for the control of the gas and air flow rates. Since they are highly nonlinear, good control performance cannot be expected for the use in open-loop control, a feedback loop is designed for each valve.

As the signals given by controller are gas flow signal and air flow signal, the gas flow and air flow signals must be converted to the valve opening. Then, an expert control algorithm is devised to obtain the relationship by classifying valve characteristics. The expert rules are given as follows:

R_1: **IF** ($E \leqslant -70\%$), **THEN** increase u by 30%.

R_2: **IF** ($E \geqslant -70\%$) AND ($E < -30\%$), **THEN** increase u by 20%.

R_3: **IF** ($E \geqslant -30\%$) AND ($E < 30\%$), **THEN** fuzzy model is used.

R_4: **IF** ($E \geqslant 30\%$) AND ($E < 70\%$), **THEN** decrease u by 20%.

R_5: **IF** ($E \geqslant 70\%$), **THEN** decrease u by 30%.

In the above rules, E is the error of gas flow rate in percent of full-scale gas flow and u is the value opening. When the error is in the range of $[-30\%, 30\%)$, the fuzzy model between the valve opening and the flow is used. The inputs of the fuzzy model are the flow deviation and its change rate, and the output is the increment of the valve opening. The fuzzy rules are shown in Table 6.5.

Table 6.5 Rules of fuzzy control for flow

EC	E							
	$(-\infty, -20)$ (%)	$[-20, -10)$ (%)	$[-10, -5)$ (%)	$[-5, 0)$ (%)	$[0, 5)$ (%)	$[5, 10)$ (%)	$[10, 20)$ (%)	$[20, +\infty)$ (%)
$(-\infty, -5)$	0	0	0	1	2	3	5	8
$[-5, -2)$	−1	0	0	0	1	2	3	5
$[-2, 0)$	−2	−1	0	0	0	1	2	3
$[0, 2)$	−3	−2	−1	0	0	0	1	2
$[2, 5)$	−5	−3	−2	−1	0	0	0	1
$[5, +\infty)$	−8	−5	−3	−2	−1	0	0	0

Table 6.6 Relationship between valve opening and feedback value

Given valve opening	30	35	40	45	50	55	60	65	70	75	80
Feedback value 1	34.6	38.3	42.5	46	50	53.6	57.6	61.9	65.2	69.6	73.5
Feedback value 2	34.5	38.5	42.4	46.4	50.3	54.3	58.2	62.2	66.1	70.1	74

Because there are some differences between a given value of the valve opening and its actual detection value, the valve opening needs to be corrected by linear compensation. To establish the relationship model between the given value of the valve opening and its actual detection value, an experiment was first carried out. The results are shown in Table 6.6.

In the table, Feedback value 1 is the actual detection value of the value opening when the valve is gradually throttled, and Feedback value 2 is the actual detection value of the value opening when the valve is gradually opened. The results show that the relationship is linear when the given value of the valve opening is about in the range between 10 and 90%. The actual detection value is close to the given value when the given value is 50%. The actual detection value is larger than the given value when the given value is less than 50%. And the actual detection value is smaller than the given value when the given value is larger than 50%. The error becomes large as the given value of the value opening moves away from 50%.

Analyzing the data in Table 6.6 reveals the relationship between the actual detection value and the given value. In order to ensure that the actual detection value of the valve opening is the same as the given value, a correction value of the control input is given as follows:

$$y = \begin{cases} 0, & 0 < x < 10.8, \\ \dfrac{x - 10.8}{0.79}, & 10.8 \leqslant x \leqslant 89.8, \\ 100, & x > 89.8, \end{cases} \tag{6.46}$$

where x is the preset given value of the valve opening and y is the actual given value after compensating.

6.4.3 Implementation and Results of Actual Runs

Industrial control software is an important part of an industrial process control system. It is responsible for the managements, the decision-makings, and the control tasks. It implements the functions of the collection, processing, and display of data, and the human–computer interaction. The quality of software directly affects the stability of the control system and product quality. Thus, software is a vital point of industrial control.

A control platform was built by combining the original distributed control environment with the InTouch configuration software. The furnace combustion optimization and control system was established on the platform. The control software was written in Visual C++, which is compatible with the platform. The OPC technology was used to seamlessly connect the control software and the PLC control system so that programs performs the control of the equipment of the process through the PLC system. This makes the control software a inseparable part of the intelligent control system of the entire CSP soaking furnace combustion process. The system hardware structure, the system data flow, the design of system functions, and key technologies are explained below. And some real run results are also presented.

1. Hardware Structure of Intelligent Control System

The system hardware structure is shown in Fig. 6.21. The system has two servers and two PLC controllers. The servers are independent with each other. One is a master server for monitoring and control, and the other is a slave server.

The control system of the soaking furnace was implemented with Siemens PLC control systems and InTouch configuration software (Wonderware Company). Simatic NET first carries out data communication with the PLCs in the site through industrial Ethernet. Then it carries out dynamic data exchange with an I/O server,

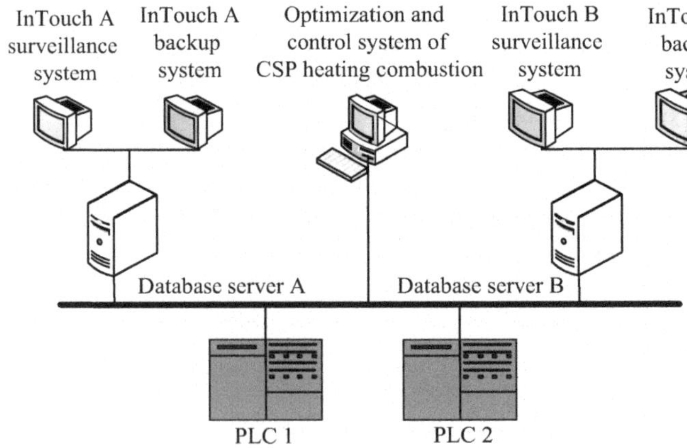

Fig. 6.21 System hardware structure

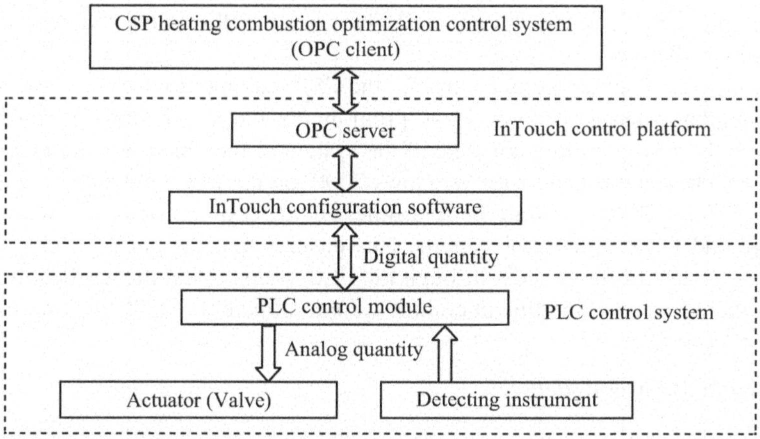

Fig. 6.22 Data flow in control system

which is a component of Wonderware. Finally, it performs monitoring and control of the process using the InTouch configuration software.

Simatic NET V6.0 is an open, multicomponent communication system at an industrial control level provided by Siemens Ltd. It connects PLCs, hosts, working stations, and personal computers to a network to perform communication between them. Simatic NET has a variety of different communication networks to adapt to different kinds of need for industrial automation, for example, Industrial Ethernet, AS-I, PROFIBUS, and PROFIBUS-PA, which meet German or international communication standards. A Simatic NET system contains a transmission medium, network accessories, transmission apparatus, protocols, and services. It connects PLCs and computers to communication processors (CP modules) on a LAN network.

Wonderware process visualization configuration software has three main programs: InTouch Application Manager, WindowMaker, and WindowViewer. InTouch Application Manager creates and manages applications. WindowMaker provides a powerful graphical tool to browse and configure InTouch applications. And WindowMaker is the development environment of InTouch; animated touch-sensitive display windows are easily be created by object-oriented graphics in this development environment. These windows are connected to industrial I/O systems and other Microsoft Windows application programs. WindowViewer displays graphics windows created in WindowMaker.

2. Design of Software Module Based on Visual C++

It is easy to obtain the data flow (Fig. 6.22) from the system hardware structure (Fig. 6.21).

Data communication is based on the OPC technology. The main functions are collecting real-time data from the PLC control system, processing of the data, and sending the data to the PLC control system. It also delivers control inputs produced by control algorithms to the PLC control system to perform the process control and to implement the seamless connection between the PLC control system and the application software.

The intelligent optimization control for the CSP soaking furnace combustion process is implemented as follows. First, programs in the application software were written in the configuration language of the system or the object-oriented C++ language, which is compatible with Windows 2000, on the PLC control configuration platform. The OPC technology built seamless connection between the application software and the PLC control system. This ensures that the application software performs control of the process through the PLC system. And the application software is included in the intelligent control system of the entire CSP soaking furnace combustion process.

3. System Software Structure

According to the design requirements and the characteristics of the process, we divided the system software into two parts: configuration software and application software. The configuration software provide human-friendly interface operating functions, such as setting control state and parameter, real-time state monitoring, and the historical data analysis. The setting of the control state and parameters provides a reference for the staff to decide whether or not to use the automatic control system for the temperature.

The application software attains data communication, establishment of the furnace temperature model, and the nonlinear compensation of valves. Fuzzy control algorithms were packaged in the class of Cfuzzy, and the design of the temperature controller is packaged in CLGCSPDlg. The application software performs measuring and control of the temperature and valves with different sampling time periods. While the temperature control is carried out with a large sampling time period, the valve control is carried out with a short sampling time period.

Modular design was used in the design process of the software to improve the maintainability of the programs and the security of data. The system software structure is shown in Fig. 6.23.

4. Running Results

The intelligent control system for the CSP soaking furnace combustion process is a two-level system composed of two process computers and Siemens PLCs. This system, which was written in the compatible Visual C++ language, carries out the temperature optimization and control, the real-time monitoring of the process, and comprehensive information management.

A control algorithm performs the optimal control of air–gas ratio and the optimal temperature control. When the gas caloric value, the pressure, and the load change, the air–gas ratio of the soaking furnace combustion process deviates from the optimal state. The intelligent control system for the CSP soaking furnace combustion process automatically optimizes the air–gas ratio and controls the air–gas ratio to the optimal state on the basis of the furnace temperature. This improves the stability of the furnace temperature and the response speed of the system. Figure 6.24 shows a temperature–time response randomly selected from the historical database for 120 min.

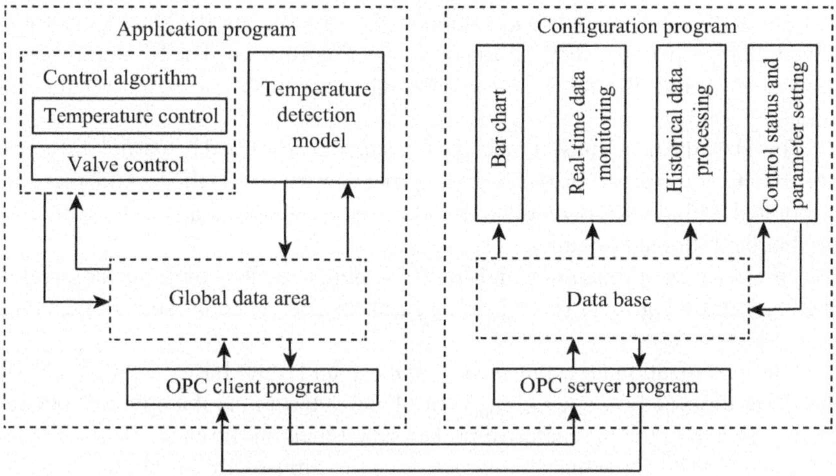

Fig. 6.23 System software structure

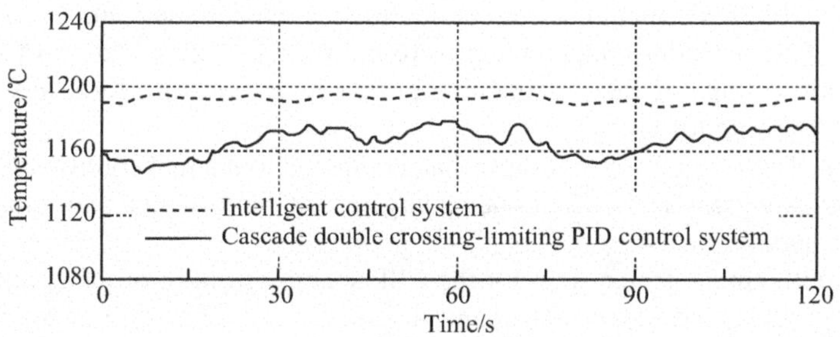

Fig. 6.24 Running result

It is clear from Fig. 6.24 that the range of the fluctuations in the temperature for the cascade double cross-limiting PID control system is [−40, 40] °C. And the range was reduced to [−10, 10] °C by the intelligent optimization and control system. Even though the fluctuations in the temperature occur, the temperature was quickly adjusted to the setting value.

6.5 Conclusion

The integrated model has been established for predicting the billet temperature, the optimization, and control system for the regenerative heating furnace, and the intelligent control system for the CSP soaking furnace of the compact strip production has been designed. The main points of the integrated model for the billet temperature prediction are as follows:

(1) The RBF-RNN models was established for predicting the zone temperatures and the billet temperature distribution of the heating furnace. The SL algorithm was employed to enable the model to adapt to changes in the process in a real-time fashion.

(2) Based on the heat transfer mechanism, a model described by partial differential of the billet temperature in the heating process was established. And then, it is incorporated with the RNN model to build the integrated model with high precision to predict the billet temperature.

We introduce an optimization and control system to control the zone temperatures of the regenerative pusher-type reheating furnace. The characteristics of the control system are as follows:

(1) The optimization and control system using an HPSO-based adaptive FNNDC method was designed for decoupling control and determining the optimal zone temperature settings of the regenerative pusher-type reheating furnace.

(2) An FNNDC for which the structure is determined by FCM clustering was employed to control the zone temperatures, and the weight parameters of the FNNDC are adjusted using the HPSO algorithm.

(3) The RBF-RNN model to predict the furnace temperatures and the integrated model to estimate billet temperatures were built for the optimization of the zone temperature settings of the regenerative pusher-type reheating furnace combined with the HPSO algorithm.

(4) Actual runs of the system show that, compared to manual control, this control method saves a great deal of fuel, yields large economic benefits, and improves the quality of steel products.

The intelligent control system for the CSP soaking furnace of the compact strip production has been designed. The control system has the following features:

(1) Based on the characteristics of the combustion process of the CSP soaking furnace, the structure of the intelligent control system was first devised. The system contains fuzzy expert control, the air–gas ratio self-optimization control, and the valve linear compensation control.

(2) The fuzzy and expert controller used the switched fuzzy controller and the expert controller to ensure that the soaking furnace temperature is within a given range and stabilizes the gas flow and the calorific value.

(3) The air–gas ratio self-optimization controller was designed to provide a suitable air–gas ratio to avoid the loss of iron scale of billets as well as smoking phenomenon caused by the fluctuations in the gas calorific value and pressure.

(4) The valve controller using fuzzy expert control method and linear compensation was designed to determine the amount of the valve openings according to the gas and air flow.

(5) Actual runs of the system show that our control strategy is effective and improves the accuracy of the air–gas ratio and the characteristics of the gas and air valves.

The combustion process is BIBO stable. Practical strategies, such as restricting the control inputs to specific ranges, were employed to ensure the BIBO stability of the control system. However, investigation of the stability of the whole control system is theoretically meaningful and will be carried out in the future. Furthermore, since a recurrent FNN (RFNN) can be used to implement a complicated control law, it is now receiving a great deal of attention in process control (e.g., [22]). Exploring its use for the temperature control of a reheating furnace and comparing it with an FNNDC is of great significance and will be carried out in the future.

References

1. Chen WH, Chung YC, Liu JL (2005) Analysis on energy consumption and performance of reheating furnaces in a hot strip mill. Int Commun Heat Mass Transf 32(5):695–706
2. Hu YK, Tan CK, Broughton J, Roach PA, Varga L (2018) Nonlinear dynamic simulation and control of large-scale reheating furnace operations using a zone method based model. Appl Therm Eng 135:41–53
3. Steinboeck A, Graichen K, Kugi A (2011) Dynamic optimization of a slab reheating furnace with consistent approximation of control variables. IEEE Trans Control Syst Technol 19(6):1444–1456
4. Steinboeck A, Wild D, Kugi A (2013) Nonlinear model predictive control of a continuous slab reheating furnace. Control Eng Pract 21(4):495–508
5. Suzuki M, Katsuki K, Imura J, Nakagawa J, Kurokawa T, Aihara K (2014) Simultaneous optimization of slab permutation scheduling and heat controlling for a reheating furnace. J Process Control 24(1):225–238
6. Mayr B, Prieler R, Demuth M, Moderer L, Hochenauer C (2017) CFD analysis of a pusher type reheating furnace and the billet heating characteristic. Appl Therm Eng 115:986–994
7. Prieler R, Mayr B, Demuth M, Holleis B, Hochenauer C (2016) Prediction of the heating characteristic of billets in a walking hearth type reheating furnace using CFD. Int J Heat Mass Transf 92:675–688
8. Wang ZJ, Shao C, Chai TY (1999) Application of multivariable technique in temperature control of reheating furnaces. In: Proceedings of IEEE international conference on control applications, pp 29–32
9. Wang J, Liu YW, Sundén B, Yang R, Baleta J, Vujanović M (2017) Analysis of slab heating characteristics in a reheating furnace. Energy Convers Manag 149:928–936
10. Kim YI, Moon KC, Kang BS, Han C, Chang KS (1998) Application of neural network to the supervisory control of a reheating furnace in the steel industry. Control Eng Pract 6(8):1009–1014
11. Sibarani H, Samyudia Y (2004) Robust nonlinear slab temperature control design for an industrial reheating furnace. Comput Aided Chem Eng 18:811–816
12. Liu YT, Lin YY, Wu SL, Chuang CH, Lin CT (2016) Brain dynamics in predicting driving fatigue using a recurrent self-evolving fuzzy neural network. IEEE Trans Neural Netw Learn Syst 27(2):347–360
13. Li S, Li YM (2014) Nonlinearly activated neural network for solving time-varying complex sylvester equation. IEEE Trans Cybern 44(8):1397–1407
14. Qin S, Xue XP (2015) A two-layer recurrent neural network for nonsmooth convex optimization problems. IEEE Trans Neural Netw Learn Syst 26(6):1149–1160
15. Zhang D, Li H (2008) A stochastic-based FPGA controller for an induction motor drive with integrated neural network algorithms. IEEE Trans Ind Electron 55(2):551–561
16. Cabrera JA, Castillo JJ, Carabias E, Ortiz A (2016) Evolutionary optimization of a motorcycle traction control system based on fuzzy logic. IEEE Trans Fuzzy Syst 23(5):1594–1607
17. Chatterjee A, Chatterjee R, Matsuno F, Endo T (2008) Augmented stable fuzzy control for flexible robotic arm using LMI approach and neuro-fuzzy state space modeling. IEEE Trans Ind Electron 55(3):1256–1270

18. Li SH, Fairbank M, Johnson C, Wunsch DC, Alonso E, Proano JL (2014) Artificial neural networks for control of a grid-connected rectifier/inverter under disturbance, dynamic and power converter switching conditions. IEEE Trans Neural Netw Learn Syst 25(4):738–750
19. Haddoun A, Benbouzid MEH, Diallo D, Abdessemed R, Ghouili J, Srairi K (2008) Modeling, analysis, and neural network control of an EV electrical differential. IEEE Trans Ind Electron 55(6):2286–2294
20. San PP, Ling SH, Nuryani N, Nguyen HT (2014) Evolvable rough-block-based neural network and its biomedical application to hypoglycemia detection system. IEEE Trans Cybern 44(8):1338–1349
21. Ijjina EP, Chalavadi KM (2016) Human action recognition using genetic algorithms and convolutional neural networks. Pattern Recognit 59:199–212
22. Lin FJ, Shen PH, Yang SL, Chou PH (2006) Recurrent radial basis function network-based fuzzy neural network control for permanent-magnet linear synchronous motor servo drive. IEEE Trans Magn 42(11):3694–3705
23. Toliyat HA, Wlas M, Krzemiriski Z (2008) Neural-network-based parameter estimations of induction motors. IEEE Trans Ind Electron 55(4):1783–1794
24. Lin FJ, Shen PH (2006) Adaptive fuzzy-neural-network control for a DSP-based permanent magnet linear synchronous motor servo drive. IEEE Trans Fuzzy Syst 14(4):481–495
25. Wang TC, He X, Huang TW (2016) Complex dynamical behavior of neural networks in circuit implementation. Neurocomputing 190:95–106
26. Das AK, Subramanian K, Sundaram S (2015) An evolving interval type-2 neurofuzzy inference system and its metacognitive sequential learning algorithm. IEEE Trans Fuzzy Syst 23(6):2080–2093
27. Dora S, Suresh S, Sundararajan N (2015) A sequential learning algorithm for a spiking neural classifier. Appl Soft Comput 36:255–268
28. Li S (2001) Comparative analysis of backpropagation and extended Kalman filter in pattern and batch forms for training neural networks. Proc IEEE Int Jt Conf Neural Netw 1(1):144–149
29. Liao YX, Wu M, She JH (2006) Modeling of reheating-furnace dynamics using neural network based on improved sequential-learning algorithm. In: Proceedings of IEEE international conference on control applications, pp 3175–3181
30. Shen YL, Ju W, Yu C, Li SY (2010) Normalized decoupling control for high-dimensional MIMO processes forapplication in room temperature control HVAC systems. Control Eng Pract 2:652–664
31. Kennedy J, Eberhart RC (2002) Particle swarm optimization. In: Proceedings of IEEE international conference on neural networks, pp. 1942–1948
32. Azimifar A, Payan S (2016) Enhancement of heat transfer of confined enclosures with free convection using blocks with PSO algorithm. Appl Therm Eng 10:79–91
33. Chen SM, Chiou CH (2015) Multiattribute decision making based on interval-valued intuitionistic fuzzy sets, PSO techniques, and evidential reasoning methodology. IEEE Trans Fuzzy Syst 23(6):1905–1916
34. Rubaai A, Young P (2016) Hardware/Software implementation of fuzzy-neural-network self-learning control methods for brushless dc motor drives. IEEE Trans Ind Appl 52(1):414–424
35. Chatterjee A, Pulasinghe K, Watanabe K, Izumi K (2005) A particle-swarm-optimized fuzzy-neural network for voice-controlled robot systems. IEEE Trans Ind Electron 52(6):1478–1489
36. Ho SL, Yang SY, Ni GZ, Wong HC (2006) A particle swarm optimizaion method with enhanced global search ability for design optimizations of electromagnetic devices. IEEE Trans Magn 42(4):1107–1110
37. Lin CJ, Hong SJ (2007) The design of neuro-fuzzy networks using particle swarm optimization and recursive singular value decomposition. Neurocomputing 71(1–3):297–310
38. Javidrad F, Nazari M (2017) A new hybrid particle swarm and simulated annealing stochastic optimization method. Appl Soft Comput 60:634–654
39. Tan KC, Goh CK, Mamun AA, Ei EZ (2008) An evolutionary artificial immune system for multi-objective optimization. Eur J Oper Res 187(2):371–392

40. Wang ZJ, Chai TY, Guan SP, Shao C (1999) Hybrid optimization setpoint strategy for slab reheating furnace temperature. In: Proceedings of american control conference, pp 4082–4086
41. Zhang RD, Tao JL (2018) A nonlinear fuzzy neural network modeling approach using an improved genetic algorithm. IEEE Trans Ind Electron 65(7):5882–5892

Index

M. Wu et al., *Intelligent Optimization and Control of Complex Metallurgical Processes*,
Engineering Applications of Computational Methods 3,
https://doi.org/10.1007/978-981-15-1145-5

The manufacturer's authorised representative in the EU is Springer Nature Customer Service Centre GmbH, Europaplatz 3, 69115 Heidelberg, Germany. If you have any concerns regarding our products, please contact ProductSafety@springernature.com

Printed and bound by CPI Group (UK) Ltd, Croydon, CR0 4YY
25/09/2026
02222901-0004